Kinetics, Transport, and Structure in Hard and Soft Materials

T0179289

Kinetics, Transport, and Structure in Hard and Soft Materials

Peter F. Green
University of Michigan

CRC Press
Taylor & Francis Group
Boca Raton   London   New York

CRC Press is an imprint of the
Taylor & Francis Group, an **informa** business

CRC Press
Taylor & Francis Group
6000 Broken Sound Parkway NW, Suite 300
Boca Raton, FL 33487-2742

First issued in paperback 2019

© 2005 by Taylor & Francis Group, LLC
CRC Press is an imprint of Taylor & Francis Group, an Informa business

No claim to original U.S. Government works

ISBN-13: 978-1-57444-768-2 (hbk)
ISBN-13: 978-0-367-39297-0 (pbk)

Library of Congress Card Number 2004062073

Library of Congress Cataloging-in-Publication Data

Green, Peter. F.
    Kinetics, transport, and structure in hard and soft materials / Peter F. Green.
        p. cm.
    Includes bibliographical references and index.
    ISBN 1-57444-768-8
    1. Materials science. 2. Transport theory. I. Title.

TA403.G675 2004
620.1'1--dc22

          2004062073

Visit the Taylor & Francis Web site at
http://www.taylorandfrancis.com

and the CRC Press Web site at
http://www.crcpress.com

To Yvett, Ashley, and Robyn

Preface

Transport phenomena play a fundamental role in a diverse range of chemical, biological, and physical processes. The connection between the mechanisms of transport of atomic or molecular entities that occur in a diverse range of hard and soft materials (metals, polymers, inorganic network glasses, and ionic crystals) and structure is discussed in this book. *Kinetics, Transport, and Structure in Hard and Soft Materials* is intended primarily as a text for senior-year undergraduates and first-year graduate students in materials science and engineering, chemical engineering, chemistry, physics, and related fields. While many topics in the book are covered at sufficient depth that new researchers in the field will find the discussions of value, aspects of this book, particularly the early stages of each chapter, are discussed at a sufficiently basic level that advanced undergraduate students will find the material instructive.

Graduate students who work on materials-related topics for a thesis or dissertation come from a diverse range of departments that include materials science, physics, chemistry, and virtually all areas of engineering. Such students develop expertise related to one particular class of materials associated with their thesis research. In recent years, our society has experienced a paradigm shift, wherein materials that were originally associated with certain applications are now routinely used where they might not have been envisioned for use years earlier. Examples include polymers as the "active" material components in devices and sensors, inorganic network glasses serving a structural (and not just aesthetic) role in buildings, and various types of organic–inorganic hybrid materials as structural elements in motor vehicles. Indeed, materials-related challenges that engineers and scientists face in a technological or scientific environment are cross-cutting and interdisciplinary, requiring a strong foundation that encompasses classes of materials and basic science.

Textbooks on the topic of kinetics and transport processes typically fall into four categories. 1) Typically, courses on kinetics taught in many materials science departments primarily emphasize the diffusion and kinetics of phase transformations in metals. 2) In the second category, solutions to the diffusion equation subject to various boundary conditions are discussed. The book by Crank, *The Mathematics of Diffusion*, is one of the best-known examples. Such books, though important, do not provide the reader with information about mechanisms of transport. 3) In the third category, diffusion and reactions are examined primarily in liquid systems. The latter is often taught in chemical engineering departments. 4) Finally, many textbooks on transport phenomena emphasize a continuum picture of transport; the connection to structure is absent. These are typically found in chemical and mechanical engineering departments. While *Kinetics, Transport, and Structure in Hard and Soft Materials* is not necessarily intended to replace those books, its intent is to educate a broad cross-section of graduate students in issues regarding transport processes in materials and their connection to materials structure.

A few years ago, while preparing a syllabus for a graduate course for students in chemical engineering, chemistry, mechanical engineering, and materials science, I was faced with an intrinsic challenge: How do I maintain the interest of this diverse collection of students? After discussions with many of my colleagues, my strategy was to emphasize the fundamentals and discuss the connection between the structure and mechanisms of transport in different classes of materials. The book also includes a discussion of physical processes, such as pattern formation, which includes phase separation (spinidal decomposition) and instabilities that develop at moving fronts leading to dendritic formation in a wide class of systems (polymers, ice, metals). This book does not examine electronic transport processes, as this is a topic covered in solid state physics courses.

This text is divided into four parts. The fundamentals of diffusional transport, "tools," are discussed in Part I. This information establishes the foundation for subsequent discussions of mechanisms of transport in crystalline materials (metals, semiconductors, ionic crystals) and in structurally disordered materials in Parts II and III, respectively. Phenomena that include spinodal decomposition, Mullins-Sekerka instabilities, and other types of instabilities that lead to morphological evolution (pattern formation) facilitated by long-range collective motions of structural entities are discussed in Part IV.

The prerequisites for *Kinetics, Transport, and Structure in Hard and Soft Materials* are basic courses on ordinary differential equations and thermodynamics or physical chemistry.

Acknowledgments

This text in many ways reflects my own personal journey, which began with studying physics and materials science. During my graduate studies at Cornell University during the early 1980s, I first developed an interest in mechanisms of transport in various classes of materials. The environment there was highly conducive to interdisciplinary research. I developed an even deeper appreciation for diffusion in polymers due largely to the influence of my Ph.D. mentor, Edward J. Kramer. Later, at Sandia National laboratories, I became interested in dynamics in inorganic network glasses and topics such as spinodal decomposition largely due to the influence of colleagues in the ceramics and polymers divisions. Upon arriving at the University of Texas at Austin, I became interested in the topic of instabilities due largely to my appointment in chemical engineering. Funding for my research by the National Science Foundation and the Robert A. Welch Foundation played a pivotal role in maintaining an active research program in various aspects of kinetics and transport.

During the preparation of this book, I benefited from the advice and direction of a number of colleagues, particularly Llewellyn Rabenberg, Venkat Ganesan, Tom Truskett, Gyeong Hwang, Isaac Sanchez, Ralph Colby, David Sidebottom, Ranko Richert, and Mark Ediger. Collectively, they devoted their time to reading chapters throughout the book. The students who have taken my graduate course on this topic during the past three years used various versions of chapters throughout the book and provided important feedback. To this end, I wanted to thank Shreyas Rajasekhara, Brian Besancon, Jamie Kropka, and Luciana Meli, who deserve special thanks for proofreading the final versions of certain chapters of the book. I also want to thank John Kieffer, Bruce Clemens, and Paulo Feirreirs for valuable discussions on some aspects of the topics discussed in this book.

Finally, and perhaps most important, this book would not have been possible without the willing encouragement, patience, and cooperation of my wife Yvett and daughters, Ashley and Robyn. I am most fortunate to have them in my life, and I am truly indebted to them for their understanding. They gave up a great deal, including the most recent summer and Christmas vacations, so that I could complete the book. It is to them that I dedicate *Kinetics, Transport, and Structure in Hard and Soft Materials.*

Contents

Part I

Tools: Elements of Diffusional Transport

"Tools" are developed in this first part of the book in order to provide a foundation for topics on mechanisms of atomic, or molecular, transport in materials covered in the remainder of the text. While the goal of Chapter 1 is to establish a phenomenological foundation of diffusional transport, Chapter 1 begins with an elementary discussion of statistical mechanics. This establishes the framework for a discussion of the Maxwell-Boltzmann distribution function, which is then used to calculate basic (average) properties of a system of noninteracting particles. Fick's 1st and 2nd laws are then introduced and solved for some very common cases involving mass transfer. Chapter 2 sets the stage for a molecular picture, described further in later chapters; the phenomenon of Brownian motion is introduced. *Random, statistically fluctuating, and incessant motions of a particle in a medium typify the phenomenon of Brownian motion.* Part I is concluded with a discussion of correlation functions, the structure factor and common experimental techniques used to study diffusion in condensed matter.

1

Elements of Transport in Systems of Noninteracting Particles and the Phenomenology of Diffusion

1.1 Introduction

Transport phenomena play a fundamental role in a diverse range of chemical, biological, and physical processes. Examples of long-range diffusional transport processes include the migration of electronic charge carriers, which are necessary for the operation of emissive displays; the transport of ions necessary for the operation of electrochemical energy storage devices, such as batteries; and the migration of large macromolecules in spatially restricted environments, such as the translocation of DNA across bacterial membranes. Morphological features (phases of differing chemical composition, and/or varying atomic or molecular organization and different size distributions, etc.) of materials profoundly influence material properties, ranging from magnetic, optical, and electronic to corrosion and mechanical properties. Annealing a material generally induces long-range atomic and molecular transport processes, which facilitate microstructural evolution. The growth of various crystalline phases of materials during annealing is controlled by atomic or molecular diffusion processes. The spatial distribution of dopants in semiconductors, which controls device performance, is determined by atomic diffusion properties. Interdiffusion between semiconductor multilayer films that make up quantum well heterostructures (components of high-speed and high-frequency digital and analog devices) influences the optical and structural properties of the heterostructures and, therefore, device performance. Clearly, the impact of diffusional transport processes on our everyday lives is profound.

The center of mass transport of an atom or molecule in a material is intimately connected to the spatial arrangement of its neighboring constituents and to its interactions with them. For crystalline materials, such as metals, the mechanism by which an atom hops from one site to another within the crystal is largely dictated by symmetries of the spatial arrangements of the atomic constituents (crystal structure) and by defects associated with the arrangement. The hopping rate is determined, in part, by the available thermal energy and by the local symmetry and defect population of the environment.

Figure 1.1 illustrates one mechanism, a vacancy mechanism, by which an atom located at site #7 migrates throughout a two-dimesional lattice. The atom may hop into the vacant site #6, as shown in the figure. It could then immediately hop back to its original location or, alternatively, another nearest neighbor atom could hop into the vacant site. This example illustrates the influence of defects (a vacancy in this case) on the diffusion process. Diverse defect mediated mechanisms of atomic transport occur in crystalline lattices, depending on the crystal symmetry and the nature of the defect population. These will be discussed in Chapters 3, 4 and 5.

In materials with structures that lack long-range order, such as entangled polymer melts, the dynamics of a long chain molecule are profoundly influenced by interactions of the chain with its neighbors. Entanglements with neighboring chains impose topological constraints on a diffusing chain such that this chain is destined to execute long-range motions along its own contour; i.e., it undergoes slithering, snake-like, motions (Fig. 1.2). Many of the unique time-dependent properties that polymers exhibit can be reconciled with this picture and will be discussed in Chapter 6.

For inorganic network glasses, such as alkali silicates (e.g., window glass), the molecular structure is characterized by a three-dimensional network of covalent bonds and by ionic bonds associated with the alkali ions. Viscous flow is accommodated by the breaking and reconstruction of bonds. The dynamics of individual cations are influenced by spatial correlations imposed on them by long-range Coulombic effects. Understanding the nature of cation migration is important for different electrochemical and sensing applications for which network glasses Chapter 7 are well suited.

The primary goal of this chapter is to provide a phenomenological description of diffusional transport in condensed media. We are initially interested in the properties of noninteracting particles. The discussion of the transport of noninteracting particles provides a natural framework to 1) introduce the topic of distribution functions, which will be used throughout this book, and 2) introduce Fick's 1st and 2nd laws, which govern the spatial and temporal evolution of species in *condensed* media. Fick's laws are solved for two simple situations: 1) steady state, time-independent, flow of particles subject to certain boundary conditions, and 2) diffusion of particles from point sources and from extended sources into surrounding media. The information developed

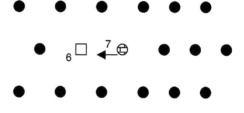

FIG. 1.1
An atom #7 can hop into an adjacent vacant site, #6. A nearby atom can hop into the site vacated by #7 or #7 can hop backwards to its original location.

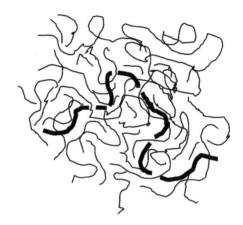

FIG. 1.2
(a) Schematic of a dense melt in which the probe chain is constrained to move along its own contour due to the topological constraints imposed by its neighbors.

in this chapter provides a context for discussions in subsequent chapters regarding *mechanisms* of diffusional transport in a diverse range of systems.

1.2 Transport of Noninteracting Particles

Of primary interest here are dynamical features of a dilute collection of a large number, N, of energetic particles enclosed within a fixed volume, V. Each particle possesses the same mass, m, and is able to translate and experiences collisions with the container and with other particles without loss of energy. In such a system it is impossible to specify the velocity or the energy of an individual particle, however it would be reasonable to inquire about average (statistical) properties, such as the average energy or average velocity. In order to answer these questions it will be necessary to calculate the relevant probability distribution functions. Other relevant properties necessary to describe the dynamics would include the flux J, the number of particles passing through a unit area per unit time, and the diffusion coefficient D. Since the particles, spaced sufficiently far apart compared to their sizes, do occasionally collide, it would be nice to know the probability of occurrence P of a collision as well as the mean free path l. Knowledge of average speeds and velocities enable calculation of these parameters (J, D, P, l), which together provide a reasonable picture of the dynamical features of this system of noninteracting particles.

1.2.1 Average Thermodynamic Properties

We are initially interested in calculating the average energy of particles located in a surrounding medium whose only influence is to provide a heat reservoir of ambient temperature T. The location of the center of mass of a molecule is $\vec{q}_i$ and its momentum is $\vec{p}_i = m\vec{v}_i$. To completely describe this large statistical system of particles in a given state (N, V, E) at an arbitrary time t, in principle one would have to consider specifying all the spatial and momentum coordinates of the particles $(\vec{q}_1(t), \vec{q}_2(t)...\vec{q}_{3N}(t), \vec{p}_1(t), \vec{p}_2(t)...\vec{p}_{3N}(t)) \equiv (\vec{q}_i(t), \vec{p}_i(t))$. Since each molecule possesses translational energy but no vibrational or rotational energy, and no interactions between molecules exist, then the energy of each molecule, E_i, is specified entirely by its kinetic energy

$$E_i = \frac{p_i^2}{2m} \qquad\qquad 1.1$$

A natural question that arises is, "What is the average energy, $\langle E \rangle$, of the system?" With this question, the following observation can be made. Throughout any given time interval the system evolves over different states and an average value of a property may be measured over a sufficiently long time interval. Alternatively, an ensemble of systems could be evaluated at a particular instant in time. An ensemble is a virtual (or mental) collection of an innumerably large number of identical systems. The ensemble average would be the time-average at thermodynamic equilibrium. To determine the average energy, we need to calculate the probability, P_i, that a particle will possess energy E_i (or reside in state i with energy E_i). Using this information the average energy may be calculated because

$$\langle E \rangle = \sum_i E_i P_i \qquad\qquad 1.2$$

To determine the probability function P_i, we begin by considering a small system A in contact with a large reservoir, A_R, at thermal equilibrium. The reservoir has an infinitely large heat capacity so the temperature, T, of the system and reservoir remains constant. This system could be a single molecule within a large body of water, an atom sitting on a lattice site, or a small sample sitting in a large oven. While the system can exchange heat with a reservoir it cannot exchange mass (N is fixed) and its volume, V, is fixed. Second, while energy is exchanged between the system and reservoir, the total energy, E_T, of the system and the reservoir remains fixed (constant),

$$E_T = E_i + E_R \qquad\qquad 1.3$$

In this equation E_R is the energy of the reservoir and E_i is that of the system. If the system comprises many particles, then there are many different ways that the energy may be distributed between the particles while obeying the constraint (Eq. 1.3). We are primarily interested in the probability that the system will possess energy E_i.

   Under the conditions described here, the appropriate ensemble would be the canonical *ensemble*. We could calculate P_i by carefully examining the statistics of ensembles or by adopting an alternative approach (which serves the purposes of this chapter). If the system resides in state i, possessing energy E_i, at time t, then the reservoir possesses energy $E_R = E_T - E_i$, and the *number of states* available to the reservoir is $\Omega(E_R) = \Omega(E_T - E_i)$. The probability, P_i, that the system possesses energy E_i is proportional to $\Omega(E_T - E_i)$,

$$P_i \propto \Omega(E_T - E_i) \tag{1.4}$$

Since $E_i \ll E_T$, we can write an approximate expression for Eq. 1.4 by expanding the logarithm of $\Omega(E_T - E_i)$ around $E = E_T$, to yield

$$\ln \Omega(E_T - E_i) = \ln \Omega(E_T) - \frac{\partial \ln \Omega}{\partial E}\bigg|_{E = E_T} E_i + \cdots \tag{1.5}$$

The higher order terms are neglected with minimal loss of accuracy because of the small magnitude of E_i in relation to E_T. The expression involving the derivative in the second term of the RHS must have units of inverse energy, and is in fact identified as

$$\beta = \frac{1}{kT} = \frac{\partial \ln \Omega}{\partial E}\bigg|_{E = E_T} \tag{1.6}$$

where k is the Boltzmann constant. This equation indicates that the number of accessible states increases with energy ($\beta > 0$). Equations 1.5 and 1.6 indicate that

$$\Omega(E_T - E_i) \approx \Omega(E_T)e^{-\beta E_i} \tag{1.7}$$

Since $\Omega(E_T)$ is constant, then Eq. 1.4 and 1.7 reveal that the probability that the system possesses energy E_i is

$$P_i \propto e^{-E_i/kT} \tag{1.8}$$

In Eq. 1.8, $e^{-E_i/kT}$ is called the Boltzmann factor, which indicates that the probability that the system will increase its energy is exponentially low. The Boltzmann factor plays an important role in a number of statistical processes. For example, it largely determines probabilities of events in thermally activated processes such as the hopping of the atom into the vacant site in Fig. 1.1 (Ch. 3).

By relying on the normalization condition, $\sum_i P_i = 1$, Eq. 1.8 becomes,

$$P_i = \frac{e^{-E_i/kT}}{\sum_i e^{-E_i/kT}} \tag{1.9}$$

The denominator of Eq. 1.9 is known as the Partition function in Statistical Mechanics,

$$Z = \sum_i e^{-E_i/kT} \tag{1.10}$$

The summation is performed over all energy (quantum) states, i.

In light of the fact that the number of particles per unit volume is large and that the system is large, then the consecutive values of the energy levels must be necessarily close. Therefore, an alternative expression for Eq. 1.9 may be considered. The number of energy levels between E and $E + dE$ is sufficiently large that E could, in principle, be treated as a continuous variable. Consequently, $P(E)dE$ would represent the probability that a system in the ensemble possesses energy between E and $E + dE$. To get an expression for $P(E)$ one would have to determine the number of states with energy in the energy range dE and this would be $g(E)dE$, where $g(E)$ is the density of states. Hence $P(E)dE \propto e^{-E/kT} g(E)dE$, and with the normalization condition,

$$P(E)dE = \frac{e^{-E/kT} g(E)dE}{\displaystyle\int_0^\infty e^{-E/kT} g(E)dE} \tag{1.11}$$

where the partition function would be specified by the denominator, $Z = \int_0^\infty e^{-E/kT} g(E)dE$.

The foregoing discussion of Partition functions is necessarily abbreviated, but it serves the purposes of this chapter. The interested reader is encouraged to consult virtually any text on Statistical Mechanics for more complete treatments of the topic.

Example 1: Average Energy, Entropy, and Pressure

With the Partition function, different average thermodynamic quantities may be determined. The average energy of the ensemble is by definition

$$\langle E(N_e, V_e, T) \rangle = \sum_i P_i E_i = \frac{\sum_i E_i(N_e, V_e) e^{-E_i/kT}}{\sum_i e^{-E_i/kT}} \tag{1.12}$$

Because $\sum \frac{E e^{-E_i/kT}}{Z} = -\sum \frac{\partial e^{-E_i/kT}/\partial \beta}{Z} = -\frac{1}{Z}\frac{\partial Z}{\partial \beta}$ the average energy of the system is

$$\langle E \rangle = -\frac{\partial \ln Z}{\partial \beta} \tag{1.13}$$

To further illustrate the point, other thermodynamic functions may be determined from knowledge of the Partition functions. An expression for the entropy may be determined by considering the differential with respect to $\langle E \rangle$,

$$d\langle E \rangle = \sum_i E_i dP_i + \sum_i P_i dE_i \tag{1.14}$$

With the use of Eq. 1.11 and some manipulation it can be shown (Problem 2) that

$$S = -k \sum_i P_i \ln P_i \tag{1.15}$$

This is an explicit expression that relates the entropy of the system to the probability that a particle possesses energy E_i. Herewith, we can also write down an expression that explicitly connects the entropy to the Partition function and to $\langle E \rangle$,

$$S = \frac{\langle E \rangle}{T} + k \ln Z \qquad 1.16$$

The Helmholtz free energy, $A = E - TS$, is readily expressed in terms of the Partition function

$$A = -kT \ln Z \qquad 1.17$$

Finally, it follows from the above that the average pressure is

$$\langle p \rangle = kT \left(\frac{\partial \ln Z}{\partial V} \right)_T \qquad 1.18$$

Further details on Partition functions may be found in virtually any text on Statistical Mechanics. An example involving a system of N noninteracting particles enclosed within a volume V is now presented in order to illustrate the utility of the Partition function.

Example 2: Equation of State for an Ideal Gas

An explicit answer for the average pressure exerted by this system of non-interacting particles is now sought. We briefly reiterate that because the particles possess no vibrational or rotational energy, and exchange only heat with the environment, their energies are specified only in terms of the kinetic energy. An expression for the Partition function for this N particle system is

$$Z_N = \sum_{i,j,k,\dots} e^{-\left(\varepsilon_i^a + \varepsilon_j^b + \varepsilon_k^c + \cdots\right)/kT} = \left(\sum_i e^{-\varepsilon_i^a/kT} \right) \left(\sum_j e^{-\varepsilon_j^b/kT} \right) \left(\sum_k e^{-\varepsilon_k^c/kT} \right) \cdots \qquad 1.19$$

where the superscripts in the exponents identify individual particles. It is noteworthy that if the system is composed of noninteracting components, then the Partition function of the system is a product of the partition functions representing each component. For the collection of gas particles of interest, each molecule may be described by the same partition function, hence the partition function for the gas, assuming that each particle is distinguishable, may be written as $Z_N = (z)^N$, where $z = \sum_i e^{-\varepsilon_i/kT}$. On the other hand, if each particle in indistinguishable, then

$$Z_N = \frac{z^N}{N!} \qquad 1.20$$

where $N!$ is associated with the number of permutations.

It is important to point out that within the classical approximation (where the energy is treated as a continuous variable, assuming that the energy spacings are small as compared to kT) the Partition function may be expressed in terms of an integral. This approximation enables calculation of the average pressure exerted by these N classical particles. In general, for a system of N components the classical partition function is

$$Z = \int \cdots \int e^{-\beta E(q_1, q_2 \dots q_{3n}, p_1, \dots p_{3N})} \frac{d^{3N}\bar{q} d^{3N}\bar{p}}{N! h^{3N}} \qquad 1.21$$

where $\frac{d^{3N}\bar{q} d^{3N}\bar{p}}{N! h^{3N}} = \frac{dr_1 dr_2 \dots dr_{3N} dp_1 \dots dp_{3N}}{N! h^{3N}}$ is the number of cells in phase space corresponding to the number of distinct states in phase space (h is Planck's constant).

Explicitly, the Partition function for an individual particle in this N-particle system is

$$z = \int_{-\infty}^{\infty} e^{-(\beta/2m)p^2} d^3\bar{p} \frac{d^3\bar{r}}{h^3} = \left(\int_{-\infty}^{\infty} e^{-(\beta/2m)p_x^2} dp_x \right) \left(\int_{-\infty}^{\infty} e^{-(\beta/2m)p_y^2} dp_y \right) \left(\int_{-\infty}^{\infty} e^{-(\beta/2m)p_z^2} dp_z \right) \frac{V}{h^3}$$

$$1.22$$

where we have taken advantage of the fact that the energy is independent of position, so $d^3\bar{q} \equiv d^3\bar{r} = dxdydz = dV$. With the use of appendix A, the integrals are readily solved and

$$Z = \frac{V^N}{N!} \left[\frac{(2\pi mkT)^{3/2}}{h^2} \right]^N \qquad 1.23$$

It follows that because the pressure depends only on the volume derivative of ln Z, then $\left(\frac{\partial \ln Z}{\partial V}\right)_T = \frac{N}{V}$. Herewith,

$$\langle p \rangle V = NkT \qquad 1.24$$

This is the equation of state for an ideal gas. This answer is, of course, not surprising considering the conditions imposed on the system. This foregoing example illustrates the utility of the Partition function.

1.2.2 Maxwell-Boltzmann Velocity Distributions

The Maxwell-Boltzmann distribution function $f(v)$ is used to calculate average dynamical properties (velocities, flux, and diffusion coefficient) of the system of particles. This distribution function is now derived; the derivation is meant to be intuitive rather than rigorous. The mean number of molecules, dN, with centers of mass between $\bar{r}$ and $\bar{r} + d\bar{r}$ and velocities between $\bar{v}$ and $\bar{v} + d\bar{v}$, simultaneously, is specified by the distribution function, $f(\bar{r}, \bar{v})$, where

$$dN = f(\bar{r}, \bar{v}) d^3\bar{r} d^3\bar{v} \qquad 1.25$$

and $r^2 = x^2 + y^2 + z^2$, $v^2 = v_x^2 + v_y^2 + v_z^2$; $d^3\vec{r} = dxdydz = dV$ and $d^3\vec{v} = dv_x dv_y dv_z$. The total number of molecules with velocity component in each direction must sum to N. Hence

$$\int_v \int_r f(\vec{r},v)d^3\vec{r}d^3\vec{v} = N \qquad\qquad 1.26$$

Again, in the absence of external forces the energy of this system does not depend on position, so $f(\vec{r},\vec{v}) = f(\vec{v})$. In fact $f(\vec{r},\vec{v}) = f(\vec{v}) = f(v)$ because the energy depends only on v^2.

In order to determine an expression $f(\vec{v})$, it should be recognized that Eq. 1.8 is now quite useful, as it is the probability that the particle possesses a particular energy. Alternatively, it maybe interpreted as the fraction of molecules that possess velocities between $\vec{v}$ and $\vec{v} + d\vec{v}$, so

$$\frac{f(v)d^3\vec{r}d^3\vec{v}}{N} \propto e^{-mv^2/2kT} d^3\vec{r}d^3\vec{v} \qquad\qquad 1.27$$

The constant of proportionality, which we designate as C, can be obtained from the normalization condition,

$$N = \int_{-\infty}^{\infty}\int_{-\infty}^{\infty} f(v)d^3\vec{r}d^3\vec{v} = C\int_{-\infty}^{\infty}\int_{-\infty}^{\infty} e^{-mv^2/2kT} d^3\vec{r}d^3\vec{v} \qquad\qquad 1.28$$

The integrals are readily solved to yield $C = (N/V)(m/2\pi kT)^{3/2}$. Finally, the expression for the Maxwell-Boltzmann distribution function is

$$f(v) = n\left(\frac{m}{2\pi kT}\right)^{3/2} e^{-mv^2/2kT} \qquad\qquad 1.29$$

where $n = N/V$. In the above equation $f(v)d^3\vec{v}$ is the mean number of particles per unit volume with velocity between $\vec{v}$ and $\vec{v} + d\vec{v}$. We now proceed to examine average velocity components and average speeds.

1.2.2.1 *Distribution of Component Velocities*

The mean velocities and the mean square velocities in different directions, $i (i = x, y$ or $z)$, are first considered. The appropriate form of the Maxwell-Boltzmann velocity distribution function must be identified in order to perform these calculations. The probability distribution function, $h(v_i)dv_i$, that enables calculation of the average velocities and mean square velocities in different directions is determined by calculating the mean number of molecules per unit volume which possess velocities that reside between v_i and $v_i + dv_i$. In the x-direction, the function is

$$h(v_x)dv_x = \int_{v_y}\int_{v_z} f(\bar{v})d^3\bar{v}$$

$$= n\left(\frac{m}{2\pi kT}\right)^{3/2}\int_{-\infty}^{\infty}\int_{-\infty}^{\infty} e^{-(m/2kT)\left(v_x^2 + v_y^2 + v_z^2\right)}dv_x dv_y dv_z \tag{1.30}$$

Upon performing the integrations, we obtain the following expression for $h(v_x)dv_x$,

$$\frac{h(v_x)dv_x}{n} = \left(\frac{m}{2\pi kT}\right)^{1/2} e^{-mv_x^2/2kT}dv_x \tag{1.31}$$

The relevant distributions for the other directions, y and z, are readily determined using the same procedure, or by inspection. It is apparent that the relationship between $f(v)$ and $h(v_i)$ is

$$\frac{f(v)d^3\bar{v}}{n} = \left(\frac{h(v_x)dv_x}{n}\right)\left(\frac{h(v_y)dv_y}{n}\right)\left(\frac{h(v_z)dv_z}{n}\right) \tag{1.32}$$

The equation representing $\frac{h(v_x)}{n}$ is a Gaussian distribution function. In terms of a variable x, the Gaussian distribution function is of the general form

$$P(x) = \left(\frac{1}{2\pi\sigma^2}\right)^{1/2}\exp\left[\frac{-(x-\langle x\rangle)^2}{2\sigma^2}\right] \tag{1.33}$$

where $\langle x\rangle$ is the average value of the variable x. The dispersion, or equivalently the standard deviation, is σ^2 where

$$\sigma^2 = \langle(x-\langle x\rangle)^2\rangle \tag{1.34}$$

This function is plotted in Fig. 1.1 for $\langle x\rangle = 0$ and $\langle x\rangle = 2$. $P(x)$ is symmetric about $\langle x\rangle$. Note, σ is sometimes called the variance. The Gaussian distribution function appears in a wide range of situations. For example, the Gaussian distribution typically represents the grade distribution for large classes. It is noteworthy that if one performs a large number of measurements of a particular physical property of a system in a laboratory and analyzes the data, σ would represent the scatter of values around the mean value of that property. We will encounter this function again when we discuss diffusion.

It is apparent from inspection of Eq. 1.31 and 1.33 that the dispersion of the component velocity $\sigma^2 = kT/m$. This result indicates that the breadth of the distribution increases with T and decreases as the mass of the particle increases. Note that σ^2 is often identified as the fluctuation of the velocity. Later we will see that if we observed a particle over a long period of time,

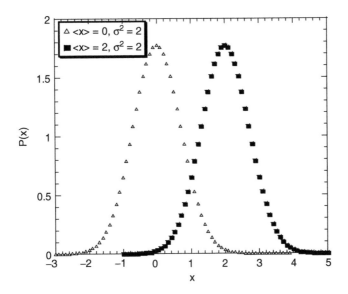

FIG. 1.3
The Gaussian distribution function $P(x)$ is plotted here for $\langle x \rangle = 0$ and for $\langle x \rangle = 2$.

its average velocity would be zero. Its velocity at a given instant, however, would not be zero. Its velocity would fluctuate about a mean value and, as the temperature increases, the fluctuations would increase. Moreover, larger particles exhibit smaller fluctuations. We will return to this issue later, in Chapter 2, because these results are relevant to the Brownian dynamics of particles.

The average component velocities are now calculated. The average velocity in the i-direction is

$$\langle v_i \rangle = \int \frac{v_i h(v_i) dv_i}{n} = 0 \qquad 1.35$$

This answer should not be surprising, as the displacement of any particle should, on average, occur with equal probability in any direction.

The mean square velocity of a particle in direction $i\,(i = x, y, z)$, is

$$\langle v_i^2 \rangle = \int \frac{v_i^2 h(v_i) dv_i}{n} = \frac{kT}{m} \qquad 1.36$$

Equation 1.36 indicates that the mean square velocity is proportional to temperature, which should not be a surprise since one expects the energy of these classical particles to increase with increasing temperature. Recall that the dispersion was also specified by $\sigma^2 = kT/m$ and is a consequence of the fact that $\langle v_i \rangle = 0$.

It is noteworthy that

$$\langle v^2 \rangle = \langle v_x^2 \rangle + \langle v_y^2 \rangle + \langle v_z^2 \rangle = \frac{3kT}{m} \tag{1.37}$$

because it implies that the total kinetic energy $\frac{1}{2}m\langle v^2 \rangle = \frac{3}{2}kT$. The kinetic energy of a particle in each direction is $kT/2$. We note, in passing, that this is the classical *equipartition theorem* which indicates that the mean value of every independent term in the quadratic expression (each corresponding to $\frac{1}{2}mv_i^2$) is $kT/2$. The implication is that the kinetic energy of a dilute gas at thermal equilibrium is proportional to its ambient temperature.

1.2.2.2 Distribution of Speeds

The mean speed and mean square speed are now discussed. In performing these calculations, it should be recalled that speed is a scalar quantity and as such is independent of the direction of motion. The calculation proceeds by asking, "What is the mean number of molecules with speeds between u and $u + du$, $F(u)du$ ($u = |\bar{v}|$)?" $F(u)du$ is determined by recognizing that

$$\frac{F(u)du}{N} = \frac{f(v)d^3\bar{v}}{N} \tag{1.38}$$

In spherical coordinates, the volume element $d^3\bar{v} = u^2\,du\,\sin\theta\,d\theta\,d\varphi$, where $0 \le u \le \infty; 0 < \theta < \pi/2; 0 < \varphi < 2\pi$. The magnitude of a given velocity vector (speed) maps out a hollow sphere and du is the thickness of this hollow sphere. Since the volume of a spherical shell of radius u and thickness du is $4\pi u^2 du$, then

$$\frac{F(u)du}{n} = 4\pi \left(\frac{m}{2\pi kT}\right)^{3/2} u^2 e^{-mu^2/2kT}\,du \tag{1.39}$$

A plot of the dependence of $F(u)$ on u is shown in Fig. 1.4. $F(u)$ increases rapidly at small values of u but decreases with increasing speed because the probability that a particle will possess a large energy is exponentially low.

The distribution function enables calculation of the average speed,

$$\langle u \rangle = \left(\frac{8}{\pi}\frac{kT}{m}\right)^{1/2} \tag{1.40}$$

revealing that the average speed, unlike the velocity, is greater than zero, as anticipated.

The mean square speed, accordingly, is

$$\langle u^2 \rangle = \frac{3kT}{m} \tag{1.41}$$

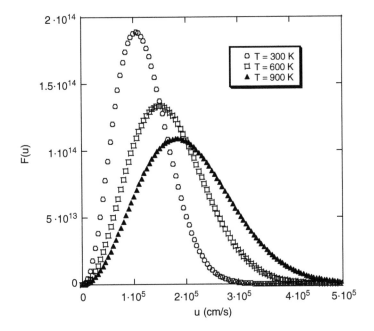

FIG. 1.4
A distribution of speeds is shown here for helium at three different temperatures. The calculation was performed using Eq. 1.39. A mean pressure of 1 atm was used and the ideal gas law
assumed to apply. The point is to illustrate that, as T increases, the breadth of the distribution
and the most probable speed increase.

This equation indicates that the total kinetic energy of a particle is $(3/2)kT$.
It is interesting to note that the average speed could have been determined
by recognizing that

$$\langle u^2 \rangle = \left\langle \left(v_x^2 + v_y^2 + v_z^2 \right) \right\rangle = \langle v^2 \rangle.$$

It might be worthwhile to briefly comment on these results in relation to a
practical issue. The speed of sound in air under standard temperature and
pressure conditions is 350 m/s. With the use of Eq. 1.39, the average speed
of a nitrogen or an oxygen molecule maybe shown to be faster (Problem 14).
These speeds are slower than a bullet from a high-caliber rifle.

1.2.3 Diffusional Transport of Noninteracting Particles

The flux, defined as the number of particles crossing a unit area per unit
time (units: mass•distance/time•volume), is important because it determines the time-dependent evolution of the concentration profile of a
diffusant in a medium. This parameter is first calculated for the collection

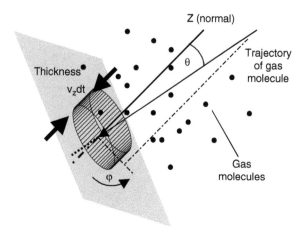

FIG. 1.5
Gas molecules crossing the area dA. Only molecules enclosed in the cylinder arrive during interval dt. The thickness of the cylinder is vdt.

of particles. Subsequently, the diffusion coefficient of these particles, which we see in the next section is connected to the flux via Fick's 1st law is calculated.

1.2.3.1 Flux of Maxwellian Particles

An imaginary plane oriented along the z-direction, as illustrated in Fig. 1.5, is now cosidered. Note that a plane with any orientation could have arbitrarily been chosen. Consider further that gas molecules impinge on an infinitesimal area dA with trajectory oriented θ with respect to the z-axis and angle φ. The mean number of molecules that cross a unit area, dA, of the plane during the interval dt is given by

$$f(\vec{v})d^3\vec{v}|dA(v_z dt)| \qquad 1.42$$

In the above equation $f(\vec{v})d^3\vec{v}$ (Section 1.2.2.1) is the average number of molecules per unit volume with velocity between $\vec{v}$ and $\vec{v}+d\vec{v}$. The expression $|dA(v_z dt)|$ represents the volume of a cylinder, whose area is dA and thickness $v_z dt$, that encloses molecules that will strike the area dA during the time interval dt. It follows that the average number of molecules that strike the area per unit time, the flux, is

$$J = \int_{v_z > 0} f(v)v_z d^3v \qquad 1.43$$

Note that the above integral is evaluated over $v_z > 0$, as $v_z < 0$ corresponds to molecules moving in the opposite direction. If we replace v_z with $u\cos\theta$ ($|\bar{v}| = u$), the flux becomes

$$J = \int_0^\infty f(u)u^3 du \int_0^{\pi/2} \sin\theta\cos\theta d\theta \int_0^{2\pi} d\varphi \qquad 1.44$$

Note that the limits over θ range from 0 to $\pi/2$; larger values of θ correspond to velocities pointing in the opposite direction ($v_z < 0$). This equation now becomes

$$J = \frac{n\langle u\rangle}{4} \qquad 1.45$$

This result is intuitive; it indicates that the flux is proportional to the number of particles per unit volume and to the average speed of the particles.

1.2.3.2 The Diffusion Coefficient and Fick's 1st Law

The diffusion coefficient of these noninteracting particles is now discussed. The analysis in this section enables the introduction of Fick's first law of diffusion. Begin by considering Fig. 1.6 which illustrates a collection of molecules crossing constant plane, $z =$ constant (gravitational effects are neglected). The number of particles at point $z + l$ (where l could be taken to be the mean free path) above the arbitrarily chosen constant plane is $c(z + l)$ and the flux of particles that travel downward is approximately $(1/6)\langle u\rangle c(z + l)$, where $\langle u\rangle$ is the average velocity of the particles in one direction. The factor of $1/6$ comes from the fact that a fraction of $1/6$ of the total number of particles, on average, moves in each of the six directions in the Cartesian coordinate system. The number of particles at $z - l$ is $(1/6)\langle u\rangle c(z - l)$. Herewith, the net flux of particles traveling in the positive z-direction is

$$J_z = (1/6)[\langle u\rangle c(z - l) - \langle u\rangle c(z + l)] \qquad 1.46$$

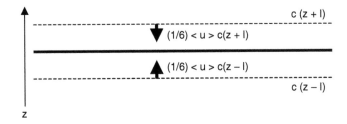

FIG. 1.6
Transport of molecules across a constant plane. The concentration is c_x, v is the velocity, and l is a mean free path.

Now $c(z)$ can be expanded using a Taylor series expansion, because l is infinitesimally small, so $c(z \pm l) = c(z) \pm \frac{\partial c}{\partial z} l \ldots$ from which it follows that

$$J_z = -(1/6)\langle u \rangle \left(2 \frac{\partial c}{\partial z} l \right)$$

$\qquad\qquad$ 1.47

This equation indicates that the flux is proportional to the concentration gradient,

$$J_z = -D \frac{\partial c}{\partial z}$$

$\qquad\qquad$ 1.48

The negative sign in Eq. 1.48 indicates that flux moves opposite the direction of the concentration gradient. In the foregoing equation, D is the diffusion coefficient,

$$D = \frac{\langle u \rangle l}{3}$$

$\qquad\qquad$ 1.49

If the mean free path is $l = \langle u \rangle \tau$, where τ is the average time between collisions of particles in the gas, then $D = (1/3)l^2/\tau$. D has units of $(distance)^2/time$. In the next section the relation between the diffusion coefficient is calculated for a particle undergoing random excursions in an arbitrary medium and shown that its mean square displacement is proportional to the product of the diffusion coefficient and the time. Note that while this equation was derived by considering a collection of noninteracting particles, the result is general and applies to a range of systems. This is Fick's first law, which will be discussed in further detail in the next section.

1.2.3.3 Collision Probabilities and the Mean Free Path

In this section, expressions are calculated for the mean free path and the average time between collisions τ in terms of molecular parameters of the system. We are initially interested in the probability that a molecule survives a collision during time t, which is denoted as $p(t)$ is initially calculated. A collision rate, ω, is defined such that ωdt is the probability that a molecule suffers a collision during the time interval between t and $t + dt$. Ultimately, the probability function of interest is the probability that a particle survives a collision for time t but experiences a collision during the subsequent interval t and $t + dt$. We define this as $P(t)dt = p(t)\omega dt$. $P(t)$ meets the normalization condition,

$$\int_0^\infty P(t)dt = 1$$

$\qquad\qquad$ 1.50

Begin by writing down the following equation for the probability that a particle survives a collision during the interval $t + dt$,

$$p(t + dt) = p(t)[1 - \omega dt]$$

$\qquad\qquad$ 1.51

Upon expanding the RHS of Eq. 1.51, $p(t+dt) \approx p(t) + [dp(t)/dt]dt$, the differential equation for $p(t)$ is

$$\frac{dp(t)}{dt} = -\omega p(t)$$ 1.52

The boundary conditions for this equation are that $p(0) = 1$ and $p(\infty) = 0$; in other words, at $t = 0$ a particle would not have collided with another, whereas at long times the particle would have suffered a collision. These considerations lead to

$$p(t) = e^{-\omega t}$$ 1.53

This equation tells us that the probability that a particle would survive a collision decreases exponentially with time.

Finally, the function of interest, the probability that a particle survives a collision for time t but collides with another during interval t and $t + dt$ is

$$P(t)dt = \omega e^{-\omega t} dt$$ 1.54

We can calculate the average time intervals between collisions,

$$\langle t \rangle = \int_0^\infty tP(t)dt = 1/\omega = \tau$$ 1.55

We now need to find an expression for ω in terms of molecular parameters. The rate at which collisions occur would be

$$\omega = n\langle \vec{V} \rangle \Sigma$$ 1.56

where n is the number of particles per unit volume (defined earlier) and Σ is the scattering cross-section, which has units of area. An indication of the probability that a scattering event will occur is provided by Σ. $\vec{V}$ is the relative velocity between two particles, and $\vec{V} = \vec{v}_2 - \vec{v}_1$.

An expression for $\langle V \rangle$ is first sought. If we consider the mean square velocity, $\langle V^2 \rangle = \langle v_1^2 \rangle + \langle v_2^2 \rangle - 2\langle v_1 \cdot v_2 \rangle$ and recognize that $\langle v_1 \cdot v_2 \rangle = 0$, then for two identical particles $\sqrt{\langle V^2 \rangle} = \sqrt{2}\sqrt{\langle v^2 \rangle} \approx \sqrt{2}\langle u \rangle$. An expression for the Σ is obtained by realizing that if we consider two spherical particles, one of diameter d_1 and other of diameter d_2, then the probability that they will collide (assuming the absence of intermolecular interactions) is $\Sigma = 2\pi(d_1/2 + d_2/2)^2$. In other words, their cross-sectional areas must overlap for a collision to occur. If $d_1 = d_2 = d$, then $\Sigma = \pi d^2$. It follows that

$$\omega = \sqrt{2}n\langle u \rangle \pi d^2$$ 1.57

The mean free path can now be expressed in terms of molecular parameters

$$l = \langle u \rangle / \omega = 1/\sqrt{2}\pi nd^2$$ 1.58

This result is intuitive in that it tells us that the mean distance between collisions is determined by the size of the molecules and by the density of molecules.

SUMMARY   It is worthwhile at this point to briefly summarize some essential details. First, it has been demonstrated how the Partition function is used to calculate average thermodynamic quantities. Second, the Maxwell-Boltzmann velocity distribution function was derived and used to calculate average dynamical properties of the system. The fact that this distribution function is Gaussian reflects the nature of the dynamics of these noninteracting particles. Calculation of the flux and the diffusion coefficient enabled introduction of Fick's 1st law. The phenomenology of diffusional transport is now discussed within the context of Fick's 1st and 2nd laws.

1.3   The Diffusion Equations: Fick's Laws

Most realistic situations involve condensed phases, where evaluating the time-dependent evolution of the spatial distribution (or time-dependent flux) of chemical species (diffusant) in such media is of particular interest. Practical concerns might include the preservation (or shelf life) of packaged foods and beverages to the protection of electronic components from corrosion due to moisture by encapsulating them within a polymeric matrix. In these cases it would be important to evaluate the time-dependent flux of the relevant gases or moisture through the packages as part of a design and reliability program. Other practical examples include microelectronic processing, with regard to the *n* or *p*-type doping of semiconductors and the carburization of iron. In each of these cases, knowledge of the time-dependent spatial distribution of the relevant chemical species is essential for processing and performance. A third class of problems involve the *welding* of metals for structural applications, or the *welding* of layers of polymers as a stage during the production of tires or for various packaging applications, or the *joining* of ceramics for use in reactors or engines. In these cases it is critical to be able to evaluate the time-dependent concentration profile of one component as it diffuses into the other.

The diffusion equations (Fick's 1st and 2nd laws, developed during the 1800s) have proven invaluable in these more general situations. Fick's second law plays a central role in evaluation of the time-dependent spatial evolution of species in a medium due to diffusion. In what follows, we begin with additional comments regarding the 1st law in order to illustrate its practical significance and its relevance toward understanding the overall phenomenology of diffusion. Fick's 2nd law is subsequently discussed.

1.3.1   Fick's 1st Law: Additional Comments

As shown earlier, the central tenet of Fick's 1st law is that the flux of particles J (units of mass•distance/time•volume) is proportional to the gradient of the concentration, $\nabla c(\bar{r},t)$. If the distribution of particles is spatially

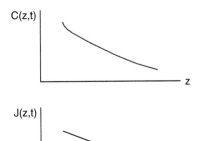

FIG. 1.7
In the absence of external driving
forces, the direction of the flux occurs
opposite to that of the concentration
gradients. The effect is to homogenize
a spatially inhomogeneous system.

inhomogeneous at time t, then in the absence of external driving forces, the particles will diffuse in order to decrease the concentration gradients (Fig. 1.7). Strictly speaking, flow is fundamentally connected to chemical potential gradients. We will address this issue in due course but for now we assume the absence of any other influences. The net flow of particles to reduce the concentration gradient is

$$\vec{J}(\vec{r},t) = -D\nabla c(\vec{r},t) \qquad 1.59$$

In general the diffusion coefficient is a tensor quantity and this is particularly important in anisotropic systems. Specifically, the equation is

$$J_i = -[D_{ij}]\frac{\partial c}{\partial x_j} \qquad 1.60$$

where the diffusion coefficient is a second-order tensor,

$$D_{ij} = \begin{bmatrix} D_{xx} & D_{xy} & D_{xz} \\ D_{yx} & D_{yy} & D_{yz} \\ D_{zx} & D_{zy} & D_{zz} \end{bmatrix} \qquad 1.61$$

Equation 1.60 may therefore be rewritten as

$$J_x = -D_{xx}\frac{\partial c}{\partial x} - D_{xy}\frac{\partial c}{\partial y} - D_{xz}\frac{\partial c}{\partial z}$$
$$J_y = -D_{yx}\frac{\partial c}{\partial x} - D_{yy}\frac{\partial c}{\partial y} - D_{yz}\frac{\partial c}{\partial z} \qquad 1.62$$
$$J_z = -D_{zx}\frac{\partial c}{\partial x} - D_{zy}\frac{\partial c}{\partial y} - D_{zz}\frac{\partial c}{\partial z}$$

This book will most often be interested in isotropic and cubic systems. In isotropic and cubic systems, the situation is less complex, all the off-diagonal terms are zero, and the diagonal terms are equal, $[D_{ij}] = D$.

1.3.1.1 Fick's 1st Law in Cylindrical Coordinates

In many situations, the geometry of the sample or the constraints (boundary conditions) on the transport process are such that use of the Cartesian coordinate system would be inappropriate. For example, the sample may possess the shape of a cylinder or a sphere. Hence it is necessary to consider Fick's laws in different coordinate systems.

Fick's first law is now considered in cylindrical coordinates. A point in the cylindrical coordinate system $P(\rho, z, \varphi)$ is related to a point in Cartesian coordinate space $P(x, y, z)$ such that $z = z$, $y = \rho\sin\varphi$, and $x = \rho\cos\varphi$; the volume element is $dV = dxdydz = \rho d\rho d\varphi dz$. The fluxes in the appropriate directions may be written

$$J_z = -D\frac{\partial c}{\partial z}$$

$$J_\rho = -D\frac{\partial c}{\partial \rho} \qquad\qquad 1.63$$

$$J_\varphi = -\frac{D}{\rho}\frac{\partial c}{\partial \varphi}$$

1.3.1.2 Fick's 1st Law in Spherical Coordinates

In the spherical coordinate system, the volume element is $dV = r^2\sin\theta dr d\theta d\varphi$ and the relation between coordinates of a point in the Cartesian and spherical coordinate systems, $P(x, y, z) = P(r, \theta, \varphi)$, is $z = r\cos\theta$, $x = r\sin\theta\cos\varphi$, $y = r\sin\theta\sin\varphi$. The fluxes in the appropriate directions are

$$J_r = -D\frac{\partial c}{\partial r}$$

$$J_\varphi = -\frac{D}{r\sin\theta}\frac{\partial c}{\partial \varphi} \qquad\qquad 1.64$$

$$J_\theta = -\frac{D}{r}\frac{\partial c}{\partial \theta}$$

We now have expressions that relate the flux of particles to the concentration gradient that are applicable to three different coordinate systems. Generally the sample geometry dictates the coordinate system that should be used. Examples in Section 1.4 will illustrate the importance of identifying the appropriate coordinate system.

1.3.2 Fick's 2nd Law

Fick's first law, while always valid within a single phase, and while useful under steady state diffusion conditions, is of limited utility. It does not provide direct information about specific time dependencies of the diffusion process. This is of obvious concern when the concentration profile in the material depends explicitly on time and on position. Fick's 2nd law should be employed to accomplish this.

To develop insight into how the spatial and temporal dependence of the concentration may be determined it is useful to consider a situation wherein the mean number of molecules per unit volume at a point varies with time (nonsteady state condition), $c = c(z,t)$. It is assumed in the following analysis that the total number of molecules in the system is conserved. Moreover, the transport process considered here occurs in one dimension; the 3-dimensional case is straightforward and is subsequently addressed. Begin by considering a thin slab of thickness dz and area A (Fig. 1.8); edge effects are ignored. If the total number of molecules is to be conserved, then the increase of the number of molecules per unit time within the slab must be equal to the difference between the total number of molecules entering one side of the slab, at location $z' = z$, per unit time and the total number of molecules per unit time exiting the other surface located at $z' = z + dz$.

This means that

$$\frac{\partial(cAdz)}{\partial t} = AJ_z(z) - AJ_z(z + dz) \qquad 1.65$$

With the use of a Taylor series expansion, this equation becomes

$$\frac{\partial c}{\partial t} dz = J_z(z) - \left(J_z(z) + \frac{\partial J_z}{\partial z} dz \right) \qquad 1.66$$

which leads to

$$\frac{\partial c}{\partial t} = -\frac{\partial J_z}{\partial z} \qquad 1.67$$

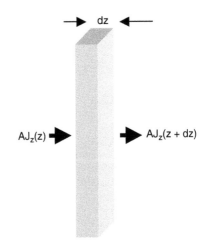

FIG. 1.8
The flow of mass across a slab of thickness dz is illustrated here. The accumulation of mass within the slab is the difference between inward and outward flow $AJ(z)dz$.

This is Fick's 2nd law; it explicitly contains the time dependence of the concentration. More generally and in three dimensions, Fick's second law may be rewritten as,

$$-\nabla \bullet \bar{j} = \frac{\partial c}{\partial t} \qquad\qquad 1.68$$

With the use of Fick's first law it is readily apparent that

$$-\nabla \bullet (-D\nabla c) = \frac{\partial c}{\partial t} \qquad\qquad 1.69$$

If the diffusion coefficient, D, is independent of concentration and location, then in one dimension Fick's second law is

$$\frac{\partial c}{\partial t} = D\frac{\partial^2 c(z)}{\partial z^2} \qquad\qquad 1.70$$

In three dimensions the 2nd law becomes

$$\nabla^2 c(\bar{r},t) - \frac{1}{D}\frac{\partial c(\bar{r},t)}{\partial t} = 0 \qquad\qquad 1.71$$

Many practical situations can be approximated reasonably well by assuming $D = constant$. Typically, if the concentration of diffusing particles is sufficiently low, wherein interactions between diffusants and interactions between the diffusants and the host environment can be ignored, then a constant D is a reasonable assumption. The concentration limits where this assumption fails will depend on the system and in some cases temperature and environmental factors.

As done for the 1st law, the 2nd law is considered in cylindrical and spherical coordinates below.

1.3.2.1   *Fick's 2nd Law in Cylindrical Coordinates*

Begin by writing down the divergence of the flux in cylindrical coordinates

$$\nabla \bullet \bar{j} = \frac{1}{\rho}\frac{\partial}{\partial \rho}(\rho J_\rho) + \frac{1}{\rho}\frac{\partial J_\varphi}{\partial \varphi} + \frac{\partial J_z}{\partial z} \qquad\qquad 1.72$$

Since the Laplacian of the concentration is

$$\nabla^2 c = \frac{1}{\rho}\frac{\partial}{\partial \rho}\left(\rho\frac{\partial c}{\partial \rho}\right) + \frac{1}{\rho^2}\frac{\partial^2 c}{\partial \varphi^2} + \frac{\partial^2 c}{\partial z^2} \qquad\qquad 1.73$$

then, in cylindrical coordinates, Fick's second law becomes

$$\frac{\partial c}{\partial t} = \frac{D}{\rho}\left[\frac{\partial}{\partial \rho}\left(\rho\frac{\partial c}{\partial \rho}\right) + \frac{\partial}{\partial \varphi}\left(\frac{1}{\rho}\frac{\partial c}{\partial \varphi}\right) + \frac{\partial}{\partial z}\left(\rho\frac{\partial c}{\partial z}\right)\right] \qquad\qquad 1.74$$

1.3.2.2 Fick's 2nd Law in Spherical Coordinates

The divergence of the flux in spherical coordinates is

$$\nabla \bullet \bar{J} = \frac{1}{r^2}\frac{\partial}{\partial r}(r^2 J_r) + \frac{1}{r\sin\theta}\frac{\partial J_\varphi}{\partial \varphi} + \frac{1}{r\sin\theta}\frac{\partial}{\partial\theta}(J_\theta \sin\theta) \qquad 1.75$$

and the Laplacian is

$$\nabla^2 c = \left(\frac{2}{r}\frac{\partial c}{\partial r} + r^2\frac{\partial^2 c}{\partial r^2}\right) + \frac{1}{r^2\sin\theta}\frac{\partial}{\partial\theta}\left(\sin\theta\frac{\partial c}{\partial\theta}\right) + \frac{1}{r^2\sin^2\theta}\frac{\partial^2 c}{\partial\varphi^2} \qquad 1.76$$

Therefore, the 2nd law in spherical coordinates is

$$\left(\frac{2}{r}\frac{\partial c}{\partial r} + \frac{\partial^2 c}{\partial r^2}\right) + \frac{1}{r^2\sin\theta}\frac{\partial}{\partial\theta}\left(\sin\theta\frac{\partial c}{\partial\theta}\right) + \frac{1}{r^2\sin^2\theta}\frac{\partial^2 c}{\partial\varphi^2} = \frac{1}{D}\frac{\partial c}{\partial t} \qquad 1.77$$

Fick's 2nd law has now been introduced. Both the 1st and 2nd laws are phenomenological and as such are devoid of information regarding the mechanism of diffusion. They are, nevertheless, of practical use and are generally applicable to a wide range of material systems, provided the appropriate diffusion boundary conditions exist. Some common examples follow; they are by no means exhaustive. The interested reader is encouraged to see the text by Crank (1975).

1.4 Simple Problems Involving Steady State Flow

Having introduced both laws, examples involving steady state, or stationary, flow across boundaries of finite thickness are now considered. As mentioned earlier, this problem is of practical significance for designing membranes, containers or packages to protect materials that are sensitive to moisture or to different gases from the environment. These equations enable the flux to be calculated directly, and knowledge of the flux enables determination of the amount of a substance that might have accumulated in a package during a specified duration of time under certain conditions.

The examples provided below involve all three coordinate systems. It will become clear that while under these conditions of stationary flow the concentration gradient ∇c across a planar layer (use of Cartesian coordinates) is constant; ∇c is nonlinear across spherical and cylindrical boundaries. Only in the case where the curvature of a boundary is such that it can be approximated as planar that the concentration gradients become constant.

1.4.1 Flow through a Planar Layer

We begin by considering the flow of particles across a membrane of thickness h, where the concentration at one end, $x = 0$, is $c = c_1$ and at the other end,

$x = h$, the concentration is $c = c_2$. Fick's 1st law in Cartesian coordinates indicates that

$$J = -D\frac{(c_2 - c_1)}{h}$$ 1.78

If $c_1 = c_2$, then there is no net flux of particles, indicating that the concentration gradient is necessary for preferential flow in one direction.

The concentration in this case varies linearly from $x = 0$ to $x = h$. This is readily seen by imposing the condition on the second law that $\partial c/\partial t = 0$ which indicates that $D\nabla^2 c = 0$. In one dimension

$$\frac{dc}{dx} = const$$ 1.79

With the above boundary conditions, the spatial dependence of composition is

$$c(x) = (c_2 - c_1)\frac{x}{h} + c_1$$ 1.80

Before going further, it might be important to realize that despite the fact that $J = 0$ when $c_1 = c_2$, the particles are constantly in motion and their behavior is characterized by local fluctuations in composition and velocity. This issue will become apparent after reading Ch. 2 where Brownian motion is discussed.

1.4.2 Steady State Flow through Nonplanar Surfaces: Cylinder

Radial flow (Fig. 1.9) of material across the interfaces of a cylindrical object (pipe, reactor, etc.) is now considered. We are interested in a hollow cylinder whose inner radius is $\rho = a$, where the concentration remains constant at c_1, and whose outer radius is $\rho = b$, where the concentration is kept constant at c_2. The stationary condition dictates that

$$\frac{1}{\rho}\frac{\partial}{\partial\rho}\left(\rho\frac{\partial c}{\partial\rho}\right) = 0$$ 1.81

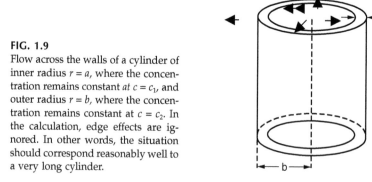

FIG. 1.9
Flow across the walls of a cylinder of inner radius $r = a$, where the concentration remains constant at $c = c_1$, and outer radius $r = b$, where the concentration remains constant at $c = c_2$. In the calculation, edge effects are ignored. In other words, the situation should correspond reasonably well to a very long cylinder.

for which the solution is

$$c(\rho) = A + B \ln \rho \qquad\qquad 1.82$$

where A and B are constants to be determined upon imposition of the boundary conditions,

$$c(\rho) = \frac{c_2 - c_1}{\ln(b/a)} \ln\left(\frac{\rho}{a}\right) + c_1 \qquad\qquad 1.83$$

In contrast to flow through a slab, where the concentration varies linearly across the thickness of the medium, the concentration exhibits a logarithmic dependence on the spatial coordinate, ρ. By extension, the concentration gradient in this case is not constant, as is the case for diffusion through a planar slab.

In fact, $c(\rho)$ only becomes approximately constant when the thickness of the layer $(b - a)$ is very small compared to a. This is readily observed when the case where $c_2 = 0$ is considered in Fig. 1.10, where $c(\rho)$ is plotted as a function of ρ/a.

1.4.3   Steady State Flow through a Spherical Interface

If flow occurs through a hollow sphere, where the inner radius is of thickness $r = a$, and the concentration is c_1, and where the outer radius $r = b$ and the concentration is kept constant at $c = c_2$, then under steady state conditions

$$c(r) = \frac{ab(c_1 - c_2)}{b - a}\frac{1}{r} + \frac{(bc_2 - ac_1)}{b - a} \qquad\qquad 1.84$$

(see Problem 29) As is the case for the foregoing situation, the concentration does not change linearly across the thickness of the sphere but varies as $1/r$.

In the aforementioned, differences between the concentration profiles under steady state flow conditions were illustrated for different geometries.

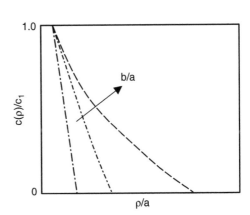

FIG. 1.10
The dependence of the concentration $c(\rho)/c_1$ is shown here as a function of the thickness of the hollow cylinder. The curvature increases as b/a increases.

In Cartesian coordinates the concentration gradient is constant across the thickness of the boundary, whereas it is not constant in spherical and cylindrical geometries. Examples involving the temporal development of the concentration profile of species subject to different boundary conditions are discussed hereafter.

1.5 Diffusion of Particles from a Point Source in One-Dimension

Fick's second law is a partial differential equation that can be solved using integral transforms. Fourier integral transforms and Laplace transforms will be used to calculate the spatial and temporal distribution of particles that diffuse into a medium from a central source. The use of integral transforms enables the conversion of a partial differential equation into a generally recognizable ordinary differential equation that can be readily solved (it is assumed that the reader is familiar with ordinary differential equations); the inverse transform of this result provides the solution of interest. Calculations will be performed in one, two, and three dimensions to illustrate subtle differences in the dynamics associated with dimensionality. A solution to the one-dimensional form of Fick's second law for the case of particles diffusing from an initial point source is first discussed.

1.5.1 Solution to Fick's 2nd Law Using Fourier Integral Transforms

The Fourier transform of a function $f(x,t)$, assuming $f(x,t)$ is well behaved and can be integrated throughout the relevant region, is

$$F(k,t) = \frac{1}{\sqrt{2\pi}} \int_{-\infty}^{\infty} f(x,t)e^{ikx}dx \qquad 1.85$$

and the inverse Fourier transform is

$$f(x,t) = \frac{1}{\sqrt{2\pi}} \int_{-\infty}^{\infty} F(k,t)e^{-ikx}dk \qquad 1.86$$

We now solve the one-dimensional diffusion equation subject to the boundary condition that $c(0) = c_0$ at $x = 0$ when $t = 0$. Stated more formally, this first boundary condition is

$$c(x,0) = c_0\delta(x) \qquad 1.87$$

where $\delta(x - a)$ is the Dirac delta function; it is equal to unity when $x = a$ otherwise it is zero; in Eq. 1.87, $a = 0$. The total amount of material present at time $t = 0$ is c_0, and in fact

$$\int_0^\infty c(x,t)dx = c_0 \qquad\qquad 1.88$$

indicating that the total amount of the diffusant remains fixed. The Fourier integral transform of $c(x,t)$ is

$$F(k,t) = \frac{1}{\sqrt{2\pi}} \int_{-\infty}^\infty c(x,t)e^{ikx}dx \qquad\qquad 1.89$$

and the integral transform of $\frac{\partial^2 c}{\partial x^2}$ is

$$F\left[\frac{\partial^2 c}{\partial x^2}\right] = -k^2 F(k,t) \qquad\qquad 1.90$$

(see Ch. 2 appendix). In addition

$$F\left[\frac{\partial c}{\partial t}\right] = \frac{\partial F(k,t)}{\partial t} \qquad\qquad 1.91$$

These transformations now permit us to rewrite Fick's second law as an ordinary differential equation,

$$\frac{\partial F(k,t)}{\partial t} + k^2 DF(k,t) = 0 \qquad\qquad 1.91$$

The general solution to this differential equation is rather straightforward and is given by

$$F(k,t) = F_0 e^{-k^2 Dt} \qquad\qquad 1.92$$

The boundary condition (Eq. 1.87) is now transformed as

$$F(k,0) = \frac{c_0}{(2\pi)^{1/2}} = F_0 \qquad\qquad 1.93$$

because the Fourier transform of $\delta(x - a)$ is $e^{-ika}/(2\pi)^{1/2}$. For the case in which $a = 0$, the Fourier transform of the boundary conditions leads to Eq. 1.93. The final stage involves calculating the inverse Fourier transform of this equation. Herewith, the inverse transform is, by definition,

$$c(x,t) = \frac{1}{(2\pi)^{1/2}} \int_{-\infty}^\infty F_0 e^{-ikx} e^{-k^2 Dt} dk \qquad\qquad 1.94$$

from which it follows that

$$c(x,t) = \frac{c_0}{\pi} \int_0^\infty e^{-k^2 Dt} \cos(kx)\,dk \qquad\qquad 1.95$$

where the relation $\cos kx = \frac{e^{kx} + e^{-ikx}}{2}$ was used. Now if you allow

$$y^2 = k^2 Dt, \qquad\qquad 1.96$$

then $dk = \frac{dy}{\sqrt{Dt}}$ and $z = \frac{x}{\sqrt{Dt}}$, so

$$c(x,t) = \frac{c_0}{\pi\sqrt{Dt}} \int_0^\infty e^{-y^2} \cos(zy)\,dy \qquad\qquad 1.97$$

This integral is solved to yield the final solution (problem 21)

$$c(x,t) = \frac{c_0}{\sqrt{4\pi Dt}} e^{-x^2/4Dt} \qquad\qquad 1.98$$

In the situation just described, the diffusion coefficient was assumed to be constant. This assumption is valid as long as the concentration of solute is sufficiently dilute, otherwise the concentration dependence of D would typically have to be accounted for. It is noteworthy that

$$P(x,t)\,dx = \frac{c(x,t)}{c_0}\,dx \qquad\qquad 1.99$$

is the probability density distribution function that describes the spatial distribution of particles undergoing one dimensional Brownian motion in a medium. The mean square displacement of a particle is readily determined from $P(x,t)$ to be

$$\langle x^2 \rangle = 2Dt \qquad\qquad 1.100$$

Earlier in this chapter, the discussion of the dynamics of the system of noninteracting particles indicated that $D = \langle u \rangle l/3$ (Eq. 1.49), which at first glance might raise a minor concern. We attempt to reconcile this by pointing out that the mean square displacement of a noninteracting particle in the x-direction could, in principle, be specified through its mean square velocity in that direction, $\xi^2 = \langle v_x^2 \rangle \langle t^2 \rangle$. Using Eq. 1.54 for $P(t)$, $\langle t^2 \rangle = 2\tau^2$ and recalling that $\langle v_x^2 \rangle = \langle v^2 \rangle/3$, then $\xi^2 = 2(\langle u \rangle^2 \tau/3)t$. If we make the approximation $\langle u \rangle^2 \approx \langle v^2 \rangle$, then it follows that $\xi^2 \approx 2Dt$. This relatively unsophisticated argument resolves this apparent discrepancy.

It is worthwhile to consider the implications of Eq. 1.98. Equation 1.98 is often called the "thin film" solution. In early experiments involving metals, very thin strips of a radioactive tracer would be placed at the surface of a metal to create a diffusion couple. The sample would subsequently be heated to allow diffusion of the radioactive element. After a sufficiently long period of time, thin strips of the sample are removed, using a lathe, and analyzed

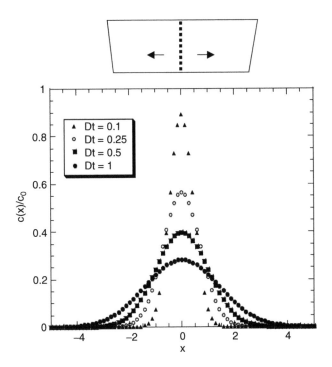

FIG. 1.11
The profile, $c(x)$, broadens with increasing Dt.

to determine the concentration based on the radioactivity. Equation 1.98 enabled subsequent determination of the diffusion coefficient. This solution (cf. Eq. 1.98) has a number of interesting features as illustrated in Fig. 1.11 and 1.12.

The concentration profile broadens with increasing time, as depicted in Fig. 1.11. A plot of $\ln(c)$ versus x^2 yields a straight line with slope $1/(4Dt)$, revealing that if t is known, D can be calculated. Moreover, the concentration at $x = 0$ decreases as $t^{1/2}$. The flux $J \propto \frac{dc}{dx} = 0$ when $x = 0$ and when $|x| \to \infty$ (Fig. 1.12). Finally the amplitude of the flux diminishes with time as the concentration profile, $c(x,t)$, broadens.

1.5.2 Solution to Fick's 2nd Law in Three Dimensions Using Laplace Transforms

Fick's 2nd law is now solved in the spherical coordinate system in 3-dimensional. A solution to this equation will be compared with the situation in one dimension. To solve this equation we will use the technique of Laplace transforms. With this technique, a partial differential equation is transformed into an ordinary differential equation for which the solution is readily recognized. The inverse transform is the desired solution.

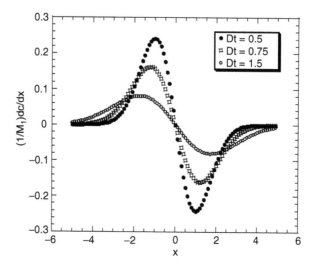

FIG. 1.12
Plots of $dc(x)/dx$ versus x are shown here for different values of Dt.

The Laplace transform of a function $f(t)$ is, by definition,

$$L[f(t)] = F(s) = \int_0^\infty e^{-st} f(t)dt \qquad 1.101$$

where $s > 0$ is a transformation variable. It is assumed that the integrand converges at large t. An inverse Laplace transform can also be performed, and it is generally unique. It is a straightforward matter to determine the Laplace transform form for simple functions. Three examples follow. The first example is the Laplace transform the function $f(t) = t^n$,

$$L[t^n] = \int_0^\infty e^{-st} t^n dt = \frac{n!}{s^{n+1}} \qquad 1.102$$

which is the well known factorial function. For the second example the function is a constant, $f(t) = c$, and the Laplace transform is c/s.

Third, the Laplace transform of a derivative is considered, and the technique of integration by parts is employed

$$L[f'(t)] = \int_0^\infty e^{-st} \frac{df(t)}{dt} dt$$

$$= e^{-st} f(t)\big|_0^\infty + s\int_0^\infty e^{-st} f(t)dt \qquad 1.103$$

$$= sF(s) - f(0)$$

Often it is convenient to use a table of Laplace transforms to evaluate different functions. Fick's 2nd law (in spherical coordinates), is first transformed into an ordinary differential equation which will be solved. An inverse transform is performed to obtain the final solution. Since only the radial evolution of the solute from the origin is of interest, then Fick's 2nd law may be written as,

$$\frac{1}{D}\frac{\partial c}{\partial t} = \frac{\partial^2 c}{\partial r^2} + \frac{2}{r}\frac{\partial c}{\partial r}$$ (1.104)

Since the boundary conditions are such that at time $t = 0$,

$$c(\bar{r},0) = c_0\delta(\bar{r})$$ (1.105)

it follows that the Laplace transform of the left-hand side of Eq. 1.104 is $(s/D)F$. With regard to the first term on the RHS of Eq. 1.104,

$$L\left[\frac{\partial^2 c}{\partial r^2}\right] = \int_0^\infty e^{-st}\frac{\partial^2 c}{\partial r^2}dt = \frac{\partial^2}{\partial r^2}\int_0^\infty ce^{-st}dt = \frac{\partial^2 F}{\partial r^2}$$ (1.106)

With this in mind, you can write down that the Laplace transform of Eq. 1.104 is,

$$\frac{\partial^2 F}{\partial r^2} + \frac{2}{r}\frac{\partial F}{\partial r} = \frac{s}{D}F$$ (1.107)

This equation may be rewritten as an ordinary second order differential equation

$$\frac{\partial^2 (Fr)}{\partial r^2} = \frac{s}{D}Fr$$ (1.108)

whose solution is

$$Fr = Ae^{\sqrt{(s/D)}r} + Be^{-\sqrt{(s/D)}r}$$ (1.109)

A and B are constants to be determined based on the boundary conditions. Since our boundary conditions dictate that at $t = 0$, $c = 0$ for large r, then $A = 0$, necessarily (the numerator increases at a much more rapid rate than the denominator decreases). This leads to

$$F = B\frac{1}{r}e^{-(r/\sqrt{D})\sqrt{s}} = B\frac{1}{r}e^{-k\sqrt{s}}$$ (1.110)

where $k = (r/D^{1/2})$. The constant B is determined by considering the boundary condition

$$\int_0^\infty c(r,t)4\pi r^2 dr = c_0$$ (1.111)

The Laplace transform of this boundary condition is

$$\int_0^\infty F(r,s)4\pi r^2 dr = \frac{c_0}{s}$$

1.112

Thus,

$$\int_0^\infty \left(B\frac{e^{-k\sqrt{s}}}{r} \right) 4\pi r^2 dr = \frac{c_0}{s}$$

1.113

Integration by parts reveals that the constant $B = c_0/4\pi D$ (see Problem 30). With the use of a table of Laplace transforms, the inverse Laplace transform is

$$L^{-1}[e^{-k\sqrt{s}}] = \frac{ke^{-k^2/4t}}{2\sqrt{\pi}t^{3/2}}$$

1.114

Consequently, in three dimensions the concentration profile is

$$c(r,t) = c_0 \frac{e^{-r^2/4Dt}}{(4\pi Dt)^{3/2}}$$

1.115

We are now in a position to compare a series of observations in three dimensions with those in one dimension. The length scale dependencies of $c(x, t)$ and of the flux are very similar in both dimensions as expected, see Fig. 1.13.

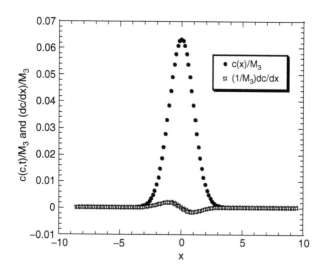

FIG. 1.13
The spatial distribution of the concentration profile and the flux are shown here for $Dt = 0.5$ in three dimensions.

The concentration profile broadens symmetrically about the origin and the flux is zero at $x = 0$ and for $x \to \infty$. On the other hand, in one dimension, the probability distribution function is given by $c(x,t)/c_0$, which also has the same shape as the concentration profile of the diffusing species. In three dimensions, the probability density function is

$$P(r,t)dr = \frac{4\pi r^2 c(r,t)}{c_0}dr \qquad 1.116$$

which differs from the one dimensional equation due to the r^2 multiplicative term. A plot of $P(r,t)$ is shown in Fig. 1.14. Whereas in one dimension $\langle x \rangle = 0$, in three dimensions $\langle r \rangle$ increases with time (Problem 23).

Using Eq. 1.116, the mean square displacement in three dimensions is readily shown to be

$$\langle r^2 \rangle = 6Dt \qquad 1.117$$

By contrast, $\langle x^2 \rangle = 2Dt$ in 1 dimension. In two dimensions, it can be shown that

$$\langle \rho^2 \rangle = 4Dt \qquad 1.118$$

It is now evident that the root mean square (RMS) displacement of a particle is determined by an important length scale, $k\sqrt{Dt}$, where the value

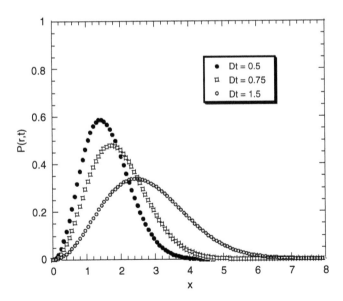

FIG. 1.14
The probability density distribution functions are plotted for different times for diffusion in three dimensions.

of k depends on the dimensionality. This $\sqrt{t}$ dependence of the dynamics is typical of particles undergoing a random walk, as discussed further in Chapter 2 (often this time dependence is identified as the so-called Fickian diffusion process). Moreover, the most probable location of a particle can be obtained by maximizing the appropriate distribution functions and in one dimension this occurs at $\langle x \rangle = 0$, but this is not the case in two and three dimensions. Why?

1.6 Concentration Profile Due to a Spatially Extended Initial Source, $f(x')$

In the foregoing section the time-dependent evolution of the concentration profile, $c(x,t)$, of a diffusant of concentration c_0 initially located at the plane $x = 0$ was determined. A similar calculation was performed in three dimensions for a solute c_0 initially concentrated at $r = 0$. In most practical situations the initial distribution is not concentrated at a plane or a point. The concentration profile, $f(x')$, is *distributed* throughout an extended region defined by x'. In principle, the time-dependent concentration profile of interest would be due to a profile that is the sum of a large number of individual planar sources (cf. Eq. 1.98) that would constitute $f(x')$. Naturally this sum of infinitesimally thin layers leads to the following solution for $c(x,t)$,

$$c(x,t) = \left(\frac{1}{4\pi Dt}\right)^{1/2} \int_{-\infty}^{\infty} f(x')e^{-(x'-x)^2/4Dt}dx' \qquad 1.119$$

As a reality check, the boundary condition $f(x') = c_0\delta(x')$ is considered, whereby the solute is concentrated in the plane $x' = 0$. In this situation Eq. 1.98 is readily recovered because $\delta(x')$ is nonzero, and is equal to unity, only when $x' = 0$.

1.6.1 Diffusion from a Semi-Infinite Source

For the second example a constant source of material, $f(x) = c_0$, located throughout the region $x > 0$ at time $t = 0$ is considered; $f(x) = 0$ for $x < 0$ (Fig. 1.15a). This situation could correspond to the doping of a semiconductor wafer with dilute concentrations of another element (the wafer is located in the vapor phase of the element), or to a polymer film absorbing moisture from its environment or for the carburization of steel. For the above equation to be valid the diffusion coefficient of the diffusant should be independent of concentration. This condition is typically met if diffusant does not react with the sample or change the structure of the sample and if the diffusant particles do not interact with each other. In fact, these conditions are typically achieved if the concentration of the diffusant is sufficiently low.

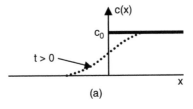

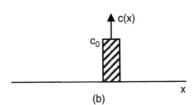

FIG. 1.15
(a) Initial concentration distribution
$f(x') = c_0$ for $x > 0$ and $f(x') = 0$ for $x < 0$;
(b) the initial concentration profile is
located between $-h < x < h$.

With the foregoing boundary conditions, Eq. 1.119 becomes

$$c(x,t) = \left(\frac{1}{4\pi Dt}\right)^{1/2} \int_0^\infty c_0 e^{-(x'-x)^2/4Dt} dx' \qquad 1.120$$

By relying on the transformation $z = \frac{x'-x}{(4Dt)^{1/2}}$, the following solution is obtained

$$c(x,t) = \frac{c_0}{2}\left\{1 + erf\left[\frac{x}{(4Dt)^{1/2}}\right]\right\} \qquad 1.121$$

where

$$erf\left[\frac{x}{(4Dt)^{1/2}}\right] = \frac{2}{\sqrt{\pi}}\int_0^{x/(4Dt)^{1/2}} e^{-z^2} dz \qquad 1.122$$

is the error function. This solution could also have been obtained using the
method of Fourier integral transforms described in 1.5.1 (see Problem 19). If
the sample thickness is finite, the solution would be valid as long as the
diffusion distance is sufficiently long yet small compared to the sample
thickness and, of course, if D is independent of composition.

1.6.2 Diffusion from a Finite Source of Thickness 2h

If the source covers only a finite location, between $-h < x < h$ (Fig. 1.15b) at
$t = 0$, then it is readily shown using Eq. 1.119, with the boundary conditions, that

$$c(x,t) = \frac{c_0}{2}\left\{erf\frac{h-x}{\sqrt{4Dt}} + erf\frac{h+x}{\sqrt{4Dt}}\right\} \qquad 1.123$$

The profile broadens symmetrically about the origin. This solution is more
appropriate than Eq. 1.98 for a thin layer of material of thickness $2h$ placed

at the center of two semi-infinite layers. The reader should solve the related problem of a film of thickness h diffusing in one direction into a medium of semi-infinite thickness.

The diffusion coefficient is extracted from the experimental concentration profile $c(x)$, which may be measured using one of a number of available techniques, by comparing $c(x)$ with the theoretical profile obtained by solving the diffusion equation subject to the appropriate boundary conditions. From an experimental perspective, the diffusing species are typically labeled so that they may be identified separately from the host environment. One of the oldest methods used to determine $c(x)$ is the radio tracer technique. A primary limitation of this technique is that the spatial (depth) resolution is poor; it is on the order of many microns. This means that samples have to be processed over sufficiently long periods of time to allow diffusion to occur over an appropriate distance before D may be determined. Typical diffusion coefficients in materials may range from 10^{-4} to 10^{-17} cm^2/s, indicating that for a diffusion distance of 10 microns, the diffusion time scales required may vary from tens of seconds to ~10^{28} seconds. Obviously the radioactive tracer technique is restricted to measuring only the faster diffusion rates.

Other techniques such as Rutherford Backscattering Spectrometry (RBS) or Secondary Ion Mass Spectrometry (SIMS) yield information about the concentration profile with depth resolutions on the order of nanometers. In RBS, a monoenergetic beam (MeV energy range) of particles is directed at a sample. A fraction of the projectiles are backscattered and the backscattered particles provide information about the depth distribution and composition of the target atoms from which they were backscattered. In SIMS, the beam, typically composed of heavier ions of lower energy than RBS, sputters atoms from the target and the ejected ions are analyzed and the concentration profiles determined. With regard to the use of RBS and SIMS, the diffusants do not necessarily have to be labeled, particularly if they are sufficiently different from the host. The interested reader is referred to references at the end of this chapter.

1.6.3 Desporption/Absorption of a Species from a Sample of Finite Dimensions

In Section 1.4, stationery (steady-state) solutions to the diffusion equations were considered. For this example, the time-dependent evolution of the concentration profile of a species diffusing within a planar sample of thickness h is considered (edge effects are ignored). The sample may have absorbed the species from the environment or it may be in the process of desorbing material into the environment. Regardless of the situation, the only constraints are that the sample of interest is of finite thickness, h, and D is constant. It will be shown that at long times the solution we obtain is identical to the stationary solution obtained in Section 1.4.

The solution to the diffusion equation may be obtained using the standard separation of variables technique where the solution is a product of a function of x, $X(x)$ and a function of time, $T(t)$,

$$c(x,t) = X(x)T(t) \qquad 1.124$$

The following boundary conditions apply

$$c(0,t) = c_1 \quad (i.e.:t \geq 0)$$

$$c(h,t) = c_2 \quad (i.e.:t \geq 0) \qquad 1.125$$

$$c(x,0) = c_0 \quad 0 < x < h \quad (t = 0)$$

The relevant equation for $c(x,t)$ may be shown to be (see Problem 19)

$$c(x,t) = c_1 + (c_2 - c_1)\frac{x}{h} + \frac{2}{\pi}\sum_{n=1}^{\infty}\frac{c_2\cos n\pi - c_1}{n}\sin\frac{n\pi x}{h}e^{-D(n\pi)^2 t/h^2}$$

$$+ \frac{4c_0}{h}\sum_{n}^{\infty}\sin\frac{(2n+1)\pi x}{h}e^{-D((2n+1)\pi)^2 t/h^2}$$

$$1.126$$

This solution is readily reconciled with the stationary solution as follows. The flux of the diffusing species develops over time and subsequently reaches its steady state value for $t \to \infty$. Upon invoking this condition, together with the boundary condition $c(x,0) = 0$, it follows that $c = c_1 + (c_2 - c_1)x/h$ and Eq. 1.80 is recovered. The flux may be obtained with the use of Fick's 1st law.

1.6.4   Permeation Experiments

It is worthwhile to consider a specific case where a gas diffuses into a membrane at an interface located at $x = 0$ and diffuses out the other side located at $x = h$ (Fig. 1.16). It is assumed here that at time $t = 0$, the sample is devoid of the gas, $c_0 = 0$, and at the interface $x = h$ the gas is readily evaporated so $c_2 = 0$. The concentration at $x = 0$ is maintained at c_1. The flux of material that diffuses out of the interface at $x = h$ is given by the first law,

$$J = -D\frac{\partial c(x,t)}{\partial x}\bigg|_{x=h} \qquad 1.127$$

Upon substitution of Eq. 1.126 into Eq. 1.127 and integrating with respect to time the total amount of material that diffused out of the interface at $x = h$ in time t is

$$A(t) = hc_1\left\{\frac{Dt}{h^2} - \frac{2}{\pi^2}\sum_{n=1}^{\infty}\frac{(-1)^n}{n^2}e^{-Dn^2\pi^2 t/h^2} - \frac{1}{6}\right\} \qquad 1.128$$

Kinetics, Transport, and Structure in Hard and Soft Materials

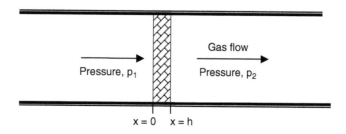

$$x = 0 \quad x = h$$

FIG. 1.16
A gas enters a membrane at $x = 0$ and exits at $x = h$. The concentration at $x = 0$ is maintained at c_1, whereas the initial concentration within the membrane and at $x = h$ is zero. The pressures to the left and to the right of the membrane are p_1 and p_2, respectively.

As $t \rightarrow \infty$, Eq. 1.128 becomes a straight line,

$$\frac{A(\infty)}{hc_1} \approx \frac{D}{h^2}t - \frac{1}{6} \qquad\qquad 1.129$$

thereby providing a convenient way to extract D. Figure 1.17 shows a typical profile of $A(t)$ as a function of Dt/h^2; at long times it is linear.

1.6.5 Time-Dependent Fluxes: Weight Gain Experiments

This particular example is presented primarily to illustrate an alternate procedure that may be used to extract the diffusion coefficient in lieu of directly measuring $c(x,t)$. It involves measuring the change in mass of the sample as a function of time in the vapor environment of a solute. The case where the concentration of solute in the sample is c_0 is considered.

The total amount of material (mass, $M(t)$) crossing the plane $x = 0$ and absorbed per unit time by the sample is

$$\frac{dM(t)}{dt} = J = -D\frac{\partial c(x,t)}{\partial x}\bigg|_{x=0} \qquad\qquad 1.130$$

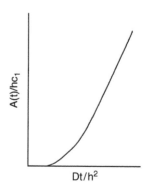

FIG. 1.17
The time dependence of $A(t)$ is shown here.

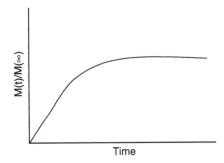

FIG. 1.18
The increase in mass of the sample
with increasing time.

The mass of substance absorbed by the sample at time t is

$$\frac{M(t)}{M(\infty)} = 1 - \frac{8}{\pi^2} \sum_{n=0}^{\infty} \frac{e^{-D(2n+1)^2 \pi^2 t / h^2}}{(2n+1)^2} \qquad 1.131$$

where $M(\infty) = h[(c_1 - c_2)/2 - c_0]$ is the mass of the sample at long times. If an experiment in which, say an elastomer, is immersed in a vapor environment, the time dependence of $M(t)$ would be similar to that illustrated in Fig. 1.18. The data in this figure could be analyzed to determine the diffusivity.

1.7 Concluding Remarks

Fick's laws enable the spatial and temporal development of the equilibrium concentration profile, $c(r,t)$, of a diffusant, subject to specified boundary conditions, to be determined. The examples discussed heretofore involved situations in which the diffusion coefficient remained independent of concentration and of position; the concentration dependence of diffusion will be dealt with in the section on interdiffusion which appears later in Ch. 10. The diffusion coefficient was determined by comparing the experimental and theoretical concentration profiles.

Specifically for the absorption of gases, the diffusion coefficient may be extracted from the time-dependent mass uptake in weight gain experiments, as described earlier. In other cases, the quantity of species that crosses a layer of material (membrane) is of particular interest. In this case, a sensor is placed at the other side of a membrane to determine the amount of the species that crosses the barrier. In this case the flux (more specifically the permeation coefficient in Problem 31) is of particular interest.

Apart from measuring concentration profiles or measuring weight uptake, a third class of experiments relies on the existence of local compositional fluctuations that characterize the dynamics. With regard to the use of these techniques, the scattering intensity of light, or neutrons, provides information about the structural evolution of the system and this enables determination of the diffusion coefficient. Scattering techniques are described in the next chapter where fluctuations are discussed. Finally, diffraction techniques may be employed. Specifically, alternating A/B/A/B ... layers of material of well defined thickness are use to create a multilayered structure from which a beam is diffracted. The diffracted beam reflects the thickness of the layers and is sensitive to interdiffusion between the layers.

The discussions in this chapter revealed an important connection between the root mean square displacement, the diffusion coefficient and the time, $\langle r^2 \rangle \propto \sqrt{Dt}$. The $\sqrt{t}$ dependence of the RMSD is a signature of long-range center of mass diffusion, often identified as Fickian diffusion. The significance of the $\sqrt{t}$ dependence of the RMS displacement will become evident in the next chapter when Brownian motion is discussed (other time-scale sub-Fickian dynamical processes are discussed in subsequent chapters). To this end, aspects of the discussion which appeared in the earlier sections of this chapter regarding the dynamics of a system of noninteracting particles have in fact provided the foundation for the future discussions on fluctuations and Brownian dynamics.

1.8   Problems

1. Using the probability distribution function

$$P(t)dt = \omega e^{-\omega t} dt$$

   a) show that $\int_0^\infty P(t)dt = 1$
   b) calculate $\langle t \rangle$ and $\langle t^2 \rangle$

2. Starting with the following equation,

$$\langle E(N_e, V_e, T) \rangle = \frac{\sum_i E_i(N_e, V_e)e^{-E_i/kT}}{\sum_i e^{-E_i/kT}}$$

   (a) Show that

$$d\langle E \rangle = -\left(kT \sum_i \ln P_i + \ln Z \right)dP_i + \sum_i P_i \left(\frac{\partial E_i}{\partial V} \right)_N dV$$

   (b) In addition show that this equation becomes

$$d\langle E \rangle = \frac{d\left(\sum_i P_i \ln P_i \right)}{\beta} + \langle p \rangle dV$$

where $-p_i dV = dE_i$ is the pV-work done on the system to increase the volume by dV and

$$\langle p \rangle = \sum_i p_i P_i$$

(c) Finally, show that $S = \frac{\langle E \rangle}{T} + k \ln Z$

3. Determine the heat capacity and entropy for the system of noninteracting particles using the Partition function. In addition, compare your answers with the predictions using classical thermodynamics.

4. Show that the integral $I(0) = \int_0^\infty e^{-ax^2} dx = \frac{1}{2}(\frac{\pi}{a})^{1/2}$

Hint: Take the square of the integral and transform to polar coordinates. In addition, solve $I(n) = \int_0^\infty e^{-ax^2} x^n dx$ for $n = 2$ and $n = 4$

5. Using the Maxwell-Boltzmann Distribution function, calculate the fraction of molecules with x-component of velocity between

$$\pm n \left(\frac{2kT}{m} \right)^{1/2} \quad \text{for } n = 1 \text{ and } n = 2.$$

6. Prove that the full width at half maximum for a Gaussian function

$$P(x) = \left(\frac{1}{2\pi\sigma^2} \right)^{1/2} \exp\left[\frac{-x^2}{2\sigma^2} \right] \quad \text{is } 2\sigma.$$

7. Using the result that expresses the connection between the average energy, $\langle E \rangle$, and the partition function, Z,

$$\langle E \rangle = -\frac{\partial \ln Z}{\partial \beta},$$

show that $-\frac{\partial \langle E \rangle}{\partial \beta} = \sigma_E^2 = \langle (E - \langle E \rangle)^2 \rangle$, where σ_E^2 is the square of the standard deviation of the energy.

8. Consider a system of harmonic oscillators. The energy of a harmonic oscillator is $E = h\nu$, where h is Planck's constant, ν is the frequency, and k is Boltzmann's constant. If $h\nu \ll kT$, show that $\sigma_E = kT$. (see Problem 7).

9. Show that $\langle v_x \rangle = 0$.

10. Calculate $\langle v_x^2 \rangle$ and show that $\langle (\Delta v_x)^2 \rangle = \langle v_x^2 \rangle$

11. Show that $\int_0^\infty F(u) du = n$

12. Show that $\frac{f(v)d^3 \bar{v}}{n} = \frac{h(v_x)dv_x}{n} \frac{h(v_y)dv_y}{n} \frac{h(v_z)dv_z}{n}$

13. The mean number of particles with speeds between u and $u + du$ is

$$F(u)du = 4\pi n \left(\frac{m}{2\pi kT} \right)^{3/2} u^2 e^{-\frac{mv^2}{2kT}} du$$

Calculate the following quantities:

  a) the mean speed u,
  b) the mean square speed $\langle u^2 \rangle$ and
  c) the most probable speed.
  d) Plot $F(u)/n$ versus $u/(2kT/m)^{1/2}$ and identify the mean speed, the root mean square speed and the most probable speed on the graph.

14. Compute the average speeds of nitrogen and oxygen molecules at $T = 300\ K$ and compare them with the speed of sound. Use the Maxwell Boltzmann distribution.

15. The pressure on a wall is the force per unit area, $\langle p \rangle = \langle F \rangle / dA$. The particles traveling to the left and to the right bombarding the wall contribute to the pressure on the wall. The force due to particles traveling in the z-direction, $v_z > 0$ (say, to the right)

$$\langle F_{z>0} \rangle = \int_{v_z > 0} f(v)v_z dA(m\bar{v})d^3\bar{v}$$

In the other direction, $v_z < 0$.

$$\langle F_{z<0} \rangle = \int_{v_z < 0} f(v)v_z dA(m\bar{v})d^3\bar{v}$$

The net force, $\langle F \rangle$, is given by the difference between the two forces,

$$\langle F \rangle = \langle F_{z>0} \rangle - \langle F_{z<0} \rangle = m \int_{-\infty}^{\infty} f(v)v_z^2 dA d^3\bar{v}$$

  (a) Show that the average pressure exerted on the wall is $\langle p \rangle = nm\langle v_z^2 \rangle$.
  (b) Show that $\langle p \rangle = nkT = \frac{NkT}{V}$. (The ideal gas law).
  (c) Why does this answer make sense?

16. Solve Fick's second law (in one dimension) using Fourier integral transforms subject to the boundary condition that

$$f(x) = \begin{cases} 0 \text{ for } x < 0 \\ c_0 \text{ for } x > 0 \end{cases} \quad t > 0.$$

17. In the presence of a driving force, Fick's first law is given by

$$J = -D\partial c/\partial x + c\langle v \rangle$$

where $\langle v \rangle$ is a drift velocity. Show that Fick's second law can now be written as

$$\frac{\partial c}{\partial t} = D\frac{\partial^2 c}{\partial x^2} - \langle v \rangle \frac{\partial c}{\partial x}$$

18. Using the following equation

$$c(x,t) = \left(\frac{1}{4\pi Dt}\right)^{1/2} \int_{-\infty}^{\infty} f(x')e^{-(x'-x)^2/4Dt}dx'$$

where $f(x')$ is the initial concentration distribution (at $t = 0$), determine the solutions for the following two sets of boundary conditions

a) $c = 0$ for $x < 0$ and $t = 0$

$$c = c_0 \text{ for } x \geq 0 \text{ and } t = 0$$

b) $c = c_0$ for $x > 0$ and $t = 0$

$$c = c_1 \text{ for } x < 0 \text{ and } t = 0$$

19. The concentration profile $c(x,t)$ equation 1.124,

$$X(x)T(t) = \sum_{k=0}^{\infty}(A_k \sin \lambda_k x + B_k \cos \lambda_k x)e^{-\lambda_k^2 Dt}$$

If the boundary conditions are such that

$$c(0,t) = c_1$$
$$c(h,t) = c_2$$
$$c(x,0) = c_0 \quad 0 < x < h$$

Determine $c(x,t)$, eqn 1.126.

20. Consider the removal of water vapor from a polymer film of thickness h. Initially, the concentration of vapor is uniform throughout the sample,

$C = C_0$ for $0 < x < h$ at $t = 0$. $C = 0$ at $x = 0$ and at $x = h$ for $t > 0$.

a) Using the method separation of variables, show that the concentration profile, $c(x,t)$, is given by:

$$c(x,t) = \frac{4c_0}{\pi}\sum_{j=1}^{\infty}\frac{1}{2j+1}\sin\left(\frac{2j+1}{h}\pi x\right)\exp\left[-\left(\frac{(2j+1)\pi}{h}\right)^2 Dt\right]$$

b) Compare this solution with the error function solutions for various times (early and late) and comment on your results.

c) Determine the average composition using the relation

$$\overline{c(t)} = \frac{1}{h}\int_{0}^{h}C(x,t)dx$$

21. Starting with

$$c(x,t) = \frac{c_0}{\pi\sqrt{Dt}} \int_0^\infty e^{-y^2} \cos(zy)dy$$

show that

$$c(x,t) = \frac{c_0}{\sqrt{4\pi Dt}} e^{-x^2/4Dt}$$

22. Solve the diffusion equation (2nd law) in two dimensions and determine the mean square displacement of a particle.

23. Calculate the most probable location of a particle in one, two, and three dimensions. Discuss your results in relation to the mean square displacement in each direction.

24. Imagine that you have a job at applied materials doing microelectronic processing. You are working on a project that requires you to diffuse indium into silicon. The specifications are that at a depth of 0.001 cm beneath the surface the concentration of indium must be one-half its value at the surface. You decide that the best way to accomplish this is to heat the sample in the presence of indium vapor at 1600°C. How long will it take to accomplish this?

25. A rubber sample of thickness 0.3 cm and mass 800 gm was placed in a humid environment for many hours. The data below shows the increase in mass of the sample as a function of time. Use this data to determine the diffusion coefficient.

Time (hrs)	Uptake (mg)
1.0000	2.0000
2.0000	4.0000
3.0000	7.0000
4.0000	8.0000
5.0000	10.000
6.0000	12.000
7.0000	14.000
8.0000	14.000
9.0000	15.000
10.000	20.000
11.000	21.000
12.000	22.000
3.0000	23.000
15.000	23.500
16.000	24.000
19.000	23.500
20.000	25.000
29.000	24.000
40.000	23.000
42.000	23.500
44.000	23.500

26. A thin film of radioactive copper is electroplated at the end of a copper cylinder. After annealing at a high temperature for 20 hours, the specimen was sectioned and the activity determined in each section,

Activity (counts/min/mg)	Average distance from end (10^{-2} cm)
5,012	1
3,980	2
2,512	3
1,414	4
525	5

Determine the diffusion coefficient of the tracer.

27. A 4 mm thick sheet of nickel has 6 at% silicon dissolved in it. The sheet is sandwiched between two infinitely thick slabs of nickel and heated to 800°C. The diffusion coefficient of Si in Ni is 6.8×10^{-9} cm²/s at this temperature. Calculate the total amount of Si that diffused out of the center of the sheet after 12 hours.

28. Using Eq. 1.124 show that $\frac{M(t)}{M(\infty)} = 1 - \frac{8}{\pi^2}\sum_{n=1}^{\infty}\frac{e^{-D(2n+1)^2\pi^2 t/h^2}}{(2n+1)^2}$

29. Consider flow of a substance across a spherical shell, of inner radius a and outer radius b. If the concentration at $r = a$ is c_1, and at $r = b$ and the concentration is kept constant at $c = c_2$, then show that under steady state conditions

$$c(r) = \frac{ab(c_1 - c_2)}{b - a}\frac{1}{r} + \frac{(bc_2 - ac_1)}{b - a}$$

Under what conditions does $c(r)$ exhibit a linear dependence on r?

30. Starting with the boundary condition,

$$\int_0^\infty \left(B\frac{e^{-k\sqrt{s}}}{r}\right)4\pi r^2 dr = \frac{c_0}{s}$$

Show that $B = c_0/4\pi D$.

31. When considering flow across a membrane of thickness h, the concentration within the membrane is linear, $c(x) = (c_2 - c_1)\frac{x}{h} + c_1$ Show that this equation may be derived subject to the boundary conditions that at $x = 0$, $c = c_1$ and at $x = h$, $c = c_2$ for $t \geq 0$. In many practical situations involving permeation, the experiment is designed such that the pressures of the gas as $x < 0$ and $x > h$ are easily measured. If the solubility of the gas in the membrane is S, then the concentration of the gas within the membrane is $c = Sp$, where p is the pressure within the material. In the pressure of the gas is p_1 and p_2, in the regions $x < 0$ and $x > h$, respectively, show that $J = P\frac{(p_1 - p_2)}{h}$ where the permeation coefficient $P = DS$.

1.9   References

R.K. Pathria, *Statistical Mechanics*, Pergamon Press, Oxford, UK, 1986.

J. Crank, *The Mathematics of Diffusion*, Oxford University Press, 1975.

F. Reif, *Fundamentals of Statistical and Thermal Physics*, McGraw Hill, New York, 1965.

Tyn. Mynt, *Partial Differential Equations of Mathematical Physics*, North Holland, NY, 1980.

Donald A. McQuarrie, *Statistical Mechanics*, University Science Books, 2000.

Handbook of Modern Ion Beam Analysis, edited by J.R. Tesmer and M. Nastasi, Materials Research Society Press, Pittsburgh, PA, 1995.

Secondary Ion Mass Spectrometry: Principles and Applications, J.C. Vickerman, A. Brown and N.M. Reed, Oxford University Press, 1989.

1.10   Appendices

1.10.1   Integrals

$$I(n) = \int_0^\infty e^{-ax^2} x^n dx = \frac{1}{2}\Gamma\left(\frac{n+1}{2}\right)a^{-(n+1)/2}$$

where the Gamma function possesses values of $\Gamma(1)=1; \Gamma(1/2)=\pi^{1/2}; \Gamma(n)=(n-1)!$ and $\Gamma(n+1)=n\Gamma(n)$.

$$I(0) = \int_0^\infty e^{-ax^2} dx = \frac{1}{2}\left(\frac{\pi}{a}\right)^{1/2}$$

$$\Gamma(n) = \int_0^\infty e^{-x} x^{n-1} dx$$

1.10.2   Fourier Integral Transforms of Derivatives

a) The integral transform of $F[f''(t)]$

We are interested in a function, $f(t)$, that is integrable and differentiable. It is further assumed that the function converges at $|t| \to \infty$ The Fourier transform of $f'(t)$ is

$$F[f'(t)] = -ikF[f(t)] \qquad \text{A.1}$$

because

$$F[f'(t)] = \frac{1}{\sqrt{2\pi}} \int_{-\infty}^{\infty} f'(t)e^{ikt}dt$$

$$= \frac{1}{\sqrt{2\pi}} \left[f(t)e^{ikt}\Big|_{-\infty}^{\infty} - ik \int_{-\infty}^{\infty} f(t)e^{ikt}dt \right] = -ikF[f(t)] \qquad \text{A.2}$$

Generally it can be shown that $F[f^{(n)}(t)] = (-ik)^n F[f(t)]$

b) *The inverse transform of a product of two functions, $H(k) \bullet F(k)$*

The convolution theorem states that

$$\frac{1}{\sqrt{2\pi}} \int_{-\infty}^{\infty} H(k) \bullet F(k)e^{-ikx}dk = \frac{1}{\sqrt{2\pi}} \int_{-\infty}^{\infty} f(x-\xi)h(\xi)d\xi \qquad \text{A.3}$$

2

Brownian Motion

2.1 Introduction

Under equilibrium conditions, the dynamics of a dilute concentration of particles of microscopic dimensions immersed in a liquid at a temperature T are random, and if the average velocity of a particle is measured over a sufficiently long time interval, for example in the x-direction, it would be zero, $\langle v_x \rangle = 0$. This is a consequence of the fact that the particle can move in any direction with equal probability. Whereas the average velocity is zero, the velocity of the particle at a given instant is typically not zero because fluctuations of the velocity occur and are specified by $\langle (\Delta v_x^2) \rangle = \langle v_x^2 \rangle = \frac{kT}{m}$. These fluctuations increase with temperature, and more massive objects experience smaller fluctuations. The random, statistically fluctuating, and incessant motions of the particles in the liquid typify the phenomenon of Brownian motion.

What are believed to be the first well-documented observations of this phenomenon were made in 1828, by an English Botanist, Robert Brown, after whom the effect is named. Brown made careful observations of the motions of pollen grains in water using an optical microscope. He reported that the motions of the pollen grains were incessant and that their behavior could not be reconciled with currents in the fluid or with evaporation. We now know that the dynamics of these particles manifest the random incessant bombardment by the molecules in the liquid.

If measurements of the displacements of a tiny particle in a liquid during fixed time intervals were to be performed, a distribution function that characterizes its dynamics could be constructed. Specifically, two parameters would be of interest: 1) the magnitude and direction of the displacement of a particle, Δx, during each fixed interval Δt; and 2) the number, n, of occurrences of such displacements. A plot of n (Δx) versus Δx, assuming that the experimental conditions are appropriate, would be Gaussian (see Problem 1)! The phenomenon of Brownian motion is observed in colloidal suspensions, smoke molecules in air, and a host of other situations.

It would be hard to fully appreciate microscopic mechanisms of diffusional transport in materials, the subject of Chapters 3 through 8, without understanding the phenomenon of Brownian motion. The diffusion of a particle within a medium is influenced by its interactions with neighboring particles that induce correlations in its location and its dynamics. To this end, two functions are introduced: one of which is the time autocorrelation function, which provides a measure of the extent to which the value of a dynamical variable at time t' is correlated by its value at an earlier time t. The second is the structure factor, which provides information regarding the structural organization of particles within the system. The Langevin analysis, which describes the effects of incessant forces on the dynamics of the Brownian particle from molecules in the medium, is subsequently introduced.

This chapter begins with an analysis of the random walk problem and a further discussion of distribution functions in order to provide insight into the time dependencies for the particle mean square displacements discussed in Chapter 1. Moreover, this provides a context for the discussions of correlation functions, structure factor, and the Langevin analysis. The discussions of these functions provide a foundation for the final topic covered in this chapter, scattering methods, used to study the structure and dynamics of materials.

2.2 The Random Walk Problem

2.2.1 Binomial Distribution Function

Begin by considering the one-dimensional motion of a particle, initially located at the origin $x = 0$ at time $t = 0$. It is assumed that the particle undergoes N independent displacements, each of magnitude L. After a large number of steps, N, the particle could reside within a range of locations, between $-NL$ to NL. The probability that the particle would reside at an extreme location is, or course, vanishingly small. We are interested in the probability, $W_N(m)$, that after N steps the particle is located at the point $x = mL$. In this model, n_r steps are taken to the right and n_l steps are taken to the left, so $n_l + n_r = N$ and $m = n_r - n_l$. The probability that a step is taken to the right is p_r and to the left the probability is p_l. The probability that n_r steps are taken to the right in any sequence and n_l steps are taken to the left in any sequence is the product of the probabilities of all steps, $p_r^{n_r} p_l^{n_l}$. Note, however, that there are many ways that N steps can occur with n_r steps to the right and n_l to the left. The number of ways by which this can be accomplished are $\frac{N!}{n_r! n_l!}$. Herewith

$$W_N(n_r) = \frac{N!}{n_r! n_l!} p_l^{n_l} p_r^{n_r} \qquad\qquad 2.1$$

Since $W_N(n_r)$ is a probability distribution function, it must satisfy the normalization condition,

$$\sum_{n_r = 0}^{N} W_N(n_r) = 1 \qquad\qquad 2.2$$

Using Eq. 2.1 and 2.2,

$$\sum \frac{N!}{n_r!(N-n_r)!}p_r^{n_r}p_l^{(N-n_r)} = (p_r + p_l)^N = 1 \qquad 2.3$$

because $p_r + p_l = 1$. Equation 2.3 is the well-known binomial theorem. If one had to ask what the average number of steps that the particle makes to the right was, $\langle n_r \rangle$, then one would surmise that the answer was the product of the total number of steps, N, and the probability that a step was taken to the right,

$$\langle n_r \rangle = Np_p \qquad 2.4$$

This result may be verified using Eq. 2.1,

$$\langle n_r \rangle = \sum_{n_r=0}^{N} n_r W_N(n_r) = \sum_{n_r=0}^{N} \frac{N!}{n_r!(N-n_r)!}p_r^{n_r}p_l^{(N-n_r)}n_r \qquad 2.5$$

The solution becomes apparent by realizing that Eq. 2.5 can be rewritten as

$$\langle n_r \rangle = \sum_{n_r=0}^{N} \frac{N!}{n_r!(N-n_r)!}\left(p_r \frac{\partial(p_r^{n_r})}{\partial p_r}\right)p_l^{(N-n_r)} \qquad 2.6$$

$$= p_r \frac{\partial}{\partial p_r}\left[\sum_{n_r=0}^{N} \frac{N!}{n_r!(N-n_r)!}p_r^{n_r}p_l^{N-n_r}\right]$$

Since the term within square brackets is $(p_r + p_l)^N$, Eq. 2.4 follows. The average number of steps taken to the left is

$$\langle n_l \rangle = Np_l \qquad 2.7$$

From Eq. 2.4 and 2.7, it follows that

$$\langle n_l \rangle + \langle n_r \rangle = N \qquad 2.8$$

which is an intuitive result. The average value of the net displacement, $m = n_l - n_r$, is

$$\langle m \rangle = \langle n_r - n_l \rangle = \langle n_r \rangle - \langle n_l \rangle = N(p_r - p_l) \qquad 2.9$$

The dispersion of n_r is

$$\langle (\Delta n_r)^2 \rangle = Np_r p_l \qquad 2.10$$

and that of $m = n_r - n_l$ is

$$\langle (\Delta m)^2 \rangle = 4Np_r p_l \qquad 2.11$$

2.2.2   One-Dimensional Random Walk: Diffusion

A connection between the mean square displacement and the time is now sought. Equation 2.1 can be rewritten in terms of the net displacement, $m = n_l - n_r$, (to the right). As $N = n_r + n_l$, then $n_r = 1/2(N - m)$ and $n_l = 1/2(N + m)$. Upon substitution, Eq. 2.1 becomes

$$W_N(m) = \frac{N!}{[(N+m)/2)]![(N-m)/2]!} p_r^{(N-m)/2} p_l^{(N+m)/2} \qquad 2.12$$

If the probability of a step taken to the left is equal to that of a step taken to the right, then $p_r = p_l = 1/2$ (since $p_r + p_l = 1$) and Eq. 2.12 can be rewritten as

$$W_N(m) = \frac{N!}{[(N+m)/2)]![(N-m)/2]!} \left(\frac{1}{2}\right)^N \qquad 2.13$$

It is readily confirmed that

$$\langle m \rangle = 0 \qquad 2.14$$

and that

$$\langle m^2 \rangle = N \qquad 2.15$$

This equation indicates that the root mean square displacement $\langle m^2 \rangle^{1/2} = N^{1/2}$. Moreover, since $x = mL$, then $\langle x \rangle = 0$, a result that is not unexpected because occurrences to the right occur with equal probability as occurrences to the left. The mean square displacement of a particle performing a random walk in one dimension is

$$\langle x^2 \rangle = NL^2 \qquad 2.16$$

revealing that the root mean square displacement is proportional to $N^{1/2}$.

We can account for the time dependence in an ad hoc fashion, for now, by specifying that if each jump occurs between time interval τ^*, then after time t the number of jumps is $N = t/\tau^*$. This will be dealt with formally in Section 2.4. Therefore, the root mean square displacement of the particle after time t is

$$x^2 = 2(L^2/2\tau^*)t \qquad 2.17$$

where $L^2/2\tau^*$ is the diffusion coefficient, $D = L^2/2\tau^*$. That $x \propto t^{1/2}$ is indicative of dynamics characterized by a large number of statistically random and independent events. This is the same result obtained for a particle diffusing in one dimension (see Chapter 1).

2.2.3   The Gaussian Distribution Function

It turns out that for $N \gg m$, Eq. 2.13 becomes the Gaussian distribution. This is illustrated using Stirling's approximation for moderately large N,

$$\ln N! = N\ln N - N + 1/2\ln 2N\pi + \cdots \qquad 2.18$$

which leads to the following

$$\ln W_N(m) \approx N \ln N - N - N \ln 2 -$$

$$\left[\frac{N+m}{2} \ln\left[\frac{N}{2}\left(1+\frac{m}{N}\right)\right] - \frac{N+m}{2} + \frac{N-m}{2} \ln\left[\frac{n}{2}\left(1-\frac{m}{N}\right)\right]\right] - \frac{1}{2}\ln 2N\pi \qquad 2.19$$

The terms involving $\log(1 \pm x)$ can be expanded into a Taylor series $\pm x - x^2 + \dots$, where $x = m/N$, since $N \gg m$. Equation 2.19 can now be simplified

$$\ln W_N(m) \approx -\frac{1}{2}\ln 2\pi N - \frac{m^2}{2N} \qquad 2.20$$

and, upon manipulation and substituting $x = mL$, the Gaussian distribution function is evident

$$W_N(x) = \left(\frac{1}{2\pi NL^2}\right)^{1/2} e^{-\frac{x^2}{2NL^2}} \qquad 2.21$$

From inspection, $\langle x \rangle = 0$ and $\langle x^2 \rangle = NL^2$, as expected. If x is treated as a continuous variable, then Eq. 2.21 can be written in the form of a probability density function,

$$P(x)dx = \frac{1}{\sqrt{4\pi Dt}} e^{-\frac{x^2}{4Dt}} dx \qquad 2.22$$

It is noteworthy that despite the approximations, the normalized distribution $P(x)dx$ is unity,

$$\int_{-\infty}^{\infty} P(x)dx = 1 \qquad 2.23$$

Further, the mean square displacement is

$$\langle x^2 \rangle = 2Dt \qquad 2.24$$

This is an important result which indicates that the root mean square displacement, $\sqrt{\langle x^2 \rangle}$, of a particle undergoing Brownian motion is proportional to $\sqrt{Dt}$. The factor of 2 is unique to one-dimensional dynamics. In three dimensions, the same basic equation is valid, except that the factor is 6, as we saw earlier. We note, parenthetically, that while this calculation was conducted for a single particle, in reality it applies to an ensemble of particles that are initially concentrated at the origin and that diffuse outward with time.

The Gaussian distribution function has appeared in a number of situations thus far in Chapters 1 and 2. It appears in other very familiar cases, such as random errors in experiments; it also describes the grade distribution for a sufficiently large class of college freshmen, among other things. The fact that this distribution appears under such diverse circumstances suggests that a "common thread" exists among the variables that characterize these otherwise unrelated events.

The *central limit theorem* of statistics provides a rationale for observations where the likelihood that variable x_i would possess a particular value is random (i.e., the specific value of a property measured at a given time, or the direction of a moving particle, in the absence of a driving force, at a given instant, or the roll of a die are random). Such events would be characterized by a Gaussian distribution. If each variable occurs with probability $p(x_i)$, where $\int p(x_i)dx_i = 1$, and if each random variable is independent, so $\langle x_i x_j \rangle = \langle x_i \rangle \langle x_j \rangle (i \neq j)$, and if σ_i, the variance of $p(x_i)$, is such that $\langle x_i^2 \rangle - \langle x_i \rangle^2 = \sigma_i^2$, then if the number of variables in the system is large then the probability density function should be Gaussian. With this in mind, it should not be surprising that this probability density distribution function characterizes the behavior of such diverse, otherwise unconnected, phenomena.

2.2.3.1   Poisson Distribution Function

Having discussed the Gaussian distribution function, it is worthwhile to make a slight detour to discuss another common distribution function, the Poisson distribution function, which is also a special case of the binomial theorem. The Poisson distribution function is generally valid when an event is rare. For example, if $p_r \to 0$, then $N \gg n_r$ and Nn_j. If we now consider the binomial distribution function and further note that if we let $\lambda = Np_r$, then with further manipulation the Poisson distribution follows from Eq. 2.1 (see Problem 7),

$$P(n) = \frac{e^{-\lambda}\lambda^n}{n!} \tag{2.26}$$

2.3   Correlation Functions

Heretofore, we have discussed distribution functions, which are useful tools for analyzing the equilibrium dynamics of particles. Time-dependent correlation functions play a central role in the analysis of transport processes. These functions quantify the extent to which two dynamic properties of a system are correlated over a period of time. The analysis of data from experiments of dynamics, such as light scattering and neutron scattering, rely on time correlation functions.

Once again, we consider a dilute gas confined within a container. The molecules constantly bombard the walls of the container, and the pressure on the

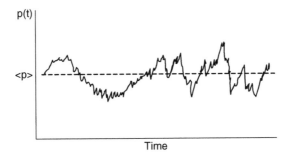

FIG. 2.1
Time dependence of the pressure in a container. It fluctuates about an average value $\langle p \rangle$.

wall fluctuates rapidly because of the incessant bombardment by individual molecules. In fact, $p(t)$ resembles a noise pattern, as illustrated in Fig. 2.1.

The noise pattern fluctuates about a mean value, $\langle p \rangle$. If we had to determine the actual pressure in the container, this task would be accomplished by taking readings at different time intervals, provided the intervals are sufficiently long compared with the time scale of the fluctuations, and subsequently averaging the measured values. The average pressure could therefore be written in terms of a time average

$$\langle p \rangle = \lim_{\Lambda \to \infty} \frac{1}{\Lambda} \int_0^{\Lambda} p(t)dt \qquad 2.27$$

where Λ is the time. Note that the initial time is necessarily arbitrary, so it is set equal to zero for convenience. The only restriction is that Λ should be sufficiently large. At equilibrium, the time average is expected to be equal to the ensemble average, which is the essence of the ergodic hypothesis of statistical mechanics.

In the above experiment we might examine values of the pressure that are taken during a sufficiently short interval, t'. For the moment, consider intervals that are short compared with the time associated with the fluctuations. If $t' \to 0$, then $p(t')$ is approximately equal to $p(0)$; at time $t = 0$, the pressure is $p(0)$. As the interval increases, the difference between the pressures measured at time $t = 0$ and $t = t'$ will increase. In the limit where the interval $t' \to \infty$, $p(0)$ is independent of $p(t')$. In other words, $p(0)$ and $p(t')$ are uncorrelated if the measurements are conducted sufficiently far apart in time. The extent to which the value of the pressure measured at time t is related to its value at later time t' is determined by the *auto-correlation* function of the pressure, which is by definition

$$\langle p(0)p(t') \rangle = \lim_{\Lambda \to \infty} \frac{1}{\Lambda} \int_0^{\Lambda} p(0)p(t')dt \qquad 2.28$$

This argument is in fact applicable to any dynamical variable, $A(t)$. Henceforth, the remainder of the discussion will involve $A(t)$ instead of $p(t)$. Further information regarding the time dependence of the autocorrelation function of $A(t)$ might be obtained by considering the following. If we evaluated events during a short time interval such that $t' \to 0$, then $\langle A(0)A(t')\rangle = \langle A(0)A(0)\rangle = \langle A^2\rangle$ (again we took $t = 0$ as the initial time). This, of course, follows from the fact that if observations are made during very close time intervals, the values of the variables would be similar. If $t' \to \infty$, that is, if observations are made at sufficiently large time intervals, then $A(0)$ and $A(t')$ are uncorrelated, which means that

$$\lim_{t'\to\infty}\langle A(0)A(t')\rangle = \langle A(0)\rangle\langle A(t')\rangle = \langle A\rangle^2 \qquad 2.29$$

The autocorrelation function decays and it can be shown that it is always true that $\langle A^2\rangle \geq \langle A(0)\rangle\langle A(t')\rangle = \langle A\rangle^2$. In fact for any initial time t and later time as $t' \to \alpha\, t'$, $\langle A^2\rangle \geq \langle A(t)\rangle\langle A(t+t')\rangle = \langle A\rangle^2$. In some situations it is known that the decay between the value of a property at long times is exponential, so

$$\langle A(0)A(t')\rangle = \langle A\rangle^2 + (\langle A^2\rangle - \langle A\rangle^2)e^{-t'/\tau}$$
$$= \langle A\rangle^2 + \sigma^2 e^{-t'/\tau} \qquad 2.30$$

where $\sigma^2 = \langle (A(t) - \langle A\rangle)^2\rangle$ is the dispersion of A (note that there is no specific stipulation that the decay need be monotonic; it could oscillate, as shown later). In Eq. 2.30, τ is the relaxation time. It characterizes the time scale beyond which the value $A(t')$ measured at time t' is no longer correlated with its value measured at earlier time $t = 0$. The time dependence of the autocorrelation function is illustrated in Fig. 2.2, where the decay is evident.

Autocorrelation functions play an important role in the analysis of dynamic properties. In a light scattering experiment, for example, the scattered intensity is determined by an autocorrelation function of the scattered field. If the

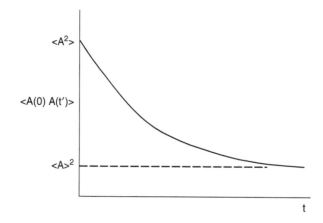

FIG. 2.2
Time dependence of the auto-correlation function for a dynamical variable.

sample is a collection of molecules, the decay of this autocorrelation function contains information about the dynamics of a molecule. In fact, the diffusion coefficient of a particle is determined by the autocorrelation function of the particle velocity. These are discussed in Sections 2.4 and 2.5.

Finally, the autocorrelation function refers to the time correlation function of the same physical quantity. A *cross-correlation function* is a time correlation function between two physical dynamical quantities A and B, $\langle A(0)B(t')\rangle$,

$$\langle A(0)B(t')\rangle = \lim_{\Lambda \to \infty} \frac{1}{\Lambda} \int_0^\Lambda A(t)B(t+t')dt \qquad 2.31$$

With this, the discussion of structure is initiated in the next section.

2.3.1 Pair Correlation Functions and the Static Structure Factor

Information regarding the spatial organization of the atomic or molecular entities that constitute the system (material, liquid, gas) is critical as it is intimately connected to dynamics. The discussion of time correlation functions and distribution functions in previous sections enables a meaningful discussion of structure and the connection between structure and dynamics in this section.

A system of N identical particles enclosed in a volume V at temperature T is considered. Since N is large it is not possible to specify the location of each particle exactly. However, the problem is readily formulated in terms of statistics and a natural question that needs to be answered is, "What is the probability that particle 1 is specifically located at position $\vec{r}_1$ within volume element $d^3\vec{r}_1$ and particle 2 at position $\vec{r}_2$ within volume $d^3\vec{r}_2$ etc.?" This is given by the probability density function

$$P^{(N)}(\vec{r}_1,\vec{r}_2...,\vec{r}_N)d^3\vec{r}_1 d^3\vec{r}_2...d^3\vec{r}_N = \frac{\int_p e^{-\beta E_N(\vec{r}_1,\vec{r}_2...\vec{r}_N)}d^3\vec{r}_1...d^3\vec{r}_N d\vec{p}_1...d\vec{p}_N}{Z}$$
$$= \frac{e^{-\beta U_N(\vec{r}_1,\vec{r}_2...\vec{r}_N)}d^3\vec{r}_1 d^3\vec{r}_2...d^3\vec{r}_N}{Z^{(N)}} \qquad 2.32$$

where Z is the classical partition function, $Z^{(N)} = \int_V e^{-\beta U_N}d^3\vec{r}_1 d^3\vec{r}_2 ... d^3\vec{r}_N$ $(d^3\vec{r}_i = dx_i dy_i dz_i)$ is the *configuration* integral and U_N is the potential energy (see Problem 13).

The *n*-particle density, $\rho^{(n)}(\vec{r}_1,\vec{r}_2...,\vec{r}_n)$, is the probability that any molecule is located at $\vec{r}_1$ within $d^3\vec{r}_1$, any molecule is located at $\vec{r}_2$ within $d^3\vec{r}_2$, and that any molecule would be located at position $\vec{r}_n$ within $d^3\vec{r}_n$ regardless of the configurations of the remaining molecules. The distribution function representing the *n*-particle density is

$$\rho^{(n)}(\vec{r}_1,\vec{r}_2...,\vec{r}_n) = \frac{N!}{(N-n)!}P^{(n)}(\vec{r}_1,\vec{r}_2,\vec{r}_3,...\vec{r}_N) \qquad 2.33$$

where $P^{(n)}(\vec{r}_1,\vec{r}_2,\vec{r}_3,...\vec{r}_N) = \frac{\int e^{-\beta U_N}d^3\vec{r}_{n+1}...d^3\vec{r}_N}{Z^{(N)}}$ is the probability density associated with the notion that particle 1 is located at position $\vec{r}_1$ within $d^3\vec{r}_1$ and particle

2 at position $\vec{r}_2$ within $d^3\vec{r}_2$, etc. The prefactor on the RHS of the equation arises from the fact that there are N ways to arrange the first particle, $(N-1)$ ways for the second and $(N-n)$ ways for the nth particle. The normalization condition is, of course $\int \rho^{(n)}(\vec{r}_1,\vec{r}_2...,\vec{r}_n)d^3\vec{r}_1...d^3\vec{r}_n = \frac{N!}{(N-n)!}$.

2.3.2 Single Particle Density Distribution Function

As an example one might ask what is the probability that any molecule would be located at position $\vec{r}_1$ within the volume element $d^3\vec{r}_1$? This is determined by the single particle density distribution function, $\rho^{(1)}(\vec{r}_1)d^3\vec{r}_1$. The integral

$$\frac{1}{N}\int \rho^{(1)}(\vec{r}_1)d^3\vec{r}_1 \qquad\qquad 2.34$$

is evaluated for an isotropic fluid where the density is uniform, $\rho^{(1)}(\vec{r}_1)=\rho$ and

$$\frac{1}{N}\int \rho^{(1)}(\vec{r}_1)d\vec{r}_1 = \frac{1}{N}\int \rho dV = 1 \qquad\qquad 2.35$$

Had the calculation been performed for a crystal then the potential, U_N, would reflect the periodicity of the crystal.

2.3.3 Pair Distribution Function

The correlation function $g^{(n)}(\vec{r}_1,\ldots,\vec{r}_n)$ for the system is defined in terms of the density distribution function,

$$\rho^{(n)}(\vec{r}_1,\ldots,\vec{r}_n)=\rho^n g^{(n)}(\vec{r}_1,\ldots,\vec{r}_n) \qquad\qquad 2.36$$

The correlation function describes the correlations between the locations of the molecules in the system. In the absence of correlations, this function $g^{(n)}$ becomes unity. The pair correlation ($n=2$) function $g^{(2)}(\vec{r}_1,\vec{r}_2)=g(r)$ is of particular significance because it provides an indication of the probability that a second molecule is located within a volume $d^3\vec{r}$ provided there is a molecule located at a distance r away. This function, which can be measured experimentally using light scattering, depends on the distance between molecules $r=|\vec{r}_{12}|=|\vec{r}_2-\vec{r}_1|$ and is related to the local density. Hence Eq. 2.36 becomes

$$\rho^{(2)}(\vec{r}_1,\vec{r}_2)=\rho^2 g^{(2)}(\vec{r}_1,\vec{r}_2)=\rho^2 g(r) \qquad\qquad 2.37$$

The number of molecules between r and $r+dr$ can readily be calculated using $g(r)$, in fact

$$\int_0^\infty \rho g(r)4\pi r^2 dr = N-1 \qquad\qquad 2.38$$

The answer $N-1$ is an indication of the fact that the molecule at the center is not counted. For a collection of spherical molecules, $g(r)$ is a damped oscillating function of position, as illustrated in Fig. 2.3. In this fig. $g(r)$ is

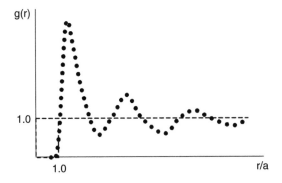

FIG. 2.3
The radial distribution function is plotted as a function of distance from a central molecule.

plotted as a function of r/a, where a is the diameter of a molecule. Note that $g(r)$ is zero as r approaches zero because of the molecule at the origin. As the distance from this central molecule increases, $g(r)$ eventually approaches a value of 1 indicating the absence of correlations with the location of the central molecule.

A significant number of measurements of dynamic processes in materials are performed using light or neutron scattering experiments. In such experiments, one parameter of primary importance is the static structure factor $S(\vec{q})$. $S(\vec{q})$ is determined by the radial distribution function $g(r)$, a shown below. The local density of a real system is not constant throughout space and is characterized by local fluctuations. The structure factor is the auto-correlation function of the Fourier components of the local density,

$$S(\vec{q}) = \frac{\langle \rho(\vec{q})\rho(-\vec{q}) \rangle}{N} \qquad \qquad 2.39$$

where

$$\rho(\vec{q}) = \int e^{-\vec{q}\cdot\vec{r}} \rho(\vec{r}) d^3\vec{r} \qquad \qquad 2.40$$

is the Fourier transform of the local density, $\rho(\vec{r})$. The local density at $\vec{r}$ is

$$\rho(\vec{r}) = \sum_{i=1}^{N} \delta(\vec{r} - \vec{r}_i) \qquad \qquad 2.41$$

Moreover, because

$$\langle \delta(\vec{r} - \vec{r}_1) \rangle = \frac{\int \delta(\vec{r} - \vec{r}_1) e^{-U_N(\vec{r}_1,\vec{r}_2...\vec{r}_N)/kT} d^3\vec{r}_1 d^3\vec{r}_2 ... d^3\vec{r}_N}{Z^{(N)}}$$

$$= \frac{\int e^{-U_N(\vec{r}_1,\vec{r}_2...\vec{r}_N)/kT} d^3\vec{r}_2 ... d^3\vec{r}_N}{Z^{(N)}} \qquad \qquad 2.42$$

it follows from the definition of the density distribution function (Eq. 2.33) that the average density at $\vec{r}$ is

$$\langle \rho(\vec{r}) \rangle = \rho^{(1)}(\vec{r}) = \left\langle \sum_{i=1}^{N} \delta(\vec{r} - \vec{r}_i) \right\rangle = \rho \qquad 2.43$$

For a homogeneous fluid, the structure factor (Eq. 2.39) is

$$S(\vec{q}) = \frac{1}{N} \left\langle \sum_{i=1}^{N} \sum_{j=1}^{N} \int e^{-q(\vec{r}-\vec{r}')} \delta(\vec{r} - \vec{r}_i) \delta(\vec{r}' - \vec{r}_j) d^3\vec{r} d^3\vec{r}' \right\rangle$$

$$= 1 + \frac{1}{N} \left\langle \sum_{i=1}^{N} \sum_{\substack{j=1 \\ (j \neq i)}}^{N} \int e^{-q(\vec{r}-\vec{r}')} \delta(\vec{r} - \vec{r}_i) \delta(\vec{r}' - \vec{r}_j) d^3\vec{r} d^3\vec{r}' \right\rangle \qquad 2.44$$

Since (Problem 18)

$$\rho^{(2)}(\vec{r}_1, \vec{r}_2) = \left\langle \sum_{i=1}^{N} \sum_{j=1}^{N} \delta(\vec{r} - \vec{r}_i) \delta(\vec{r}' - \vec{r}_j) \right\rangle \qquad 2.45$$

the expression which provides a connection between the structure factor the radial distribution (cf. Eq. 2.37) follows,

$$S(\vec{q}) = 1 + \rho \int e^{-i\vec{q} \cdot \vec{r}} g(r) d^3\vec{r} \qquad 2.46$$

where $\vec{r} = \vec{r}_2 - \vec{r}_1$. This equation indicates that the structure factor is the Fourier transform of $g(r)$. It is customary to define a new total correlation function

$$h(r) = g(r) - 1 \qquad 2.47$$

Which indicates that $h(r) \to 0$ as $g(r) \to 1$.

With regard to the question of structure, the important point that should be emphasized here is that scattering experiments measure the intensity which is determined by the structure factor. In a typical X-ray scattering experiment of a crystalline material, a series of sharp peaks would comprise the structure factor, reflecting the underlying periodicity of the structure. In the case of an isotropic liquid the structure factor would be a strongly damped oscillating function with increasing wave vector $\vec{q}$. The dampening reflects the fact that the locations of molecules away from the central molecule are not influenced by the presence of that molecule. In fact in the limit where q approaches 0, $S(0)$ provides information about the isothermal compressibility of the system. The essential features of a scattering experiment that measures the dynamic structure factor, $S(\vec{q}, t)$, are described in Section 2.5 of this chapter.

2.4 Langevin Analysis

Heretofore, we have provided a cursory description, at best, of the Brownian process with the random walk analysis. Having concluded the discussion of correlation functions we may now proceed with a reasonable analysis that accounts for the effect of the surrounding molecules that constitute the liquid on the particle dynamics. The analysis developed by Langevin is now described.

The number of bombardments per second that a Brownian particle experiences in a liquid is enormous, $\sim 10^{21}$, therefore it is not possible to analyze the dynamics in terms of discrete events. Since each collision influences the dynamics of the particle, then at any given instant, the velocity of the particle is not equal to the average velocity, and instead fluctuates about its mean value. The interaction of a Brownian particle with its environment can be described in terms of a "random" force, regarded as arising from two contributions. The first, $f(t)$, is associated with an "average" viscous drag force. The second component, $\bar{F}(t)$, reflects fluctuations that are rapid compared with the time scale of the fluctuations of the particle velocity. $\bar{F}(t)$ is therefore not correlated on time scales comparable to the relaxation time associated with the dynamics of the particle. $\bar{F}(t)$ is necessarily a random ("noise") fluctuating force and the ensemble average (or long time average) of $\bar{F}(t)$ is $\langle \bar{F}(t) \rangle = 0$. Nevertheless, note that if the random force did not exist the particle would come to rest due to the viscous drag forces (Stokes law). Conversely, if the random force were too large, then the kinetic energy of the particle would increase. To this end, there exists a connection between the random force and the frictional drag force. This connection is made by what is known in Statistical Mechanics as a *fluctuation dissipation theorem*, discussed below in this section.

With this in mind we proceed with an equation of motion for a particle of mass m. Using Newton's 2nd law, the equation has two contributions,

$$m\frac{d\bar{v}}{dt} = \bar{f}(t) + \bar{F}(t) \qquad\qquad 2.48$$

The viscous drag is associated with a friction factor $\zeta = 1/B$, where B is the mobility. Equation 2.48 may therefore be rewritten as

$$m\frac{d\bar{v}}{dt} = -\frac{\bar{v}}{B} + \bar{F}(t) \qquad\qquad 2.49$$

where the negative sign indicates that the effect of the viscous drag force is to slow down the particle. The drag is assumed to be governed by Stokes law, which stipulates that the frictional force (or viscous drag) which a spherical particle of radius a experiences in a medium with a viscosity of η is $\zeta = 6\pi a\eta$ ($\zeta = 1/B$), assuming nonslip boundary conditions. The foregoing equation (Eq. 2.49) is referred to as a stochastic differential equation because $\bar{F}(t)$ is a random force. Hence the solution to this equation is the probability that the particle will have a velocity $\bar{v}$ at time t subject to the boundary conditions that at time $t = 0$ its velocity is $\bar{v}(0)$. At sufficiently long times the particle is at equilibrium with its surroundings and the velocity distribution is Maxwellian (Chapter 1), independent of its initial velocity. A solution to Eq. 2.49 is

$$\bar{v}(t) = \bar{v}(0)e^{-(\zeta/m)t} + \int_0^t e^{-(\zeta/m)(t-t')}\bar{F}(t')dt' \qquad 2.50$$

The dynamics of the particle are characterized by a relaxation time τ, where the relaxation time $\tau = mB$. Upon taking the ensemble average (note that $\langle F(t)\rangle = 0$) it is evident that the drift velocity (Eq. 2.50) approaches zero rapidly

$$\langle v(t)\rangle = v(0)e^{-\frac{t}{\tau}} \qquad 2.51$$

In other words, the effect of the viscous drag is irreversible and dissipative.

2.4.1 Velocity Autocorrelation Function

The other dynamical variable of interest is the autocorrelation function of the velocity,

$$\langle \bar{v}(0)\bar{v}(t)\rangle = \langle \bar{v}(0)\bar{v}(0)\rangle e^{-(\zeta/m)t} + \int_0^t e^{-(\zeta/m)(t-t')}\langle \bar{v}(0)\bar{F}(t')\rangle dt' \qquad 2.52$$

The second term in this equation is zero since the initial velocity and the random force are not correlated, i.e.: $\langle \bar{v}(0)\cdot \bar{F}(t')\rangle = 0$

$$\langle \bar{v}(0)\bar{v}(t)\rangle = \frac{3kT}{m}e^{-t/\tau} \qquad 2.53$$

To obtain this equation we also took advantage of the *equipartition theorem* (Chapter 1), $\langle \bar{v}(0)\bar{v}(0)\rangle = \langle v^2\rangle = 3kT/m$. This is the autocorrelation function for the velocity.

2.4.2 Mean Square Velocity

The other dynamic property of interest is the mean square velocity,

$$\langle \bar{v}(t)\cdot \bar{v}(t)\rangle = \langle v^2(0)\rangle e^{-2t/\tau} + e^{-2t/\tau}\int dt'\int dt'' e^{-(t'+t'')/\tau}\langle F(t')\cdot F(t'')\rangle \qquad 2.54$$

The relaxation times associated with the fluctuating forces are vanishingly small compared to those that characterize the fluctuations of the velocity, so

$F(t')$ and $F(t'')$ could be approximated as two random, independent, variables when $t' \neq t''$. The implication is it may be assumed that

$$\langle F(t') \cdot F(t'') \rangle = K\delta(t' - t'') \qquad 2.55$$

where K is a constant. It may be shown (see Problem 12) that $K = 6kT/\tau m$ and that Eq. 2.54 becomes

$$\langle \bar{v}(t)\bar{v}(t) \rangle = \frac{3kT}{m} + \left(v^2(0) - \frac{3kT}{m} \right) e^{-2t/\tau} \qquad 2.56$$

which indicates that the mean square velocity is approaches $3kT/m$, the *equipartition* value at long times.

With the value of K, Eq. 2.55 may be rewritten as $\langle F(t') \cdot F(t'') \rangle = 6\zeta kT\delta(t' - t'')$, which is a statement of a *fluctuation dissipation theorem*. In this situation the theorem provides a connection between the random fluctuating force that influences the particle velocity with the dissipative forces that act to slow down the particle.

2.4.3 Mean Square Displacement

The mean square displacement, which in this case represents fluctuations in the location of the particle that occur at equilibrium ($\langle x \rangle = 0$), is now calculated. We will accomplish this by considering the one-dimensional form of Eq. 2.49 because it is convenient. By multiplying both sides of the equation by x, we obtain

$$mx \frac{d}{dt}\left(\frac{dx}{dt} \right) = -\frac{x}{B}\frac{dx}{dt} + xF(t) \qquad 2.57$$

where $v = dx/dt$. Equation 2.57 may be rewritten (Problem 19) as

$$\frac{d\langle x^2 \rangle}{dt} = \frac{2kT\tau}{m}e^{-t/\tau} + 2\frac{kT}{m} \qquad 2.58$$

which is readily solved for $\langle x^2 \rangle$ subject to the boundary conditions: at time $t = 0$, $x = 0$ and $dx/dt = 0$

$$\langle x^2 \rangle = 2kTBt + ekTB\tau(e^{-t/\tau} - 1) \qquad 2.59$$

The solution, Eq. 2.59, describes two limiting physical situations regarding the behavior of the mean square displacement of a particle undergoing Brownian motion. The first concerns the very early stage behavior, where $t \ll \tau$. Here the exponential term can be expanded and

$$\langle x^2 \rangle \cong \frac{kT}{m}t^2 \qquad 2.60$$

which indicates that $\langle x^2 \rangle = \langle v^2 \rangle t^2$. Physically this result reveals that at short times, $t \ll \tau$, the particle moves with a constant thermal velocity $v = (kT/m)^{1/2}$, which is not surprising.

The real impact of the environment occurs at long times, $t \gg \tau$, where

$$\langle x^2 \rangle \cong 2kTBt \qquad\qquad 2.61$$

2.4.4   Stokes–Einstein Equation

Clearly, the main message from the foregoing discussion is that at sufficiently long times, the mean square displacement of the particle whose dynamics are characterized by random fluctuating motions is proportional to t. Note that this is the same result we obtained from the one-dimensional random walk analysis where $\langle x^2 \rangle = 2Dt$. Here we identify the diffusion coefficient of the particle as

$$D = kTB \qquad\qquad 2.62$$

This is the well known Einstein relation that provides a connection between the diffusion coefficient and the mobility of a particle undergoing Brownian motion. In essence, the particle experiences a frictional resistance ζ as it migrates throughout the liquid in response to the thermal energy, kT. In fact, Eq. 2.61 can be rewritten as

$$\langle x^2 \rangle = \frac{kT}{3\pi\eta a}t \qquad\qquad 2.63$$

which indicates that as the viscosity of the medium increases the mean square displacement is reduced.

2.4.5   Nernst-Einstein Equation

The influence of an external force such as an electric field on a charged particle in this medium is now considered. The effect of the electric field is to impart a force of eE, where E is the electric field and e is the charge, on the particle. Under these conditions, Eq. 2.49 (again considering one dimension) becomes,

$$m\frac{dv}{dt} = eE - \frac{v}{B} + F(t) \qquad\qquad 2.64$$

Under steady-state conditions, $dv/dt = 0$. By taking the long-time average of this equation, we arrive at the result that $eE = v/B$. With this result and Eq. 2.62, we obtain what is often referred to as the *Nernst-Einstein* relation,

$$D = \frac{kT}{e}\mu \qquad\qquad 2.65$$

which provides a direct connection between the diffusion coefficient and the "mobility" (now $\mu = eB$) of the charged particle. The Nearnst-Einstein equation plays a central role in a variety of electrochemical processes. We will revisit the Nearnst-Einstein equation later in Chapter 7 where the transport of ionic species in glasses is considered. At that point, a more general form of the equation is introduced.

2.5 Light Scattering: Measurement of Diffusion

The molecules in air scatter light elastically as well as inelastically. The elastic scattering, also known as Rayleigh scattering, of sunlight from air is largely responsible for the blue sky. With regard to inelastic scattering, light interacts with the molecules and experiences a shift in frequency (energy); this is the basis of Raman scattering. As we saw earlier, auto-correlation functions play a central role at describing the dynamics of a variety of systems. In this section we describe a light scattering experiment that provides information about the dynamics of particles undergoing Brownian motion in a medium. In a light scattering experiment, a laser source provides a monochromatic beam of light of constant intensity. The scattered intensity is determined by an auto-correlation of the scattered field and is measured by a detector. For dynamic light scattering (DLS) experiments the scattered photons emanating from the sample are counted and their temporal dependence analyzed. If the sample is a collection of molecules, the auto-correlation function contains information about their dynamics. In static light scattering, routinely used in many laboratories to measure structure, size, and shape, the average intensity of light scattered, $I(q)$, with a given polarization is measured for various values of wave vector q. In addition to DLS, inelastic light scattering measurements are also used to study dynamics. Experiments such as Brillouin scattering, which will not be discussed here, is associated with the scattering of light from phonon modes within the material.

Consider the scattering of light from a sample, Fig. 2.4. The incident electric field of a plane wave, you may recall from your freshman Physics course, is specified by

$$E_i(\vec{r},t) = E_0 e^{i(\vec{k}_i \bullet \vec{r} - \omega_i t)} \qquad 2.66$$

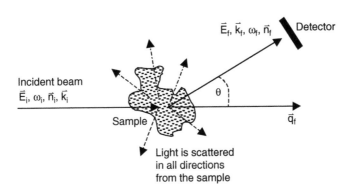

FIG. 2.4
Schematic of a light scattering experiment. The parameters that characterize the incident beam are $\vec{E}_i, \omega_i, \bar{n}_i, \vec{k}_i$ and for the scattered beam the parameters are $\vec{E}_f, \vec{k}_f, \omega_f, \bar{n}_f$.

where E_0 is the magnitude of the field, $\bar{k}_i$ is the wave vector, of magnitude $k = |\bar{k}| = 2\pi/\lambda$ (radians/distance), pointing in the direction of the field, λ is the wavelength of light in the medium. The angular frequency is ω (radians/second). The subscript i identifies this as the incident field.

2.5.1 The Scattered Field

The scattered electric field contains information about the dynamics of the molecules. We consider scattering from a dilute collection of molecules. The incident light induces a time-dependent dipole moment in each molecule. This follows from the fact that any molecule is characterized by a polarizability (which is generally anisotropic) and the magnitude of the induced dipole moment is the product of the polarizability of the molecule and the electric field, $\bar{\mu}(\bar{r},t) = \tilde{\alpha}(\bar{r},t)\bar{E}(\bar{r},t)$, where the polarizability, $\tilde{\alpha}$, is generally a tensor.

The polarizability may be written in terms the sum of an average value and a fluctuating component, due to the species in the medium

$$\tilde{\alpha}(\bar{r},t) = \langle\alpha\rangle + \delta\alpha_{is}^j(\bar{r},t) \qquad\qquad 2.67$$

where

$$\delta\alpha_{is}(\bar{r},t) = \sum_{j=1}^{N'} \alpha_{is}^j \delta(\bar{r} - \bar{r}_j(t)) \qquad\qquad 2.68$$

In these equations N' is the number of molecules illuminated by the beam and the superscript refers to the jth molecule. Recall that the local density is also specified in a manner similar to that of the fluctuating dipole (cf. Eq. 2.41); $\rho(\bar{r},t) = \sum_{j=1}^{N'} \delta(\bar{r} - \bar{r}_j(t))$. In essence, fluctuations in the local density are reflected in fluctuations of the dipoles.

It is a well-documented phenomenon that time-dependent fluctuating dipoles give off radiation. This means that the electric field that arrives at the detector is determined by the polarizability of the molecules. The scattered electric field that arrives at the detector a distance R away ($R \gg d$ (sample dimension) $\gg \lambda$) is the sum of all the fields from the elements in the volume illuminated by the beam

$$E(\bar{r},t) = \frac{E_0\omega_i^2}{cR^2} e^{i(\bar{k}_s\bar{R}-\omega_s t)} \delta\alpha_{is}(\bar{q},t) \qquad\qquad 2.69$$

where

$$\delta\alpha_{is}(\bar{q},t) = \int e^{i\bar{q}\bullet\bar{r}} \delta\alpha_{is}(\bar{r},t) d^3\bar{r} \qquad\qquad 2.70$$

The wave vector $\bar{q} = \bar{k}_i - \bar{k}_f$ and $q^2 = |\bar{k}_f - \bar{k}_i|^2 = k_f^2 + k_i^2 - 2k_f k_i \cos\theta$. Since the wavelength remains the same after scattering (assuming quasi elastic scattering) $|\bar{k}_i| = |\bar{k}_f|$, and $q = 2k_i \sin\frac{\theta}{2}$. This implies that for visible light q is very small because λ is hundreds of nanometers. The implication is that the resolution,

ΔD, at which dynamical, and structural, information can be obtained is in principle limited since $\Delta D \sim 1/q$. However, if the scattering angle, θ, is chosen to be very small ($\sim$ few degrees), then the dimension can be increased. In fact, the resolution can be on the order of a micron with the appropriate values of q and λ. X-rays and neutrons, which possess smaller wavelengths, can probe smaller length scales and it is possible with modern instruments for the length scales probed by visible light and X-rays to overlap.

The scattering intensity spectrum, or the spectral density, of a dynamical property $A(t)$ is by definition

$$I_A(\omega) = \frac{1}{2\pi} \int_{-\infty}^{\infty} e^{i\omega t} \langle A^*(0)A(t)\rangle dt \qquad 2.71$$

where A^* is the complex conjugate of the function A. The spectral density of the scattered electric field is therefore

$$I_E(\omega_f) = \frac{1}{2\pi} \int_{-\infty}^{\infty} e^{i\omega t} \langle E_s^*(R,0)E_s(R,t)\rangle dt \qquad 2.72$$

This result indicates *that the scattering intensity is the Fourier transform of the autocorrelation of the scattered electric field.*

2.5.2 Scattering from a Dilute Collection of Molecules

In typical experiments, samples may be liquids, gases, or polymeric mixtures, mixtures of colloidal particles or other types of complex fluids. Optically transparent samples are typically required because they readily scatter light. The primary requirement for use of this technique is that the particles (molecules, etc.) much be smaller than the wavelength of light. The intensity of the field that arrives at the detector is, based on Eqns. 2.69 and 2.72,

$$I(\vec{q},\omega) \propto \int_{-\infty}^{\infty} \langle \delta\alpha_{is}(\vec{q},0)\delta\alpha_{is}(\vec{q},t)\rangle e^{-i\omega t} dt \qquad 2.73$$

where $\omega = \omega_f - \omega_i$.

If the system is dilute, and the molecules are only weakly interacting, then the scattered wave is a superposition of all the waves scattered by each of the N' particles *illuminated* by the beam and

$$\delta\alpha_{if}(\vec{q},t) = \sum_{j=1}^{N'} \alpha_{if}^j(t)e^{i\vec{q}\cdot\vec{r}(t)} \qquad 2.74$$

The scattered intensity $I(\vec{q},t)$ can be rewritten such that it is proportional to $F(\vec{q},t)$

$$F(\vec{q},t) = \langle \varphi^*(\vec{q},0)\varphi(\vec{q},t)\rangle \qquad 2.75$$

where $F(\vec{q},t)$ the dynamic structure factor, whose importance will become evident momentarily and

$$\varphi(\vec{q},t) = \sum_{j=1}^{N'} e^{i\vec{q}\bullet\vec{r}_j(t)} \qquad\qquad 2.76$$

Note that N' is time dependent, $N' = N'(t)$, since molecules enter and depart from the scattering volume. $\varphi(\vec{q},t)$ may be rewritten in terms of the local density (Eqs. 2.40 and 2.4.1)

$$\varphi(\vec{q},t) = \int_{V'} \sum_{j=1}^{N} e^{i\vec{q}\bullet\vec{r}_j(t)} \delta(\vec{r}-\vec{r}_j)d^3\vec{r} \qquad\qquad 2.77$$

In scattering experiments it is convenient to define a parameter $b_j(t)$ which possesses values of 0 or 1; $b_j(t) = 1$ when a molecule resides within the scattering volume V and zero otherwise. Hence

$$\varphi(\vec{q},t) = \sum_{j=1}^{N} b_j(t) e^{i\vec{q}\bullet\vec{r}_j(t)} \qquad\qquad 2.78$$

It follows that

$$F'(\vec{q},t) = \left\langle \sum_{j=1}^{N} b_j(0) b_j(t) e^{i\vec{q}\cdot[\vec{r}_j(t)-\vec{r}_j(0)]} \right\rangle \qquad\qquad 2.79$$

For noninteracting spherically symmetric molecules, the scattering intensity is

$$I(\vec{q},t) \propto F'(\vec{q},t) = \langle N \rangle F(\vec{q},t) \qquad\qquad 2.80$$

The function $F(\vec{q},t)$ is known as the *dynamic structure factor in light scattering experiments* or, in neutron scattering experiments, the *intermediate scattering function*.

2.5.3 Measurement of Diffusion

The scattering function, defined in terms of the Fourier transform of $F(\vec{q},t)$, is

$$S(\vec{q},\omega) = \frac{1}{2\pi} \int_{-\infty}^{\infty} e^{-i\omega t} F(\vec{q},t)dt \qquad\qquad 2.81$$

With regard to the connection to diffusion, a time dependent radial distribution function $G(\vec{r},t)$ (the well-known van Hove space-time correlation function) is related to the scattering function such that

$$S(\vec{q},\omega) = \int_{-\infty}^{\infty} e^{i(\vec{q}\cdot\vec{r}-\omega t)} G(\vec{q},t)dtd^3\vec{r} \qquad\qquad 2.82$$

and

$$G(\vec{r},t) = \frac{1}{(2\pi)^3} \int\limits_{-\infty}^{\infty} e^{i(\vec{q}\cdot\vec{r}-\omega t)} S(\vec{q},\omega) d\omega d^3\vec{q} \qquad 2.83$$

it may be shown (Problem 23) that

$$F(\vec{q},t) = \int G(\vec{r},t) e^{-i\vec{q}\cdot\vec{r}} d^3\vec{r} \qquad 2.84$$

and

$$G(\vec{r},t) = \frac{1}{(2\pi)^3} \int e^{-i\vec{q}\cdot\vec{r}} F_s(\vec{q},t) d^3 q \qquad 2.85$$

If the particles undergo Brownian motion $G(\vec{r},t)$, the Van-Hove function is a solution to the diffusion equation,

$$\frac{\partial G}{\partial t} = D\frac{\partial^2 G}{\partial x^2} \qquad 2.86$$

By taking the Fourier integral transform of this equation, we arrive at

$$\frac{\partial F(q,t)}{\partial t} = -q^2 D F(q,t) \qquad 2.87$$

The solution to this equation

$$F(q,t) = e^{-q^2 Dt} \qquad 2.88$$

Note further that the Fourier transform of the Gaussian function is

$$F(q,t) = e^{-q^2 \langle r^2(t)\rangle/6} \qquad 2.89$$

Based on the derivation in appendix 1, $\langle r^2(t)\rangle = 2t\int_0^\infty \langle v(0)v(t')\rangle dt'$. This result indicates that the diffusion coefficient is determined by the velocity autocorrelation function

$$D = \frac{1}{3}\int\limits_0^\infty \langle \vec{v}(0)\vec{v}(t)\rangle dt \qquad 2.90$$

This result is sometimes referred to as a Green-Kubo relation. The dynamic structure factor may be written as

$$F_s = e^{-t/\tau} \qquad 2.91$$

where $\tau = 1/(q^2 D)$ is the relaxation time. In frequency domain, $F(\omega,q)$ is Lorenzian,

$$F(\omega,q) = \frac{1}{\pi}\frac{q^2 D}{\omega^2 + (q^2 D)^2} \qquad 2.92$$

Scattering techniques (light, X-rays, and so forth) are powerful tools that are (routinely) used to study dynamic properties of materials. In later chapters we will revisit this topic when we discuss interdiffusion and spinodal decomposition in concentrated mixtures. The foregoing discussion was not meant to be exhaustive but meant to illustrate the use of correlation functions for the study of dynamics. The reader is referred to other references on the topic, located at the end of the chapter.

2.6   Problems for Chapter 2

1. Consider the following data that describe the Brownian motion of a spherical particle in a liquid. The displacement, Δx, of the particle during a specified time interval (2 seconds) is determined. The frequency at which each displacement is observed was determined from the observations. The data are shown below.

Δx (nanometers)	N, frequency
$< -5.5 \times 10^{-3}$	0
$-5 \pm 15 \times 10^{-3}$	1
$-4 \pm 15 \times 10^{-3}$	2
$-3 \pm 15 \times 10^{-3}$	15
$-2 \pm 15 \times 10^{-3}$	32
$-1 \pm 15 \times 10^{-3}$	95
$0 \pm 15 \times 10^{-3}$	111
$1 \pm 15 \times 10^{-3}$	87
$2 \pm 15 \times 10^{-3}$	47
$3 \pm 15 \times 10^{-3}$	8
$4 \pm 15 \times 10^{-3}$	5
$5 \pm 15 \times 10^{-3}$	0
$> 5.5 \times 10^{-3}$	0

Data taken from Pathria, 1980

   a) Using this data determine the diffusion coefficient of the particle.
   b) If the viscosity of the liquid is $\eta = 10^{-2}$ poise, $T = 300$ K, the radius of the particle is 4×10^{-5} cm, determine the diffusion coefficient.
   c) Please comment on the results obtained from (a) and (b).
2. Using the following distribution function

$$P_N(n_r) = \frac{N!}{n_r! n_l!} p_r^{n_r} p_l^{n_l}, \text{ show that } \sum_{n_r=0}^{N} P_N(n_r) = 1$$

3. Show that $\overline{n_r} = N p_r$

4. Show that $\overline{n_r} + \overline{n_p} = N$

5. Show that the relative width of the Gaussian distribution is

$$\frac{\left[\overline{(\Delta n_r)^2}\right]^{1/2}}{\overline{n_r}} = \frac{1}{\sqrt{N}}, \text{ provided that } p_r = p_l = 1/2.$$

6. If $m = n_r - n_l$, calculate the dispersion of m.

7. Derive the Stokes-Einstein equation from the Green-Kubo relation

8. Let us consider the motion of a molecule in a gas. The molecule moves in such a manner that it makes unit displacements of distance L between collisions. These displacements occur in any direction with equal probability. What is the mean square displacement $\langle R^2 \rangle$ of the molecule after N steps?

9. A person loads a bullet in one chamber of a revolver, leaving the other five chambers of the cylinder empty. The player then spins the cylinder, aims at an object, and pulls the trigger.

   a) What is the probability that the gun fires if the trigger is pulled N times?

   b) What is the probability that the person does not fire the gun after $(N - 1)$ tries but is successful on the Nth try?

   c) What do you believe is the mean number of times that this person gets to pull the trigger in order to fire the revolver?

10. Using Eq. 5 in Appendix 2 show that for $t \ll \tau, \langle \Delta R^2(t) \rangle = \langle V^2 \rangle t^2$

11. Show that the friction coefficient in the Stokes-Einstein equation can be expressed in terms of the velocity autocorrelation function

12. Show that for a Brownian particle, $\langle \vec{v}(t)\vec{v}(t) \rangle = \frac{3kT}{m} + (v^2(0) - \frac{3kT}{m})e^{-2t/\tau}$.

13. Starting with

$$P^{(N)}(\vec{r}_1, \vec{r}_2 \ldots, \vec{r}_N)d^3\vec{r}_1 d^3\vec{r}_2 \ldots d^3\vec{r}_N = \frac{\int_p e^{-\beta E_N} d^3\vec{r}_1 \ldots d^3\vec{r}_N d\vec{p}_1 \ldots d\vec{p}_N}{Z}$$

show that $P^{(N)} = \frac{e^{-\beta U_N} d^3\vec{r}_1 d^3\vec{r}_2 \ldots d^3\vec{r}_N}{Z^{(N)}}$

14. Show that for an ideal gas $\rho_N^{(2)}(\vec{r}_1, \vec{r}_2) = \rho^2(1 - \frac{1}{N})$.

15. The Poisson distribution function is generally valid when an event is rare. Using the following approximations, $p_r \rightarrow 0$, $N \gg n_r$, and $N \approx n_l$, show that Eq. 2.1 becomes

$$P(n) = \frac{e^{-\lambda}\lambda^n}{n!} \qquad \qquad 2.27$$

16. Starting with the static density-density correlation function

$$G(\vec{r}) = \frac{1}{N} \int \langle \rho(\vec{r}' + \vec{r}) \rho(\vec{r}') \rangle d^3\vec{r}' \text{ show that } G(\vec{r}) = \rho g(\vec{r}) + \delta(\vec{r})$$

17. Show that $F(\vec{q}, 0) = S(\vec{q})$.

18. Starting with the fact that $\langle \delta(\vec{r} - \vec{r}_1) \rangle = \frac{\int e^{-\beta U_N} d^3\vec{r}_2 \dots d^3\vec{r}_N}{Z^{(N)}}$, where

$$Z^{(N)} = \int e^{-\beta U_N} d^3\vec{r}_1 d^3\vec{r}_2 \dots d^3\vec{r}_N$$

show that $\rho^{(2)}(\vec{r}_1, \vec{r}_2) = \langle \sum_{i=1}^{N} \sum_{j=1}^{N} \delta(\vec{r} - \vec{r}_i) \delta(\vec{r}' - \vec{r}_j) \rangle$.

19. Show that Eq. 2.57 may be rewritten as

$$\frac{d}{dt} \left\langle x \frac{dx}{dt} \right\rangle = \frac{-1}{\tau} \left\langle x \frac{dx}{dt} \right\rangle + \left\langle \left(\frac{dx}{dt} \right)^2 \right\rangle$$

where $\tau = Bm$ *and* $\langle xF(t) \rangle = \langle x \rangle \langle F(t) \rangle = 0$ (recall $F(t)$ and x are uncorrelated and that $\langle F(t) \rangle = 0$). Relying on the fact that $\langle (dx/dt)^2 \rangle = \langle v_x^2 \rangle = kT/m$, show that

$$\frac{d^2 \langle x^2 \rangle}{dt^2} + \frac{1}{\tau} \frac{d \langle x^2 \rangle}{dt} = \frac{2kT}{m}$$

20. Consider a solution that contains a collection of noninteracting particles. The potential energy of a the particles if $U = mgz$. The probability that a particle will be found at height z is specified by the Boltzmann factor, e^{-mgz}.

   a) What is the equilibrium concentration of particles at height z?

   b) The particles move under the influence of a viscous drag force $-\zeta \vec{v}$ (in the z-direction it is $-\zeta v_z$). The equation of motion is $md^2z/dt^2 = -mg - \zeta v_z$. If the terminal velocity of the particles is mg/ζ, what is the flux?

   c) The total flux in the system has two contributions, $\vec{J} = \vec{J}_{Diff} + \vec{J}_g$, where the first is due to suppression of the concentration gradient and the latter is to the gravitational forces. If the particles do not segregate to the interfaces of the container, and the total flux vanishes everywhere, $\vec{J} = 0$, derive the Einstein equation, $D = kt/\zeta$.

21. Show that the scattered intensity decreases as λ^{-4} for Rayleigh scattering.

22. Starting with Eq. 2.81 show that

$$F(\vec{q}, t) = \int G(\vec{r}, t) e^{-i\vec{q} \cdot \vec{r}} d^3\vec{r}$$

23. Explain the conditions under which eqn. 2 is transformed into eqn. 3 in the appendix.

2.7 Appendix: The Diffusion Coefficient

Earlier it was mentioned that the diffusion coefficient is an autocorrelation function of the particle velocity. This is readily seen from the following. Generally, one can write the displacement of the particle as

$$r(t) = \int \bar{v}(\tau)d\tau \qquad\qquad 1$$

It follows that the mean square displacement is

$$\langle r^2(t) \rangle = \int_0^t \int_0^t \langle \bar{v}(t_1)\bar{v}(t_2) \rangle dt_2 dt_1 \qquad\qquad 2$$

which (see Problem 23) becomes

$$\langle r^2(t) \rangle = 2 \int_0^t \int_0^{t_2} \langle \bar{v}(0)\bar{v}(t_2 - t_1) \rangle dt_1 dt_2 \qquad\qquad 3$$

If we change variables using $\tau = t_2 - t_1$, then

$$\langle r^2(t) \rangle = 2 \int_0^t dt_2 \int_0^{t_2} \langle \bar{v}(0)\bar{v}(\tau) \rangle dt_2 d\tau \qquad\qquad 4$$

We can then integrate by parts

$$\langle r^2(t) \rangle = 2 \int_0^t (t - \tau) \langle \bar{v}(0)\bar{v}(\tau) \rangle d\tau \qquad\qquad 5$$

The center of mass diffusion coefficient is recovered as sufficiently long times. From a practical point of view, one should be careful that when measuring diffusion coefficients that represent long-range dynamics. The time interval needs to be sufficiently long to ensure that indeed one measures a true center of mass diffusion coefficient.

2.8 References

"Stochastic problems in Physics and Astronomy," Chanrdasekhar, S.; *Reviews of Modern Physics*, 15, 1 (1943).

J.-P Hansen and I.R. McDonald, *Theory of Simple Liquids*, 2nd ed. Academic Press, INC, CA, 1990.

Dynamic Light Scattering: Applications of Photon Correlation Spectroscopy, ed. Robert Pecora, Plenum Press, NY, 1985.

Dynamic Light Scattering, B.J. Berne and R. Pecora, John Wiley and Sons, NY, 1976.

H.J.V. Tyrrell and K.R. Harris, *Diffusion in Liquids*, Butterworths, London 1984.

R.K. Pathria, *Statistical Mechanics*, Pergamon Press, Oxford UK, 1980.

J. Crank, *The Mathematics of Diffusion*, Oxford University Press 1975.

F. Reif, *Fundamentals of Statistical and Thermal Physics*, Mcgraw Hill, New York, 1965.

Tyn. Mynt, Partial Differential Equations of mathematical Physics, North Holland, NY 1980.

Part II

Diffusion in Crystalline Materials

The center of mass transport of an atom in a material is intimately connected to the spatial arrangement of its neighboring constituents and to its interactions with them. The dynamics of a microscopic particle in a *simple* fluid are characterized by random thermal fluctuations. The geometry of the molecular constituents and the degrees of freedom afforded them by the nature of their interactions with neighbors dictates the mechanism by which molecules are able to migrate. Mechanisms of transport in materials that exhibit long-range structural order are discussed. Specifically, hopping transport mechanisms in metals, ionic crystals and semiconductors are discussed in Part II.

3

Structure, Defects and Atomic Diffusion in Crystalline Metals

3.1 Introduction

This chapter dicusses elements of atomic diffusion in crystalline lattices and is largely devoted to metals. Atomic migration plays a central role in the processing of materials and in the reliability and performance of device and sensor technologies. The societal impact is profound. Properties of materials, from magnetic, optical, and electronic to corrosion and mechanical, are strongly influenced by the microstructural features of materials. During processing, the annealing of materials induces atomic migration and the associated evolution of microstructural features. The growth of various crystalline phases of materials during annealing is controlled by atomic diffusion processes. Stresses that develop in materials during fabrication are often relieved as a result of atomic diffusion processes. Control of the spatial distribution of dopants in semiconductors, which influences device performance, is controlled by atomic diffusion properties. The optoelectronic properties of quantum well heterostructures (e.g., GaInNAs/GaAs and InGaN/GaN multilayer structures) are influenced by atomic (nitrogen and indium) diffusion across the layers. Heterostructures are essential components of high performance high speed and high frequency digital and analog devices. Solid state magnetic field sensors, for applications such as magnetic storage, are made of magnetic metallic layers. Atomic diffusion across the interfaces affects the spatial compositional profile and hence the magnetic properties. As a final example, common processes such as the rate of oxidation at interfaces are often controlled by atomic diffusion.

The diffusion coefficient of an atom in a crystal typically exhibits an Arrhenius dependence on temperature,

$$D = D_0 e^{-Q/kT} \qquad\qquad 3.1$$

Both Q and D_0 conceal information regarding the nature of the defect mediated transport mechanism. As a specific example we briefly consider diffusion via a vacancy mechanism. Q is determined by the enthalpy of formation of

FIG. 3.1
Diffusion of an atom due to a singe
mechanism (e.g., single vacancy or interstitial mechanism) is Arrhenius, as
illustrated by the thick solid line. If a
second mechanism is operable at high
T then deviations from Arrhenius behavior may occur, as indicated by the
dashed line. In polycrystalline samples, which contain a large fraction of
grain boundaries, both D_0 and the activation energy are changed. Typical
behavior is illustrated by the dotted
line; the activation energy is smaller
and D_0 is generally larger.

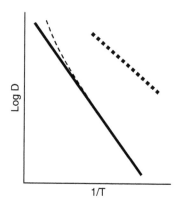

a vacancy, H_v^f, and by the enthalpy of migration, H_v^m, of an atom from one
site in the crystal lattice to a neighboring vacant site. The prefactor D_0 is
determined by the nearest neighbor atomic jump distance, natural atomic
hopping frequencies, crystal symmetry, and by entropic effects associated
with the formation of vacancies.

Defects in materials are ubiquitous and they have a profound impact on
diffusional transport processes. They influence the rate and mechanism by
which an atom migrates throughout a crystal. Defects found in metals
include point defects, vacant sites, atoms lodged in the interstices or impurities, line defects (dislocations), planar defects (grain boundaries), and voids
created by large clusters of vacancies. The presence of dislocations and grain
boundaries tend to increase the effective diffusivities of atoms. The sketch
in Fig. 3.1 illustrates the influence of defects on diffusional transport. At low
temperatures $\log D \propto 1/T$, which typically indicates a single defect mediated
mechanism of transport. The deviation from the $1/T$ dependence at high T,
represented by the broken line, is often indicative of at least one additional
mechanism of atomic transport that operates simultaneously. For example,
at low T, single vacancies may be responsible for diffusional transport,
whereas at high T the presence of divacancies would enhance the effective
diffusivities beyond that due to single vacancies. The dotted line might
represent diffusion in the same material except that it possesses a large
concentration of grain boundaries, e.g., polycrystalline sample. The essential
point is that defects control the rate and mechanism of transport and that
D_0 and Q reveal information about such processes.

A diverse range of defects typically appear during various stages of processing. They develop during materials growth. Clusters of vacancies or
interstitials can often form due to radiation damage and deformation. Dislocations and grain boundaries are also the result of mechanical deformation.

In this chapter atomic diffusion in crystalline materials is discussed. It will
be shown that the nature of the crystal structure and of the point defects
have a profound influence on the mechanism by which an atom is destined
to migrate. While the discussion will largely discuss the situation in metals,

the general ideas and concepts described in this chapter apply to other crystalline materials. Specific details regarding characteristics of atomic transport in elemental semiconductors and ionic crystals, in which defects are typically charged, are addressed in subsequent chapters. In the next Section, 3.2, crystal structure is discussed. Students in Materials Science and Engineering will already be intimately familiar with this topic and may forgo reading it and continue with Section 3.3.

3.2   Crystal Structure and Point Defects

3.2.1   Bravais Lattices

Atoms in a crystal arrange themselves on a three-dimensional periodic array of points in space that is called a lattice. The structure of the lattice is characterized by long-range order, wherein the periodic arrangement of points persists over distances many times the interatomic spacing. This arrangement is such that each point on the lattice has identical surroundings; the structure of the lattice is characterized by different symmetry conditions, including translational, rotational, inversion, and mirror symmetry. Consider, for simplicity, a two-dimensional square lattice, as shown in Fig. 3.2. The lattice possesses translational symmetry in that any point, arbitrarily chosen, can be translated in any direction to coincide with another point. If the square lattice is rotated by 90°, each point on the lattice coincides with another point. In other words, the surroundings of any point remain invariant. In fact, rotations through 180°, 270°, and 360° produce the same result. The square lattice therefore possesses four-fold rotational symmetry. The mirror symmetry condition is also obeyed by this lattice. We will not dwell further on this issue of symmetry. Nevertheless, it suffices to say that similar concepts apply to the cubic lattice (three dimensions) as well as to other periodic arrangements of points in space.

   Bravais, a French crystallographer, in 1848, proved that in fact there exist only 14 ways to arrange points in three-dimensional space (5 in two dimensions) to meet the requisite criteria that define a lattice. Herewith, there exist

FIG. 3.2
A two-dimensional square lattice is shown here. $|\bar{a}| = |b| = a$.

TABLE 3.1

Characteristics of the unit cells of the 14 Bravais lattices and associates 7 crystal systems

System	Axial lengths and angles	Bravias Lattice
Cubic	$a = b = c,$ $\alpha = \beta = \gamma = 90°$	Simple Body-centered Face-centered
Tetragonal	$a = b \neq a = c$ $\alpha = \beta = \gamma = 90°$	Simple Body-centered
Otrhorhombic	$a \neq b \neq c$ $\alpha = \beta = \gamma = 90°$	Simple Body-centered Face-centered Base-centered
Rhombohedral	$a = b = c$ $\alpha = \beta = \gamma \neq 90°$	Simple
Hexagonal	$a = b \neq c$ $\alpha = \gamma = 90°; \beta = 120°$	Simple
Monoclinic	$a \neq b \neq c$ $\alpha = \gamma = 90° \neq \beta$	Simple Base-centered
Triclinic	$a \neq b \neq c$ $\alpha \neq \beta \neq \gamma \neq 90°$	Simple

14 Bravais lattices. These lattices are organized into 7 crystal systems, cubic, tetragonal, orthorhombic, trigonal, hexagonal, monoclinic, and triclinic. The smallest group of points that possesses the same symmetry as the lattice is identified as the unit cell. The unit cell is characterized by three vectors, $\vec{a}$, $\vec{b}$ and $\vec{c}$. In Fig. 3.1, the unit cell is a square so the vectors in the x and y-directions are of magnitude $|\vec{a}| = |\vec{b}| = a$ and the angle between them is 90°.

The relative magnitudes of the vectors and the angles that characterize unit cells representing the 14 Bravais lattices and the associated 7 crystal systems are described in Table 3.1. Interestingly, the atoms of metals generally organize themselves into cubic structures, body centered cubic (BCC), face centered cubic (FCC), and hexagonal close packed (HCP). These three structures, particularly the BCC and FCC systems, will be discussed in further detail in the next section.

3.2.2   Unit Cells, Crystal Directions, and Crystal Planes

The notation [hkl] denotes a direction in a lattice and $\langle hkl \rangle$ denotes a family of directions; h, k, and l are integers. Using the diagram in Fig. 3.3, the c-direction is denoted by [001], the b-direction by [010] and the a-direction by [100]. The direction [111] passes through points in space ($\frac{1}{2}$, $\frac{1}{2}$, $\frac{1}{2}$), (1,1,1), (2,2,2), etc. The direction [112] passes through the point ($\frac{1}{2}$, $\frac{1}{2}$, 1), etc. The eight directions, [111], [11$\bar{1}$], [1$\bar{1}$1], [$\bar{1}$11], [$\bar{1}\bar{1}\bar{1}$], [$\bar{1}\bar{1}$1], [$\bar{1}$1$\bar{1}$] and [1$\bar{1}\bar{1}$]

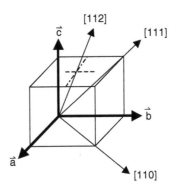

FIG. 3.3
Different directions are identified in
the diagram above.

compose the family of directions ⟨111⟩. These directions are all related by symmetry.

A BCC unit cell is shown in Fig. 3.4a. In order to illustrate salient features of the structure, it is common to rely on so-called "hard sphere" models. Specifically, atoms are imagined to be hard spheres that occupy the maximum volume possible within the unit cell, while satisfying the requisite symmetry conditions (i.e., their centers of mass coincide with lattice points). In the BCC system, the spheres touch along the ⟨111⟩ directions, whereas they necessarily do not in the ⟨100⟩ directions.

If the radius of each sphere is R and the lattice spacing is a, then based on geometrical considerations, $a = (4/\sqrt{3})R$. It is easily shown that the number of nearest neighbors (coordination number) is 8, the atomic packing fraction (fractional volume occupied by spheres) is 0.68 and the number of atoms per unit cell is 2. Note that in this structure, each atom in the unit cell is equivalent in that any atom chosen at random would serve as the center of a unit cell.

The FCC structure is a close packed arrangement of atoms, as illustrated in Fig. 3.4b. In this geometry the spheres, each of radius R, are in contact along the ⟨110⟩ directions (face diagonals) and the relation between a and R is $a = 2\sqrt{2}R$. In the FCC structure, the coordination number is 12 and the number of atoms per unit cell is 4. The atomic packing fraction is 0.74, which is the largest packing density at which spheres of equal size can be organized in three dimensions, in any geometry.

Finally, we note, parenthetically, that based on knowledge of the number of atoms per unit cell (n), the atomic weight (A), the volume per unit cell, (V_c), and Avogadro's number (N_A), the density, (ρ), of many FCC, BCC and HCP metals can be calculated with reasonable accuracy, $\rho = \frac{nA}{V_c N_A}$. Problems at the end of the chapter provide a reasonable assessment of the utility of this strategy involving hard-sphere models. The HCP unit cell is shown in Fig. 3.4(c). Like the FCC structure, the HCP structure is close packed with an atomic packing fraction of 0.74. In the HCP system, the c/a ratio is 1.633 and the coordination number is 12.

Crystal planes in the BCC and FCC systems are now briefly described. Crystal planes are designated with the notation (hkl), where, as before, h, k,

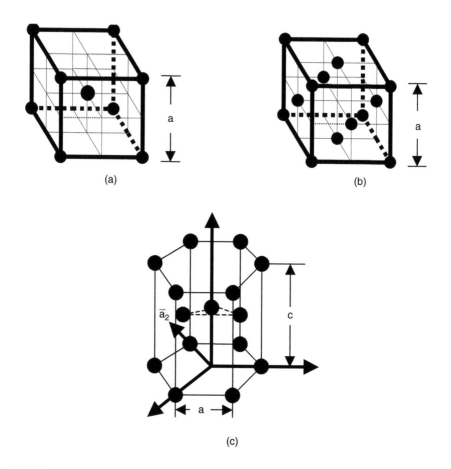

FIG. 3.4

(a) A BCC unit cell is depicted here. The lines point in the ⟨111⟩ directions (b) An FCC unit cell is shown here. The lattice spacing is a (c). The HCP unit cell is shown here. The angles between the a_i directions are 60° apart.

and l are integers. In the cubic system, (hkl) is perpendicular to $[hkl]$. Planes designated by (100) are perpendicular to the direction [100]. In Fig. 3.5, a plane that cuts the c-axis at a value of 1/2, the b-axis at a value of 1, and the a-axis at a value of 1 is designated (112). A series of parallel planes is designated $(nh\ nk\ nl)$ where n is an integer.

Generally, in order to determine the designation of a plane, the points at which the plane cuts the axis are inverted. If the inverted numbers $(h'k'l')$ are not all integers, then they are multiplied by the smallest integer possible in order to create the smallest integral values of h, k and l.

Planes that cross axes in the negative directions are designated with an over bar. For example, the plane that cuts the c-axis at $(0,0,-1)$ is designated $(00\bar{1})$. Figure 3.6 shows a series of planes designated by appropriate Miller indices. A family of planes, those related by symmetry, is designated by $\{hkl\}$.

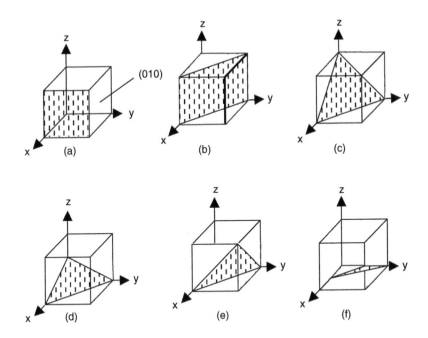

FIG. 3.5
Planes identified by Miller indicees (a) (100), (b) (110), (c) (111), (d) (112), (e) (11$\bar{1}$) and (f) (21$\bar{2}$).

For example, {111} represents the family of eight planes, (111), ($\bar{1}\bar{1}1$), (11$\bar{1}$), (1$\bar{1}$1), etc.

The following equations indicate the distance between planes (*hkl*) in three common systems. The distance between planes for the cubic system ($\alpha = \beta = \gamma = 90°$) is given by

$$d_{hkl} = \frac{a}{(h^2 + k^2 + l^2)^{1/2}} \qquad 3.2$$

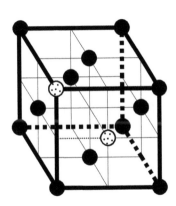

FIG. 3.6
A divacancy in an FCC unit cell is
shown here.

In the tetragonal system, $a = b$, $\alpha = \beta = \gamma = 90°$,

$$\frac{1}{d_{hkl}^2} = \left(\frac{h^2 + k^2}{a^2}\right) + \frac{l^2}{c^2}$$ 3.3

In the case of the orthorhombic system, $\alpha = \beta = \gamma = 90°$,

$$\frac{1}{d_{hkl}^2} = \frac{h^2}{a^2} + \frac{k^2}{b^2} + \frac{l^2}{c^2}$$ 3.4

and for the hexagonal system, $a = b$, $\alpha = \beta = 90°$, $\gamma = 120°$,

$$\frac{1}{d_{hkl}^2} = \frac{4}{3}\left(\frac{h^2 + hk + k^2}{a^2}\right) + \frac{l^2}{c^2}$$ 3.5

It is clear from the above that planes designated with larger values of h, k, and l are more closely spaced than those designated by smaller values of the integers (Warren, 1969).

3.2.3 Atomic Defects in Crystals

Generally, atomic defects, or point defects as they are often called, include: 1) vacant lattice sites (single or divacancies), 2) atoms lodged in interstitial sites (self-interstitials or interstitial impurities), and 3) impurity atoms, at very low concentrations, that maybe incorporated in the host (substational impurities) (Crawford and Slifkin, 1980). Point defects in the crystal lattice affect the rhestivity of metals because they can scatter conduction electrons. It is impossible to avoid them in crystals and they play a central role in atomic diffusion processes.

The fraction of vacancies in a monoelemental crystal under thermodynamic equilibrium is

$$X_v = e^{-G_v^f/kT}$$ 3.6

where G_v^f is the free energy of formation per vacancy. G_v^f is the difference between the Gibbs free energy/vacancy of a crystal with and without vacancies. Thermodynamically, it is possible to have divacancies, trivacancies, and clusters of vacancies, though the probability of finding larger clusters of vacancies decreases rapidly with increasing size. A diagram of a divacancy, which consists of two vacant nearest neighbor sites, in an FCC lattice is shown in Fig. 3.6 oriented along the $\langle 110 \rangle$ direction.

The equilibrium concentration of divacancies, as shown later in Section 3.5, is

$$X_d = \frac{z}{2}X_v^2 e^{\left(\frac{\Delta G}{kT}\right)}$$ 3.7

where ΔG is the binding energy between two vacancies and $z/2$ is the number of distinct orientations of a divacancy. A local strain field is created when a vacant site is created. The strain field associated with two nearest neighbor vacancies is smaller than that associated with two separate vacancies. This is the essence of the attraction between two vacancies that are in close proximity. Since the binding energy is typically lower than the free energy of formation, the $X_d \ll X_v$. The fraction of trivacancies is much smaller than that of divacancies,

$$X_{3v} = \lambda X_v^3 e^{\left(\frac{\Delta G_{3v}}{kT}\right)}$$

3.8

where ΔG_{3v} is the binding energy of a trivacancy and λ is the number of distinct orientations. Generally, the fraction of vacancies in a crystal would be approximated by the sum of the individual contributions,

$$X_v^T = X_v + 2X_d + 3X_{3v} + 4X_{4V} + \cdots + nX_{nv}$$

3.9

Interstitial impurities are not uncommon in metals. Small atoms like carbon, nitrogen, and hydrogen readily migrate throughout the interstices of different crystal lattices. Consider as an example a BCC unit cell. Two types of interstitials, depending on the symmetry of their environment, exist, octahedral, or tetrahedral. Octahedral sites are identified in Fig. 3.7 based on the (octahedral) symmetry of the local environment of the interstitial atom.

The centers of all the faces in the BCC unit cell are octahedral sites and so are the centers of the edges. Tetrahedral sites are depicted in Fig. 3.8. The symmetry of the local environment identifies these interstitial as such.

Self-interstitials (where a host atom resides in an interstitial location) are not common in metals due to the large strain fields associated with their existence. Typically, self-interstitials exist in metals due to external factors

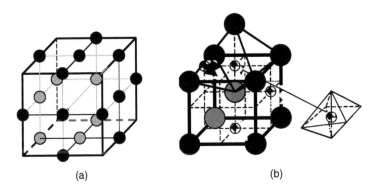

(a)   (b)

FIG. 3.7
(a) Octahedral interstitial locations are identified in this BCC unit cell; (b) the symmetry of an octahedral interstitial is illustrated here.

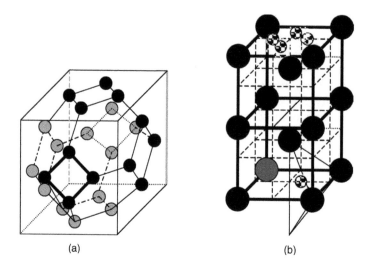

FIG. 3.8

(a) Tetrahedral interstitial sites are identified in this figure for a BCC cell; (b) the local symmetry of a tetrahedral site is identified at the bottom of this figure. The situation is identical for the other faces.

such as mechanical deformation or radiation damage. Interstitials are generally accompanied by a dilation of the local environment, opposite to the effect of vacancies. In fact, because their strain fields are opposite in sign, it is unlikely that vacancies and self-interstitials can exist in adjacent locations.

The fraction of self interstitials is given by

$$X_i = e^{-G_i^f/kT} \qquad\qquad 3.10$$

Like divacancies, it is also possible to find nearest neighbor interstitial pairs and other related defects in metals. Figure 3.9 shows examples of interstitial

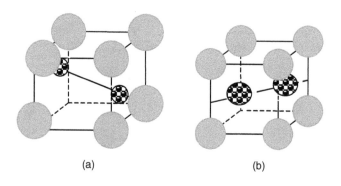

FIG. 3.9

Interstitial pairs. (a) $\langle 111 \rangle$ split "dumbbell" configuration (two interstitials surround the vacant site); (b) $\langle 110 \rangle$ split of "dumbbell."

pairs. An equation for the total interstitial concentration X_i^T, analogous to equation 3.9, may be defined. In general, the free energy of formation for an interstitial is greater than that of a vacancy, $G_i^f > G_v^f$, indicating that the fraction of self-interstitials is lower at thermal equilibrium in metals. In the next section elements of the diffusion process are discussed.

3.3 Tracer and Self-Diffusion in Crystals

Equation 3.1 shows that D depends on a prefactor, D_0, and a Boltzmann factor. The goal of this section is to determine a more detailed expression for the diffusion coefficient, thereby illustrating the factors that influence atomic diffusion in metals. The work in this section will provide the necessary framework for a subsequent discussion of various defect-mediated mechanisms of diffusion in metals.

In Chapters 1 and 2, we introduced the random walk problem and Fick's laws of diffusion. We showed that the mean square displacement of a particle during time, t, in three dimensions, is given by

$$\langle R^2 \rangle = 6Dt \qquad 3.11$$

where D is the diffusion coefficient. Through Fick's 1st law of diffusion, D, together with the concentration gradient, determines the amount of material that crosses a given cross-sectional area per unit time. In two dimensions, we showed that $\langle R^2 \rangle = 4Dt$ and in one dimension, $\langle R^2 \rangle = 2Dt$. Generally, in any dimension

$$\langle R^2 \rangle = \gamma Dt, \qquad 3.12$$

where γ would be 1/2, 1/4, and 1/6, in one, two, and three dimensions, respectively. The connection between this result and the random walk problem is addressed in the next section.

3.3.1 Random Walk in 3-D

We now discuss the Random Walk problem in three dimensions so we can later make a more direct connection to mechanisms of diffusion. We begin by considering an atom that makes N hops, each with an elementary jump vector $\vec{r}_i$. The final location of the atom with the vector

$$\vec{R}_N = \vec{r}_1 + \vec{r}_2 + \vec{r}_3 + \vec{r}_4 + \cdots \vec{r}_N = \sum_{i=1}^{N} \vec{r}_i \qquad 3.13$$

The magnitude of $\bar{R}_N$ is readily be determined by considering

$$
\begin{aligned}
\vec{R}_N \bullet \vec{R}_N = \vec{r}_1 \bullet \vec{r}_1 + \vec{r}_1 \bullet \vec{r}_2 + \vec{r}_1 \bullet \vec{r}_3 + \cdots + \vec{r}_1 \bullet \vec{r}_N \\
\vec{r}_2 \bullet \vec{r}_1 + \vec{r}_2 \bullet \vec{r}_2 + \vec{r}_2 \bullet \vec{r}_3 + \cdots + \vec{r}_2 \bullet \vec{r}_N \\
\vec{r}_3 \bullet \vec{r}_1 + \vec{r}_3 \bullet \vec{r}_2 + \vec{r}_3 \bullet \vec{r}_3 + \cdots + \vec{r}_3 \bullet \vec{r}_N \\
\cdot \\
\vec{r}_N \bullet \vec{r}_1 + \vec{r}_N \bullet \vec{r}_2 + \vec{r}_N \bullet \vec{r}_3 + \cdots + \vec{r}_N \bullet \vec{r}_N
\end{aligned}
\tag{3.14}
$$

This result can be simplified by collecting terms appropriately

$$
\begin{aligned}
\vec{R}_N \bullet \vec{R}_N &= \sum_{i=1}^{N} \vec{r}_i^2 + 2\sum_{i=1}^{N-1} \vec{r}_i \bullet \vec{r}_{i+1} + 2\sum_{i=1}^{N-2} \vec{r}_i \bullet \vec{r}_{i+2} + \\
&= \sum_{i=1}^{N} r_i^2 + 2\sum_{j=1}^{N-1}\sum_{i=1}^{N-j} \vec{r}_i \bullet \vec{r}_{j+i}
\end{aligned}
\tag{3.15}
$$

One should recall that in Eq. 3.15 the dot product

$$
\vec{r}_{i+j} \bullet \vec{r}_{i+j} = r_i r_{i+j} \cos\theta_{i,i+j}
\tag{3.16}
$$

where $|\vec{r}_i| = r_i$.

With this in mind, Eq. 3.16 can be further rewritten as

$$
\vec{R}_N \bullet \vec{R}_N = R_N^2 = \sum_{i=1}^{N} r_i^2 + 2\sum_{j=1}^{N-1}\sum_{i=1}^{N-j} r_i r_{i+j} \cos\theta_{i,i+j}
\tag{3.17}
$$

In crystalline solids with cubic symmetry, a nearest neighbor jump distance between atomic sites can be identified. Hence Eq. 3.1 can be simplified by allowing $r_i = r$. Herewith

$$
R_N^2 = Nr^2 \left(1 + \frac{2}{N} \sum_{j=1}^{N-1}\sum_{i=1}^{N-j} \cos\theta_{i,i+j} \right)
\tag{3.18}
$$

We are now in a position to consider the displacements of a large number of identical particles. By doing so, we can immediately consider an ensemble average whereby

$$
\langle R^2 \rangle = Nr^2 \left(1 + \frac{2}{N} \sum_{j=1}^{N-1}\sum_{i=1}^{N-j} \langle \cos\theta_{i,i+j} \rangle \right)
\tag{3.19}
$$

The quantity within parentheses is identified as the correlation factor f,

$$
f = 1 + \frac{2}{N} \sum_{j=1}^{N-1}\sum_{i=1}^{N-j} \langle \cos\theta_{i,i+j} \rangle
\tag{3.20}
$$

The value of f depends on crystal symmetry. It should be clear that if the motion of the atom is truly random, then $f = 1$, since in a truly random process, as we showed earlier, $\langle R^2 \rangle = Nr^2$. In reality, the dynamics are often not truly random in atomic crystals, except in the case of the so-called interstitial mechanism, which we discuss later. In general, f indicates the extent to which the direction of a hop is correlated with that of a previous hop, hence the name correlation coefficient. The correlation coefficient will further be discussed in Section 3.10 of this chapter.

For now we will compare Eqs. 3.12 and 3.19 with Eq. 3.20, whereupon it becomes clear that

$$D = \gamma \frac{N}{t} r^2 f \tag{3.21}$$

If $\Gamma_t = N/t$ is defined as the number of jumps per unit time, an expression for the diffusion coefficient in terms of the jump frequency is

$$D = \gamma \Gamma_t r^2 f \tag{3.22}$$

Γ_t will depend on 1) the activation barrier that the particle must surmount as it migrates from one point to another, on 2) the mechanism of transport and on 3) the number of nearest neighbor sites. Intuitively $\Gamma_t = z\Gamma P$, where z is the number of equivalent jumps; it is the number of nearest neighbor atomic sites available if diffusion occurs via a vacancy mechanism. Γ is the jump frequency of the atom and it is temperature dependent. P is determined by the mechanism of transport. If the atom hops via a vacancy mechanism, then P is the probability that a site is available and would be equal to the fraction of vacant sites in the crystal, $P = X_v$. The nearest-neighbor jump distance, r, is determined by atomic arrangements, atomic size, and by the mechanism of transport. Finally, as suggested earlier, f is associated with crystal symmetry and with the mechanism by which the atom traverses from one location to another. In summary, an expression for the diffusion of an atom in a crystalline lattice may be written as

$$D = \gamma z \Gamma \, Pr^2 f \tag{3.23}$$

In the subsequent sections, each of these parameters will be discussed in relation to crystal structure and migration mechanism. It then becomes clear that equation 1, specifically D_0, can possess somewhat distinct forms depending on the crystal system, defect concentration and the mechanism of transport. The reader should be forewarned to be patient. It will take some time to slowly dissect this expression for the diffusion coefficient (Eq. 3.18). In the section that follows we discuss the jump frequency.

3.3.1.1 The Jump Frequency, Γ

The parameter, Γ, describes the rate at which an atom hops from one nearest neighbor site to another. It is therefore anticipated that Γ would be a function of temperature, atomic structure, atomic vibrational frequency and the Gibbs Free energy of migration. In this section an expression for Γ is derived. The derivation is meant to be intuitive rather than quantitative.

As an atom migrates from one equilibrium site to another, it experiences repulsive interactions from its neighbors. At each equilibrium position, it vibrates at a frequency on the order of $v_D \approx 10^{13}$ Hz, the so-called Debye frequency, discussed in the next section. The thermal energy, kT, at ambient temperature is 0.025 eV while ~1.0 eV is required by the atom to surmount the barrier. The difference between kT and the activation energy indicates that the atom does not possess sufficient energy to hop to a new site and, moreover, the hopping rate is temperature dependent. It turns out that the atom occasionally acquires enough energy from phonons in the crystal to move from its current location to a nearest neighbor location. A phonon is a quantum of lattice vibrations (sound wave), and will be discussed in the next section in connection with the Debye frequency.

Consider for convenience the hop of an atom initially located at position "1" (Fig. 3.10), to location "3," a vacant site.

The probability that an atom will possess sufficient energy to hop into a new site is dictated by the Boltzmann factor, $e^{-G_m/kT}$, (Chapter 1). Therefore the rate at which a hop occurs is

$$\Gamma = v_D e^{-G^m/kT} \qquad\qquad 3.24$$

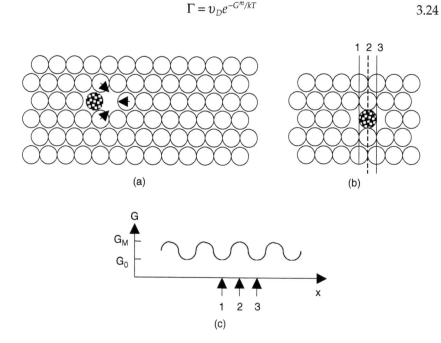

(a)   (b)

(c)

FIG. 3.10
(a) A vacancy is created in this two-dimensional lattice. The surrounding atoms relax inward after the vacant site is created. (b) Labeled atom experiences most resistance as it migrates through location 2, on its way from 1 to 3. (c) Its Free energy is maximum at this location.

where $G^m = G_m - G_0$ is the free energy change associated with migration of the atom from the bottom of the well to the top of the well and v_D is identified as the so-called Debye frequency.

3.3.2 Debye Frequency

The significance of v_D is now discussed. Collective vibrations of each of the N atoms in a crystal lead to the development of a spectrum of normal mode vibrations that range in frequency from 0 up to approximately 10^{13} Hz (cycles/sec). For example, in a one dimensional crystal, the normal mode, which possesses the shortest wavelength, is associated with the vibration of each successive atom in the opposite direction, as illustrated in Fig. 3.11a.

On the other hand, longer wavelength modes are associated groups of atoms vibrating in unison on this 1D crystal. These vibrations in a crystal are quantized. To understand the origin of the Debye frequency, it is instructive to consider the internal energy of a crystal since vibrations of atoms contribute largely to the internal energy. The energy of a mode, due to n phonons, of an elastic wave in the crystal is specified by $\varepsilon = (n+1/2)\hbar\omega$, where ω is the frequency. This expression for the energy is associated with the fact that the dynamics are similar to a quantum harmonic oscillator of frequency ω. Consequently, the expression for the energy of a phonon possesses the same functional dependence on the number of modes as the energy for photons (quanta of light).

For long wavelength dynamics, the sample may be treated as an elastic continuum. The displacement of a point in the sample, $U(r,t)$, satisfies the wave equation from elasticity theory that describes the propagation of sound waves through a medium. For a wave of amplitude U_0 traveling in direction $\vec{k}$,

$$U(\vec{r},t) = U_0 e^{i(\vec{k}\cdot\vec{r}-\omega t)} \tag{3.25}$$

where $k = |\vec{k}| = 2\pi/\lambda$ and λ is the wavelength associated with a phase velocity $u = \omega/k$ ($\omega = 2\pi v$). Of particular interest to us is a standing wave that is established within the crystal, in this case assumed to be a cube of length L,

$$U(\vec{r},t) = 2U_0 e^{-i\vec{k}\cdot\vec{r}} \cos(\omega t) \tag{3.26}$$

$k_i L = n_i \pi$, $k^2 = (\frac{\pi}{L})^2 [n_x^2 + n_y^2 + n_z^2]$. The number of standing waves with wave vector less than k, $\Omega(k)$, may be calculated by recognizing that these

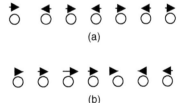

(a)

(b)

FIG. 3.11
In part a, the wavelength is equal to the interatomic spacing, whereas in part b the wavelength is longer, many atomic distances.

standing waves are enclosed within a region of a sphere of "radius" $R = (n_x^2 + n_y^2 + n_z^2)^{1/2} = Lk/\pi$. Since only positive values of k are permitted ($k_x > 0$, $k_y > 0$ and $k_z > 0$),

$$\Omega(k) = \frac{1}{8}\left[\frac{4}{3}\pi\left(\frac{Lk}{\pi}\right)^3\right] = \frac{Vk^3}{6\pi^2} \qquad 3.27$$

alternatively,

$$\Omega(\omega) = \frac{V}{6\pi^2}\frac{\omega^3}{u^3} \qquad 3.28$$

The number of modes between ω and $\omega + d\omega$ (corresponding to wave vectors between k and $k + dk$) is

$$\Omega(\omega)d\omega = 3\frac{V\omega^2}{2\pi^2 u^3}d\omega \qquad 3.29$$

In the above equation the factor of 3 is introduced to reflect the fact that $\bar{u}(\bar{r},t)$ possesses 3 polarization directions, two transverse and one longitudinal, for each wave vector $\bar{k}$. The total number of modes can't exceed $3N$, where N is the number of atoms in the crystal. To meet this requirement Debye indicated that

$$\int_0^\infty \Omega(\omega)d\omega = \int_0^{\omega_D} \Omega'(\omega)d\omega = 3N \qquad 3.30$$

where $\Omega'(\omega)d\omega = \Omega(\omega)d\omega$ for $\omega < \omega_D$ and $\Omega(\omega) = 0$ for $\omega > \omega_D$.

Debye identified a maximum normal mode frequency, ν_D, as the upper cutoff limit with lower limit being equal to zero. It is the long-wavelength normal modes that are responsible for providing sufficient energy for the transport of atoms in solids. The Debye theory enjoyed tremendous success at predicting the low temperature heat capacity of solids. Typical Debye frequencies are shown in Table 3.2 for different solids. Note that the frequencies are all on the order of $\nu_D \approx 10^{12}$ s^{-1}.

3.3.2.1 An Expression for the Tracer Diffusion Coefficient

Based on the foregoing discussion, we are now in a position to write down a somewhat more complete expression for Eq. 3.23

$$D = \gamma z\,Pr^2\,f\nu_D e^{-G^m/kT} \qquad 3.31$$

This expression is very similar to the equation describing the temperature dependence of D, Eq. 3.1. Note, however, that $G^m = H^m - TS^m$, where H^m is the enthalpy of migration and S^m is the entropy associated with the fact that the directions in which the particle may move are restricted by the symmetry of the lattice; atoms cannot hop in any arbitrary direction in space. Now we rewrite Eq. 3.31

$$D = \gamma z\,Pr^2\,\nu_D f e^{S^m/k} e^{-H^m/kT} \qquad 3.32$$

TABLE 3.2

Debye frequencies for a range of solids are tabulated here

Solid	v_D (s^{-1})
Na	3×10^{12}
K	2.08×10^{12}
Cu	6.6×10^{12}
Ag	4.68×12^{12}
Au	3.53×10^{12}
Be	2.1×10^{13}
Mg	6.1×10^{12}
Zn	5.25×12^{12}
Cd	3.6×10^{12}
Fe	8.2×10^{12}
Co	8.1×10^{12}
Ni	7.8×10^{12}
Al	8.2×10^{12}
Ge	6.1×10^{12}
Sn	5.5×10^{12}
Pb	1.8×10^{12}
Pt	4.7×10^{12}
Diamond (carbon)	3.9×10^{13}

Data taken from Kittel, 1976

In Eq. 3.32, D depends on r, the nearest neighbor (*n.n.*) jump distance, z, the number of equivalent jumps (in this case the number of *n.n* sites), f, the correlation factor, v_D, the Debye frequency, and on P, the probability that a jump would occur. The number of available nearest neighbor atoms and the nearest neighbor jump distance, r, are functions of crystal structure, as we saw earlier. Therefore it is apparent that the magnitude of the diffusivities in BCC and FCC crystals are in principle different. Moreover, both f and P depend on the defect mediated mechanism of transport.

3.4   Atomic Transport in Crystals via a Single Vacancy Mechanism

The hopping of an atom, mediated by the presence of a vacancy, is perhaps the most important of all atomic diffusion mechanisms, especially at elevated temperatures. In close packed (HCP, FCC) metals the vacancy mechanism plays an especially important role in atomic diffusion, in contrast to diffusion in the more open BCC, or diamond-like structures (discussed in the next chapter). In the vacancy mechanism for atomic diffusion, an atom is allowed to jump only if an adjacent site is vacant (Fig. 3.12).

In a simple cubic system the nearest neighbor jump vector is $\vec{r} = a \langle 100 \rangle$. At any given moment vacancies migrate throughout the crystal in random

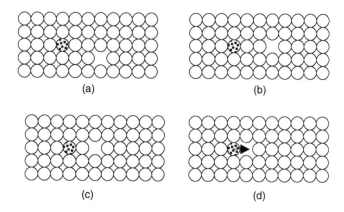

FIG. 3.12

A tracer or solute impurity atom is imagined to migrate via a vacancy mechanism. At a given instant, a vacant site is not available for it to hop into, but one is located a few lattice spacings away. The surrounding atoms can move such that the vacancy appears next to the foreign atom, allowing it to jump. This is illustrated in steps a through d.

directions, assuming the absence of external driving force. Vacancies can originate from "sources," which generally include free surfaces, grain boundaries, or dislocations. They arrive at a nearest neighbor site of an atom from random directions, as illustrated in Fig. 3.12. In light of these comments it is important to make a distinction between self diffusion and tracer diffusion processes.

It is necessary to label the diffusant in order to evaluate its migration through the medium in which it resides. In situations where it is necessary to determine the diffusion coefficient of species A throughout an A-environment, *self-diffusion coefficient*, it is typical to rely on the use of an isotope of A, because the dynamics of an A-isotope should closely mimic those of species A. However, since an isotope is used instead of the actual species A, then in this case a *tracer diffusion coefficient* is obtained. *Tracer diffusion* and *self-diffusion* are described in the next section.

3.4.1 Self-Diffusion and Tracer Diffusion via a Vacancy Mechanism

With regard to the vacancy mechanism, tracer-A is allowed to hop only if a vacant site is available next to it. The probability that a vacant site is available is provided by the concentration of vacancies, $P = X_v$ (Eq. 3.6). When the tagged atom (tracer) hops, there is a greater than random probability that it would hop back into the site just vacated. In this regard the motion of the tracer is correlated (only a fraction of its hops contribute to truly random diffusion). In order to gain information about *self-diffusion* from *tracer diffusion* the effects due to correlation and to slight differences in isotopic mass must be reconciled. The tracer diffusion coefficient is related to the self diffusion coefficient such that

$$D_v^T = f D_v^{SD} \qquad\qquad 3.33$$

which indicates that the tracer diffusion coefficient measured in an experiment is smaller than the self-diffusion coefficient because $f < 1$. In light of the foregoing, it is necessary to consider additional comments regarding the correlation coefficient in relation to the vacancy mechanism in crystals. It is clear from Eq. 3.20 that

$$f = \lim_{N \to \infty} \langle R^2 \rangle / Nr^2, \qquad\qquad 3.34$$

and that f may be interpreted as the fraction of hops that contribute to diffusion due entirely to random hopping. In other words, correlations reduce the fraction of hops that contribute to a truly random diffusion process. In fact, one can, alternatively, write the correlation factor

$$f = \frac{D_{actual}}{D_{random}} \qquad\qquad 3.35$$

To further illustrate the significance of f, we might consider, for a moment, the hopping of a tracer into a vacant site. The probability that it will hop into the vacant site is $1/z$. There is also a probability of $1/z$ that it will hop back into the original location to cancel its previous jump. This process involves 2 hops. It follows that $2/z$ is the fraction of hops that will not contribute to a random hopping process. If this is the case then

$$f \approx 1 - \frac{2}{z} \qquad\qquad 3.36$$

Values of the correlation function, calculated using Eq. 3.20, are in the second column of Table 3.3.

Based on Eq. 3.32 and 3.6, the *self-diffusion* coefficient, assuming diffusion occurs via a single vacancy mechanism, is

$$D_v^{SD} = \gamma z r^2 v_D e^{-(G_v^m + G_v^f)/kT} \qquad\qquad 3.37$$

TABLE 3.3

Comparison of correlation coefficient in the cubic system

Structure	f^*	f (Eq. 3.36)
SC	0.6531	0.67
BCC	0.727	0.75
FCC	0.782	0.83

*Calculated using a more rigorous procedure, involving Eq. 3.20.

TABLE 3.4

This table lists a series of parameters associated with vacancy formation and migration in various metals. The measurements were obtained using different techniques

Metal	$T_m(K)$	H_v^f (eV)	S_v^m/k	H_v^m (eV)	H_{2v}^m (eV)	Structure
Au	1,333	0.86–0.94	0.5–1.2	0.89	0.94	FCC
Ag	1,234	0.99	1.5	0.86	0.6	FCC
Cu	1,353	1.0	~2	1.1		FCC
Al	983	0.73	2–2.4	0.65	0.5	FCC
Pt	2,044	1.5		1.4		FCC
Pb	600	0.5		0.6		FCC
Nb	2,688	2		2.1		FCC
Mo	2,873	2.3		1.7		BCC
Fe	1,083	1.5		1.1		BCC
Na	370	0.4		0.04		BCC
Mg	924	0.7		0.7		HCP

Point Defects in Solids, Crawford and Slifkin, (Volume 1) Plenum Press, NY, 1972.

because $f = 1$ (correlation effects are absent) and $P = X_v$.

If the crystal structure is FCC then the nearest neighbor distance is $r = a/\sqrt{2}$ the jump vector is $\frac{a}{\sqrt{2}}\langle 110 \rangle$ and $z = 12$, which indicates that

$$D_v^{SD} = v_D e^{(S_v^m + S_v^f)/k} e^{-(H_v^m + H_v^f)/kT} \qquad 3.38$$

Enthalpies of migration, H^m, and enthalpies of formation, H_v^f, associated with the transport and formation of single vacancies in metals are given in Table 3.4. The magnitudes of these enthalpies are typically on the order of ~eV as opposed to many eVs. The prefactor in this case, D_0, depends on the entropy of formation of a vacancy on the lattice. It also depends on the entropy of migration, which reflects the degrees of freedom available in the limited phase space.

Thus far, the expression describing the concentration dependence of vacancies on temperature has been used to obtain an expression for the diffusion coefficient. Understanding the origins of the temperature dependence of X_v is an important topic in its own right and therefore an entire section is devoted to it.

3.5 The Equilibrium Vacancy Concentration

The equilibrium concentration of vacancies in a crystal is now calculated. Vacancies are distributed throughout crystals under conditions of thermodynamic equilibrium. In most models developed to calculate the average defect concentration in crystals, it is assumed that each defect is a statistically independent entity, so the energy associated with the creation of each defect

is additive. The interactions between defects are neglected in these calculations. This is typically not an unreasonable assumption considering that
the fraction of defects is usually small, fractions of a percent near the
melting temperature, as we see later. The combinatorial entropy of mixing
of the species on the lattice provides an important contribution to the
overall free energy of the system. It is evident that $g = G'(T,P) - G(T,P)$,
the difference between the free energy of a crystal with vacancies has two
contributions, one term describing the combinatorial entropy of mixing
and the other due to enthalpy,

$$g \approx N_v G_v^f - kT \ln \Omega_v \qquad 3.39$$

where $\Omega_v(N_v)$ is the number of ways that N_v (indistinguishable) vacancies
can be arranged on N_L sites;

$$\Omega = \frac{N_L!}{(N_L - N_v)! N_v!} \qquad 3.40$$

Note that it is assumed that the sample is composed only of single vacancies
and that they are located sufficiently far apart that they do not interact.
Since N_L and N_v are very large numbers, Stirlings approximation,
$\ln N! \approx N \ln N - N$, may be used, which leads to

$$G = X_v G_v^f + kT\{(X_v \ln X_v + (1 - X_v)\ln(1 - X_v)\} \qquad 3.41$$

where $X = N_v/N_L$ and $G = g/N_L$ is the free energy per lattice site.

If this equation is minimized to determine the equilibrium concentration
of vacancies, $\frac{\partial g}{\partial N_v} = 0$, we obtain the result that the equilibrium fraction of
vacancies, X_v, is

$$\ln X_v = -G_v^f/kT \qquad 3.6$$

where the approximation, $\frac{N_v}{N_L - N_v} \approx X_v$. With this, the derivation of the temperature dependence of the vacancy concentration is concluded. Within the
Harmonic approximation, the entropy associated with the formation of a
defect is $S^f = k \sum_i \ln(v_i^0/v_i)$ where v_i^0 is the vibrational frequency of the pure
crystal and v_i is that of crystal containing defects. For vacancies in most metals,
$S \sim 1k - 2k$, which indicates that the entropic contribution to the free energy
of formation is relatively small. Figure 3.13 illustrates the relative influence of
the entropy of mixing in relation to the free energy of formation per vacancy.
It is the interplay between the two contributions, entropy and enthalpy, that
determines the vacancy concentration at thermodynamic equilibrium. In the
next section we compare the predictions, Eq. 3.6, with actual experiments.

3.5.1 Vacancy Concentration in Crystals: Experiment versus Theory

Many experiments aimed at measuring the concentration of vacancies in
crystals exploit the fact that certain physical properties are influenced by the

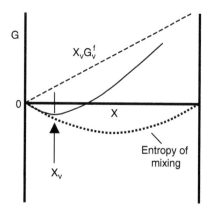

FIG. 3.13
Relative contribution of the free energy
of formation and the entropy of mixing
to the free energy in the system.

vacancy concentration. Vacancy concentrations can typically be inferred from measurements of physical properties such as volume changes, electrical resistivity changes and, in some special circumstances, heat capacity. In this regard, a clever experiment was performed by Simmons and Balluffi during the 1960s. They showed that the vacancy concentration X_v could be determined if the temperature dependence of the change in length, Δl, of a sample with cubic symmetry using dilatometry, and changes in the lattice parameter, Δa, measured simultaneously using X-rays, then

$$X_v^T = 3\left(\frac{\Delta l}{l} - \frac{\Delta a}{a}\right) \qquad 3.42$$

It is worthwhile to take a moment to examine the origins of equation 3.42. In crystals, thermal expansion is due to lattice vibrations and arises due to the anharmonicity of the interaction potential between atoms in the crystal. If the interactions were harmonic, then solids would not exhibit thermal expansion. The presence of point defects, particularly vacancies, and their increase in number with increasing temperature provides an additional contribution to the increase in the size of a sample, as measured using dilatometry, for example. Therefore, if one were to measure the change in size of the lattice parameter with temperature, then one could determine the vacancy concentration. Experimentally it has been shown that Eq. 3.15 is indeed an accurate description of the vacancy concentration in metals.

The creation of a vacancy is imagined to occur with the removal of an atom from the interior of the crystal and placing it at the free surface. When a vacancy is created within the crystal, a local distortion (strain field) in the vicinity of the vacant site occurs, wherein the atoms locally relax inward. This leads to a reduction in the local volume, Δv. The volume of a vacant site is therefore

$$v_v = v_a - \Delta v_v = \beta v_a \qquad 3.43$$

where v_v is the volume associated with a vacant site and v_a is that of an atomic site; Δv_v is the local reduction ($\beta < 1$). If we imagine that there are N_v vacant sites and N_L lattice sites, then the average volume associated with a site in this crystal is

$$\langle v \rangle = \frac{N_v}{N_L} v_v + \frac{N_L - N_v}{N_L} v_a \qquad 3.44$$

assuming the rule of mixtures. The difference between the volume per site of a perfect crystal and the crystal containing defects is,

$$\langle v \rangle - v_a = \frac{N_v}{N_L} \beta v_a + \frac{N_L - N_v}{N_L} v_a - v_a$$

$$= X_v \beta v_a - X_v v_a \qquad 3.45$$

This equation can be rewritten to yield

$$\frac{\langle v \rangle - v_a}{v_a} = \frac{\Delta v}{v} = X_v(\beta - 1) \qquad 3.46$$

If we assume that the lattice parameter increases from a to $a + \Delta a$, then

$$\frac{3\Delta a}{a} = X_v(\beta - 1) \qquad 3.47$$

The total volume of the crystal with vacancies is

$$V_c = N_v \beta v_a + N_L v_a \qquad 3.48$$

If the volume of the perfect crystal, $V = N_L v_a$, is subtracted from V_c, and assuming that the sample is a cube and that the length increases from l to Δl, then

$$\frac{3\Delta l}{l} = X_v \beta \qquad 3.49$$

A comparison of Eq. 3.42 and 3.44 leads to Eq. 3.42 for the vacancy concentration. Experiments confirm the exponential dependence of X_v on temperature in metals.

The foregoing example involved measurements of samples in which vacancies would be the primary defect. However, if a sample contained vacancies and self interstitials, then for a cubic crystal,

$$X_v^T - X_i^T = 3\left(\frac{\Delta l}{l} - \frac{\Delta a}{a} \right) \qquad 3.50$$

$3(\frac{\Delta l}{l} - \frac{\Delta a}{a}) > 0$ would imply that vacancies are the predominant defects, whereas $3(\frac{\Delta l}{l} - \frac{\Delta a}{a}) < 0$ would imply that self interstitials would be dominant.

It should be noted that self-interstitials are typically not the predominant defect under equilibrium conditions. The strain fields associated with self-interstitials typically extends beyond a single lattice position, which largely explains the large formation energies compared to vacancies.

Another method that has proven to be reasonably effective at assessing the concentration of vacancies is measurement of the resistivity (see for example, J.E. Bauerle and J.S. Koehler, "Quenched-in lattice defects in gold," Physical Review, 107, 1493, 1957). The resistivity can be written as

$$\rho \approx \sum_{j} \rho_j X_j + \rho_{lattice} \qquad\qquad 3.51$$

where ρ_j is the contribution to the resistivity due to defect j; $\rho_{lattice}$ is the contribution to the resistivity due to phonons from lattice vibrations. It is clear that sample preparation is critical in these experiments. Typically, thin wires of the sample are annealed at high temperatures where the defect concentration is high. If the sample is a pure metal then there is confidence that the dominant defect contribution is due to vacancies. One concern is that if the exterior of the sample cools at a faster rate than the interior, then plastic deformation can occur. This will lead to the creation of dislocations that act as sinks for vacancies, thereby affecting the vacancy concentration. After annealing, the sample is quenched to liquid helium temperatures. Typical quenching rates exceed 10^4 C/sec. At liquid helium temperatures, the contribution to the resistivity due to lattice vibrations is negligible and the main contributor is the vacancy concentration, believed to be retained at low T after the quench (a reasonable assumption). Resistivity experiments also confirm the theoretical prediction. The change in resistivity is shown to be $\Delta\rho = Ae^{-H_0^f/kT_Q}$, where A is a constant and T_Q is the quench temperature.

Other techniques used to determine the vacancy concentration involve measurement of the heat capacity. This can be understood by considering the Gibbs free energy of a crystal with vacancies. Knowledge of the enthalpy enables calculation of the heat capacity at constant pressure in terms of the equilibrium vacancy concentration. This method has not proven to be as reliable as the other methods. Another effective method used to determine the vacancy concentration is Positron-electron annihilation. This technique measures the concentration under equilibrium conditions at temperature T. Positrons are introduced into the sample. They are attracted by vacancies and in contrast repelled by the positive iron cores. Positrons are annihilated by electrons. However the rate of annihilation is different when the electrons are free, as opposed to being trapped by vacancies. This difference in response of the positrons provides one method by which the vacancy concentration may be determined.

Table 3.4 shows typical values of vacancy formation energies for different metals. Note that the enthalpies of formation for vacancies are typically on the order of ~1 eV.

TABLE 3.5

Values of D_0 are shown here for a limited number of metals

Metal	D_0 (cm²/sec)	Structure
Cu	0.2	FCC
Ag	0.4	FCC
Ni	1.3	FCC
Au	0.1	FCC
Pb	0.28	FCC
Mg	1.0	HCP
Nb	12	BCC
Na	0.2	BCC

From Shewmon, *Diffusion in Solids*, McGraw Hill, N.Y. 1963.

The Table that follows (Table 3.5) lists typical values for D_0 self diffusion in selected metals. Note that the prefactors all reside within a certain range of values.

3.6 Divacancies and Their Effect on Diffusion

Divacancies are present in metals, particularly at high temperatures where their contribution to the diffusivity is expected to become important. There is evidence that diffusivity exhibits deviations from Arrhenius behavior at high temperatures due, in some cases, to divacancies. Figure 3.14 shows the temperature dependence of the self-diffusion of sodium. The data deviates from an Arrhenius temperature dependence at high temperatures. This deviation suggests that more than one type of mechanism is operational at higher temperatures, whereas the lower T data is consistent with a vacancy mechanism. While the situation is not entirely clear-cut, divacancies have been implicated as being responsible for the deviation in Na. In other systems, the deviations are appropriately rationalized in terms of a phase transformation in the material. Other explanations for such deviations have also been attributed to a temperature-dependent enthalpy, though the evidence is less certain. In this section the effect of divacancies on self-diffusion is discussed. We first calculate the equilibrium concentration of divacancies in a crystal and then discuss their influence on diffusion.

Earlier we highlighted the existence of a strain field in the vicinity of a vacancy. An effective attraction between vacancies may occur, resulting in the creation of divacancies. The origin of the attraction is associated with the reduction in the free energy associated with the local strain field in the vicinity of a divacancy compared two independent single vacancies (Larger vacancy clusters often occur in metals as a result of radiation). The vacancies have to be sufficiently close for this attraction to occur.

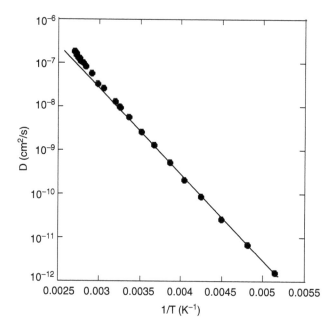

FIG. 3.14
The temperature dependence of the self-diffusion of Na is plotted here (data taken from N.N. Mundy, (1971)).

In the model that is about to be described, the total vacancy concentration is due to single and divacancies, $X_v^T = X_v + 2X_d$. The free energy difference between a perfect crystal and a crystal with single vacancies and divacancies is given by

$$g = N_v G_v^f + N_d G_d^f - kT \ln \Omega_d \Omega_v \qquad 3.52$$

where N_d is the number of divacancies. The first term represents the free energy of formation of single vacancies and the second represents that of the divacancies. Ω_v is the number of ways that N_v single vacancies can be arranged onto $N_L - N_d \alpha$ sites; α is the number of nearest neighbor sites to the divacancy lattice positions. Ω_d is the number of ways (positions/orientations) that N_d divacancies can be arranged on the remaining $(z/2)N_L - zN_v$ locations. Note that the factor of $z/2$ is needed because the divacancy has $z/2$ distinct orientations. It follows that $\Omega = \Omega_v \Omega_d$,

$$\Omega = \frac{(N_L - \alpha N_d)!}{[(N_L - \alpha N_d) - N_v]! N_v!} \bullet \frac{\left(\frac{z}{2} N_L - z N_v\right)!}{\left[\left(\frac{z}{2} N_L - z N_v\right) - N_d\right]! N_d!} \qquad 3.53$$

We can take advantage of the fact that $N_L \gg N_v$ and $N_L \gg N_d$, so, assuming that the vacancies and divacancies are formed independently,

$$\Omega = \frac{N_L!}{[N_L - N_v]!N_v!} \cdot \frac{\left(\frac{z}{2}N_L\right)!}{\left(\frac{z}{2}N_L - N_d\right)!N_d!} \qquad 3.54$$

Stirling's approximation can be used to simplify this equation

$$\ln\Omega = N_L \ln N_L + \frac{z}{2}N_L \ln\left(\frac{z}{2}N_L\right) - (N_L - N_v)\ln(N_L - N_v)$$

$$- N_v \ln N_v - \left(\frac{z}{2}N_L - N_d\right) - N_d \ln N_d \qquad 3.55$$

The condition for thermodynamic equilibrium for the relevant equation

$$v + v \rightarrow d \qquad 3.56$$

implies that

$$\mu_d - 2\mu_v = 0 \qquad 3.57$$

where $\mu_v = \frac{\partial g}{\partial N_v}$ and $\mu_d = \frac{\partial g}{\partial N_d}$. Having done so, we arrive at the result that

$$2G_v^f - G_d^f \approx -kT\left(2\ln N_L - 2\ln N_v - \ln\left(\frac{z}{2}N_L\right) + \ln N_d\right) \qquad 3.58$$

Since $X_v = N_v/N_L$ and $N_d/N_L = X_d$, then

$$X_d = \frac{z}{2}X_v^2 e^{\left(\frac{\Delta G}{kT}\right)} \qquad 3.7$$

where $\Delta G = 2G_v^f - G_d^f$ is a binding energy (reduction in free energy associated with the formation of the pair).

If diffusion occurs via single and via divacancies, we could treat both processes as independent and write the total tracer diffusion coefficient as a sum of single and divacancy contributions

$$D^T = D_v + D_d \qquad 3.59$$

This is not unreasonable, since the concentrations of each one are small. If we consider tracer diffusion to occur in an FCC lattice, then D_v is given by Eq. 3.59. On the other hand,

$$D_d = \frac{r_d^2}{6}gv_D X_d f_d e^{-G_d^m/kT} \qquad 3.60$$

where g is the number of equivalent jumps. The divacancy would migrate like a dumbell, i.e., the two vacancies move together as a pair without dissociating and associating. The divacancy can move when one vacancy moves to a new site where the distance between them remains the same. f_d is a correlation factor for divavancies. It is left as an exercise at the end of the chapter to show that at high temperatures D^T deviates from an Arrhenius temperature dependence.

As suggested in Section 3.1, in the absence of structural transitions in metals, the divacancy is often implicated as being responsible for an enhancement of self-diffusion in metals at high temperatures near the melting temperature, T_m. This, as indicated above, is largely associated with the fact that their concentrations increase appreciably near the melting temperature and they are relatively mobile. Recently, molecular dynamic simulations strongly indicate that in copper and aluminum, such an enhancement would be due to a high concentration of self-interstitials (K. Nordlund and R.S. Averback). In fact, the concentration of self-interstitials would be expected to increase near T_m. The interstitials possess a split dumbbell-like configuration. The entropic contribution is large $S_{dbell}^f \approx 15\ k$ (for a single vacancy $S_v^f = 2.3\ k$ and for a divacancy $S_d^f = 5$) but the migration enthalpy is small, $H_{dbell}^m = 0.081$ eV, compared to 0.7 eV and 0.26 eV, for single and divacancies, respectively. The high concentration of slef interstitials and their relatively large mobilities are believed to be responsible for the enhancement of self-diffusion in Cu and Al.

3.7 Diffusion of Interstitials in Crystals

The migration of interstitial impurities through lattices is associated with a migration energy per atom, H_i^m. Their dynamics are not correlated ($f = 1$) since, invariably, sites are readily available into which the atoms can hop. In an FCC lattice, a self-interstitial atom can hop from one site to another with jump vector is $a/2\langle 110\rangle$. The number of equivalent jumps (nearest neighbor sites here) is $z = 4$. These considerations would lead to a diffusion coefficient of

$$D = \left(\frac{a^2}{6}\nu_D e^{(S_i^m + S_i^f)/k}\right)e^{-(H_i^m + H_i^f)/kT} \qquad 3.61$$

In the case of an interstitial impurity

$$D = \left(\frac{a^2}{6}\nu_D e^{S_i^m/k}\right)e^{-H_i^m/kT} \qquad 3.62$$

The temperature dependence of the interstitial diffusion of carbon in three BCC metals is shown in Fig. 3.15. The differences between the slopes reflect differences in the activation energies associated with diffusional transport.

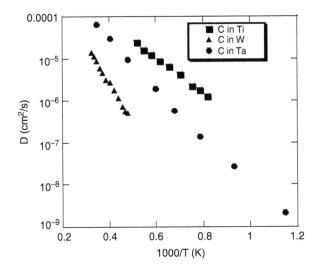

FIG. 3.15
The interstitial diffusion of C into three metals is shown here. Data taken from I.I. Kovenski, 1964.

Values for the prefactor D_0 and the migration enthalpies are shown in the table below to illustrate the magnitudes of the parameters that characterize the interstitial diffusion process.

3.8 Ring Mechanism of Atomic Diffusion

A 4-membered ring (or exchange) mechanism is illustrated in Fig. 3.16. The atoms in the plane move in the direction of the arrows to exchange places. In essence, this is a cooperative process. A 3-membered ring mechanism in the (111) plane or a 2-membered exchange process are also possible and are in fact energetically more favorable. The important attribute of a ring or exchange mechanism is that it is not mediated by defects. The ring (or exchange)

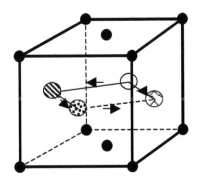

FIG. 3.16
The atoms in the middle plane can exchange positions via a 4-membered ring mechanism.

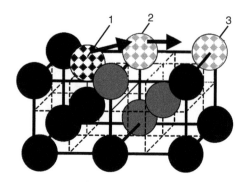

FIG. 3.17
The interstitialcy mechanism is illustrated here.

mechanism is generally not an important mechanism in metals, except in
cases where the defect concentration is high, such as near grain boundaries,
or in heavily radiation damaged materials.

There is evidence from computer simulations that in silicon the silicon
atoms can migrate via an exchange-type mechanism in this directional bonding material (Pandey). This will be discussed further in Chapter 5 on elemental semiconductors.

3.9   The Interstitialcy Mechanism of Atomic Diffusion

The interstitialcy mechanism typically involves the migration of self-
interstitials. In this mechanism, an atom on the lattice moves to an interstitial
site and displaces a lattice atom, as shown in Fig. 3.17 for a BCC lattice.
Specifically atom #1 moves in the $[1\bar{1}0]$ direction, displacing atom #2. Atom
#2 then displaces another lattice atom, say atom #3.

This mechanism can accommodate rapid transport of self interstitials in a
crystal.

3.10   Diffusion in the Presence of Impurities

Impurities are impossible to eliminate entirely from materials during processing. They can be problematic in that vacancies are often attracted to them
for different reasons. The strain field in the presence of a large substitutional
impurity is somewhat alleviated in the presence of a vacant site. This lowers
the free energy of the system. The electrostatic field in the vicinity of a vacant
site is different from a normal site, which can lead to a net attraction to a
substitutional impurity, depending on its charge state. This often becomes

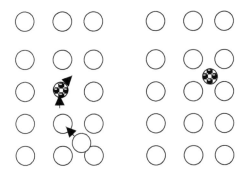

FIG. 3.18
A "kick-out" mechanism is illustrated here wherein a self-interstitial "kicks out" a substitutional impurity which subsequently becomes an interstitial impurity.

an issue in semiconductor based systems. The binding of vacancies to substitutional impurities can affect the vacancy concentration.

Generally, the presence of impurities has a profound impact on diffusion processes that occur in materials. Vacancy–impurity or impurity-substitutional atom interactions, for example, can engender multiple mechanisms of transport. Two additional mechanisms involving the diffusion of a substitutional impurity are briefly described hereafter. This is followed by a reasonably quantitative analysis of diffusion in the case where a vacancy and in interstitial impurity exhibit a strong associative interaction.

3.10.1 "Kick-Out" and Dissociative Mechanisms

This mechanism would occur when impurity atoms are present in the system. Consider the diagram in Fig. 3.18. A self-interstitial atom, which belongs to the host, displaces a substitutional *impurity* in the lattice and the displaced atom becomes an interstitial. This self-interstitial subsequently diffuses. Such a situation would arise in cases where the substitutional impurity is somewhat smaller than a host atom and would, on average, spend more time diffusing as an interstitial since it would be more energetically favorable.

In addition to the "kick-out" mechanism, a so-called *dissociative* mechanism is also probable, wherein the substitutional impurity would simply hop to an interstitial site and migrate via an interstitial mechanism because this would be a more efficient process. The "kick-out" and dissociative mechanisms are common in semiconductor based systems, as discussed later in Chapter 5.

3.10.2 Diffusion of Vacancy-Substitutional Impurity Pairs

3.10.2.1 *Concentration of Vacancies and Impurities in a Dilute Alloy*

An alloy composed on N lattice sites containing N_i randomly distributed solute atoms is considered here. In this sample a fraction, p, of the solute atoms form nearest neighbor associations with vacancies. $N_s p$ solute atoms associated with vacancies ($N_s p$ pairs) and N_v unassociated (free) vacancies. The free energy of formation of a free vacancy is G_v^f and $G_v^f + \Delta G$ is the free energy of formation of a vacancy next to an impurity (ΔG is the solute-vacancy binding energy). The free energy difference of the alloy with defects and without defects is $g = G'(P, T) - G(T, P)$,

$$g = N_v G_v^f + N_s p\left(G_v^f + \Delta G\right) - kT \ln \Omega \qquad 3.63$$

where $\Omega = \Omega_v \Omega_s \Omega_p$. The first challenge would be to write down the number of ways of setting down N_v free vacancies, Ω_v, the number of ways of placing $N_s(1 - p)$ unpaired solute atoms, Ω_s, and the number of possible locations/orientations of $N_s p$ vacancy-impurity pairs, Ω_p, on N lattice sites.

The total number of sites, N, is the sum of the number of solute atoms, N_s, the number of vacant sites N_v, the number of sites occupied by pairs $2N_s p$, the number of unassociated solute atoms $N_s(1 - p)$ and the number of host atoms, N_H; $N = N_s + N_H + N_i p + N_v$. The analysis presented here follows that of Lidiard (see Howard and Lidiard (1964), Allant and Lidiard, (1993)).

The number of ways of setting down $N_i p$ pairs on N sites is now considered. The number of ways of arranging the first pair is $z(N_H + N_s + N_s p + N_v)$ where z is the number of equivalent orientations. For the second pair, the number of sites available is reduced by two, so this number is $N - 2 = (N_H + N_s + N_s p + N_v - 2)$ and the number of sites available to the $(j - 1)$th pair is $(N_s + N_H + N_s p + N_v - 2j)$. We must account for the number of orientations for each pair. The number of ways of arranging $N_i p$ pairs on N lattice sites is

$$\Omega_p = \frac{(z)^{N_i p}}{N_i p!} \prod_{j=0}^{N_i p - 1} (N_s + N_H + N_s p + N_v - 2j) \qquad 3.64$$

At the end of this process, there remain $(N_s + N_H - N_s p + N_v)$ sites on which the free vacancies and the unpaired impurities reside. The number of ways of placing $N_s(1 - p)$ free solute atoms on $(N_s + N_H - N_s p + N_v)$ sites is

$$\Omega_s = \frac{[N_s(1-p) + N_H + N_v]!}{[N_s(1-p)]!(N_H + N_v)!} \qquad 3.65$$

Finally, the free vacancies need to be placed on the remaining $(N_s + N_H - N_s p + N_v) - (z+1)N_s(1-p)$ sites,

$$\Omega_s = \frac{[-zN_s(1-p) + N_H + N_v]!}{[N_H - zN_s(1-p)]! N_v!} \qquad 3.66$$

In the foregoing, the formation of complexes that would be due to the nearest neighbor proximity of a solute atom next to a pair or a vacancy next to a pair, etc., have been omitted. This approximation is not unreasonable since the fraction of complexes could be sufficiently low under normal conditions.

The equilibrium number of pairs is

$$\mu_p = \left(\frac{\partial G}{\partial p}\right)_{N_v, N_s} = 0 \qquad 3.67$$

leading to the fraction of vacancy-solute pairs,

$$X_p \approx zX_s e^{-(G_b^f + \Delta G)/kT} \qquad 3.68$$

This is a reasonably intuitive result that indicates that if there exists a fraction of X_s solute atoms the fraction of pairs is proportional to the Boltzmann factor with the relevant free energy of formation and the number of orientations allowed for each pair. The vacancy concentration is obtained from the relation,

$$\mu_v = \left(\frac{\partial G}{\partial p}\right)_{N_v, N_s} = 0 \qquad 3.69$$

where it is assumed that single vacancies, everywhere, are at equilibrium. The fraction of vacancies is specified by

$$X_v \cong [1 - zX_s + (z-1)X_p]e^{-G_b^f/kT} \qquad 3.70$$

Since $X_p \ll X_s$ then the concentration of free vacancies in the dilute alloy is smaller than in the pure crystal because of the solute-vacancy interaction.

3.10.2.2 Substitutional Impurity-Vacancy Pair Diffusion

The presence of impurities affects the transport of vacancies because vacancies interact with them. In fact, the vacancies and impurities exchange sites at a greater than random probability because of this interaction. If the vacancy-impurity interaction is sufficiently strong, then the rate at which a vacancy exchanges places with the impurity, Γ_2, is much larger than the rate at which it exchanges places with a host (solvent) atom, Γ_1 (i.e. $\Gamma_2 \gg \Gamma_1$). This indicates that the exchange occurs with a probability of unity and that the diffusion coefficient of the impurity is controlled by the rate at which the vacancy moves through the lattice (i.e., proportional to the very small rate at which the vacancy exchanges with the solvent atom), $D_I \propto \Gamma_1$. Since the correlation coefficient $f = \Gamma_1/\Gamma_2$, D_1 would be proportional to $f\Gamma_2$. On the contrary, if $\Gamma_2 \ll \Gamma_1$, then D_1 is proportional to Γ_2.

How would the situation change of we considered the effects of a lattice more seriously? We must first consider Eq. 3.20 so we can calculate f for cubic systems. In cubic, close packed systems, all exchanges between a vacancy and a tracer are identical; the jump distances are identical and the only difference between these exchanges is the orientation of each jump. Because of this, $\cos\theta_{i,i+j}$ does not depend on the direction of the ith jump. It depends on the average direction of the $i + j^{th}$ jump. In other words, $r_i \bullet r_{i+j}$ depends only on j. The average angle between two successive jumps in the cubic system is the same. Note further that $r_i \bullet r_{i+1} = l^2 \cos\theta$ and that for the next nearest neighbor hop, $r_i \bullet r_{i+2} = l^2(\cos^2\theta)$ (the projection along the ith jump direction) and $r_i \bullet r_{i+3} = l^2 \cos^3\theta$; in general

$$\langle r_i \bullet r_{i+j} \rangle = l^2 \langle \cos\theta_j \rangle = l^2 \langle \cos\theta_1 \rangle^j \qquad 3.71$$

This results in a simplification of Eq. 3.20

$$f = \lim_{N \to \infty} \left(1 + \frac{2}{N} \sum_{j=1}^{N-1} (N-j) \right) \langle \cos\theta \rangle^j \qquad 3.72$$

Since $(N - j)/N$ approaches unity for large N, then a further simplification occurs for Eq. 3.72. It is left as an exercise for the reader to show that Eq. 3.72 may be written as

$$f = \frac{1 + \langle \cos\theta \rangle}{1 - \langle \cos\theta \rangle} \qquad 3.73$$

A more general method for calculating f in different systems is discussed by Le Claire and Lidiard (1956).

The value of $\langle \cos\theta \rangle$ is always negative ($\theta > 90°$), indicating that $f < 1$ always for a vacancy mechanism. We might consider a two-dimensional lattice (Fig. 3.19).

In Fig. 3.19, lets assume that the impurity atom initially at position #6 makes an exchange with the vacancy and now sits at position #7. We need to calculate the probability that on the next jump, the impurity will go to location 1, 2, 3, 4, 5, or 6. The probabilities are P_1, P_2, P_3, P_4, P_5, and P_6, respectively. $\cos\theta$ is the angle between the first jump and the next jump. The average value of $\cos\theta$ is,

FIG. 3.19
A two-dimensional lattice is shown here. The impurity atom is located at position 6 after it exchanges places with the vacancy now located at position 7.

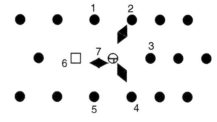

$$\langle \cos\theta \rangle = P_6\cos\theta_6 + P_5\cos\theta_5 + \cdots + P_1\cos\theta_1 \qquad 3.74$$

Since the vacancy-interstitial pair is bound, then the next hop by the vacancy is limited to positions 1, 5, or 7, only (migration to other sites would lead dissociation). If the rate at which the vacancy exchanges position with a host atom is Γ_1 and that which it exchanges with the impurity is Γ_2, as before, then the probability that it will exchange with position with the impurity on the next jump is

$$P_6 = \frac{\Gamma_2}{\Gamma_2 + 2\Gamma_1} \qquad 3.75$$

and $\langle \cos\theta_1 \rangle = -P_6$. The correlation factor is $f = \Gamma_1/(\Gamma_2 + \Gamma_1)$. It follows that the diffusion coefficient of the impurity is,

$$D_I \propto f\Gamma_2 = \frac{\Gamma_1\Gamma_2}{\Gamma_2 + \Gamma_1} \qquad 3.76$$

The foregoing case applies to the situation in which all impurities are bonded to vacancies.

It is worthwhile to mention that if the guest atom were a tracer, and it was not subject to the restriction that it had to remain a nearest neighbor of the vacant site, then the probability of moving to vacant site on its next hop (assuming it is at location #6 after a hop from location #7) is $1/z$ (in this case $z = 6$). This means that $f = 1 - 2/z$. If we relaxed this restriction and asked what is the probability that it would return after n hops then additional terms in the equation would have to be examined. Doing this for a cubic lattice provides somewhat more accurate values of f. The results are in the first column of Table 3.5. It turns out that it is probably not worth the hassle because it is hard to measure diffusion coefficients with the level of accuracy required to discern the difference.

TABLE 3.6

Parameters that characterize interstitial diffusion in some systems are shown here (Data taken from Shewmon, 1989)

Host Metal	Interstitial (solute)	D_0 (cm²/sec)	h_i^m (kcal/mol)	s_i^m/R
Ta	C	0.00061	38.5	0.73
Ta	N	0.0056	37.8	0.73
Fe	C	0.02	20.1	2.4
Fe	N	0.003	18.2	0.69
Nb	C	0.004	33.0	0.51
Nb	N	0.0086	34.9	1.3

3.11 Isotope Effects

We mentioned earlier that it is commonplace to use isotopes to study diffusion in metals. The manifested in differences in size and mass of the isotopes are actual differences in the diffusivities of the actual atom and its isotope. If the differences in activated volumes are ignored, then the ratio of the diffusivities of two isotopic analogs, α and β would be

$$\frac{D_\beta - D_\alpha}{D_\beta} = f_\alpha \left(1 - \left(\frac{M_\beta}{M_\alpha}\right)^{1/2}\right) \qquad 3.77$$

If the deformations of the host are accounted for during the diffusion process, then f_α is replaced by $f_\alpha \Delta K$. Typically, ΔK is of order unity, as opposed to 5 or 0.1 (Le Claire, A.D 1966, Franklin, W 1969). In many typical situations, D_α and D_β are not very different.

3.12 Effects of Pressure on Diffusion

We showed, hitherto, that the addition of a vacancy or an interstitial to the crystal involves a volume change. Local volume changes also accompany the migration of these entities. We can begin with Eq. 3.45 for the self-diffusion coefficient, D. Recall that D is determined by the Gibbs free energy, which includes information about the activated volumes associated with these processes. The derivative of Eq. 3.35 with respect to pressure, P, leads to

$$\frac{\partial \ln\left(D_v^{SD}/\gamma z r^2 \nu_D\right)}{\partial P}\bigg|_T = -\frac{1}{kT}\frac{\partial\left(G_v^f + G_v^m\right)}{\partial P}\bigg|_T \qquad 3.78$$

$$= -\frac{1}{kT}(V_f + V_m)$$

The partial atomic volume associated with the formation of a vacancy is V_f and the partial atomic volume associated with the migration of a vacancy, is V_m. The sum of the two is sometimes called an activation volume, $V_a = V_f + V_m$. We note parenthetically that while in metals the volume of a vacant lattice site is smaller than that of an atomic site, the opposite is true in ionic crystals. In ionic crystals, the surrounding atoms relax outward due to the Coulombic repulsion of like charges.

If the external pressure is increased, the sample will lose vacancies, as is evident from the equation below,

$$\frac{\partial \ln X_v}{\partial P} = -\frac{V_f}{kT} \qquad 3.79$$

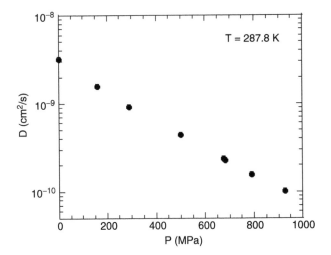

FIG. 3.20
The pressure dependence of self-diffusion in Na is shown here (Data adopted from J.N. Mundy, (1971)).

One can also argue that increasing the pressure also results in a decrease in the activated volume associated with the migration of the vacancy. Estimates of V_m and V_f indicate that $V_f \sim 0.6\ \Omega$ and $V_{fm} \sim 0.1\ \Omega$, where Ω is the atomic volume, indicating that V_m is small in comparison. The information presented here indicates that the diffusion coefficient should decrease with increasing pressure, which is observed experimentally. The data in Fig. 3.20 shows that the self-diffusion decreases appreciably with increasing pressure. It is left as an exercise to determine the activation volumes from these data.

3.13 Diffusion Near Dislocations and Grain Boundaries

Dislocations and grain boundaries act as sources and sinks for vacancies. In fact, atomic transport is known to occur rapidly in the presence of these defects. The term *short circuit diffusion* is often used to describe the enhancement of the rates of transport. Generally, diffusion in the bulk phase and along the defects is considered to occur independently.

As an example, we might consider copper (Sorensen et al. 2000). Vacancy mechanisms, interstitialcy mechanisms and ring mechanisms, involving collections of molecules, occur near grain boundaries, as suggested by simulations and experiments in this system. The rates of transport parallel and perpendicular to the boundaries are also different. The values of $D_0 \sim 10^{-6}\ \mathrm{m^2/s}$ (an order of magnitude smaller than the bulk value) and enthalpies of migration

~0.6 eV (half that of the bulk value) are obtained for vacancy and interstitialcy related mechanisms. The correlation factors associated with transport along different types of grain boundaries are also temperature dependent. This is because the concentration of vacancies near the grain boundaries is highly temperature dependent. It is known that the vacancies in different sites of the boundary are characterized by different structures and formation energies. In fact the atomic jump distances in the vicinity of grain boundaries are characterized by a distribution of lengths and jump rates, as shown by computer simulations (Kwok et al. 1984). The data in Fig. 3.21 illustrates the influence of grain boundaries on self-diffusion in a polycrystalline Cu sample compared to lattice diffusion. Data from polycrystalline samples of two different purities are shown in order to illustrate the variability of diffusion rates. The situation may be summarized as follows: the free energy of formation of a vacancy or an interstitial in the vicinity of a grain boundary or a dislocation is reduced compared to the interior of the lattice. Depending on the orientation of the grain boundary, the energies could vary. Impurities often segregate to grain boundaries and they interact with vacancies and interstitials to change the dynamics in very complex ways. The free energy of migration is also lower near grain boundaries. The average jump distances are also different from the interior of the lattice. In short, understanding the atomic migration processes near planar and line defects present major challenges because much of the behavior can be system specific and therefore defy a simple universal picture.

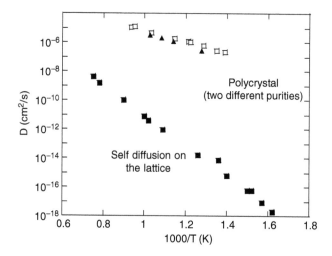

FIG. 3.21

The self-diffusion coefficient of Cu in the Cu lattice is compared with the self-diffusion in a Cu polycrystal where effects of grain boundaries are illustrated. The open squares and triangles represent samples with minor differences in purity. *Data From Diffusion, American Soc of Metals, Metals Park Ohio, 1973 and from T. Surholt and C.H.R. Herzig, Acta Mater. Vol. 45, 3817 (1997) and M.R. Sorensen, (2000).*

3.14 Final Remarks

The nature of the point defects in crystalline solids determines the mechanism of atomic transport. Defect formation energies determine the concentrations of the relevant defects in the solid and, as the concentrations exhibit an exponential dependence on these energies, small differences in formation energies lead to large differences in concentrations of the relevant species. The migration energies for most species are very similar (~0.5 eV–1.5 eV) and because the diffusion coefficient is determined by the formation and migration energies, it is reasonable that the formation energies are largely responsible for much of the differences in transport properties. It is also clear that impurities influence atomic diffusivity in materials.

The prefactors for diffusion in many solids are determined by the Debye frequency, v_D, the nearest neighbor distances, r, the coordination number, z, and the entropies of formation and migration. With the exception of the formation entropies, the other variables in the prefactor $(zr^2 f v_D)$ are comparable for virtually all materials. The migration energies also do not differ significantly from one material to another. It stands to reason that differences between the formation entropies determine the variations in the magnitude of the prefactor from one type of solid to another.

To this end, a number of empirical correlations have been made between different classes (including different crystal systems) of materials. In fact, D_0 is approximately constant for different classes of materials. Diamond cubic structures exhibit the largest prefactors (~50–100 cm^2/s), followed by alkali halides (~10), then FCC crystals (~0.5). BCC crystals exhibit wide variation but, for the most part, span the range of FCC crystals yet remain somewhat comparable to or larger than the alkali halides. The diffusion coefficient at the melting point, $D(T_m)$, is also shown to be constant, depending on the material class. In systems with the diamond cubic structure, $D(T_m) \sim 10^{-12}$ cm^2/s, whereas for alkali halides, $D(T_m) \sim 10^{-9}$ cm^2/s. It is comparable in FCC, BCC, and HCP metals (10^{-8} cm^2/s) but the BCC crystals exhibit the widest variation. The largest values are found in rare earth metals, 10^{-6} cm^2/s. The opposite trends are observed for the ratio of the activation energy to the thermal energy at T_m, Q/kT ($Q = H^f + H^m$), with the diamond cubic structure possessing the largest values (Brown and Ashby 1980).

3.15 Problems for Chapter 3

1. Compare the size of the largest octahedral interstitial atoms that can be incorporated into FCC and BCC lattices without distortion. Assume that the lattice constant is a and that the atomic radius is R.

2. An expression for tracer diffusion coefficient for a single vacancy mechanism is

$$D_v^{SD} = gv_D e^{-(G_v^m + G_v^f)/kT}$$

What is the value of g for diffusion in BCC, FCC, and simple cubic lattices?

3. Write down an expression for the tracer diffusion coefficient for a small impurity atom undergoing diffusion via an interstitial mechanism on (i) tetrahedral sites and (ii) octahedral sites in a BCC crystal.

4. Show that for a vacancy mechanism in the cubic system:

(i) $\langle r_i \bullet r_{i+j} \rangle = l^2 \langle \cos \theta_j \rangle = l^2 \langle \cos \theta_1 \rangle^j$ and (ii) show that $f = \dfrac{1 + \langle \cos \theta \rangle}{1 - \langle \cos \theta \rangle}$

5. The diffusion coefficient of a vacancy-interstitial pair may be expresses as $D_I = Kf\Gamma_2$. Determine an expression for K.

6. Derive Eq. 3.49, $\dfrac{3\Delta l}{l} = X_v \beta$

7. Estimate the average distance of separation between vacancies at the melting point of Au and of Cu. Ignore the entropy of formation.

8. Equation 3.24 could be written

$$\int_0^\infty \Omega(\omega)d\omega = \int_0^{\omega_D} V\left(\frac{\omega^2 d\omega}{2\pi^2 c_L^2} + \frac{\omega^2 d\omega}{2\pi^2 c_T^2} \right) = 3N$$

where c_L^2 and c_T^2 are the velocities of the longitudinal and transverse modes, respectively.

a) Determine an expression for ω_D.

b) If $\int_0^{\omega_D} V(\frac{\omega^2 d\omega}{2\pi^2 c_L^2}) = N$ represents the contribution of the longitudinal modes to the total number of normal modes. What is the equivalent expression for the contribution from the transverse normal modes?

c) Determine an expression for the minimum wavelength of each mode. What would you surmise would be the smallest limiting wavelength of the system.

9. Derive an equation describing the concentration of divacancies

$$X_d = \frac{z}{2} X_v^2 e^{\left(\frac{2G_v^f - G_d^f}{kT} \right)}$$

10. Determine an expression for the diffusion coefficient of a di-interstitial moving on octahedral sites of an FCC lattice. In this equation the number of orientations and the jump distance needs to be identified. The jump frequency may be identified as n_{2D} and the free energy of migration as G_m.

11. Determine the equilibrium vacancy concentration for Al and Au at one half their melting points.

12. Draw $\langle 110 \rangle$ and $\langle 111 \rangle$ split interstitials in a BCC lattice.

13. Calculate the X_v, and X_d at one half of the melting point of gold $H_v^f = 5.17$ kJ/mol, $S_v^f/k = 1$ and $\Delta G = 0.824$ kJ/mol, $T_m = 1336$ K).

14. In some solids, the dependence of the self-diffusion coefficient on temperature is not completely Arrhenius (ln D is not linear with $1/T$). There can be two reasons for this. It is possible that diffusion occurs via more than one mechanism or the enthalpy may be temperature dependent. In the latter situation (for a vacancy mechanism), the enthalpy of formation can be expanded in terms of a Taylor series expansion,

$$h(T) = h(T_0) + \alpha k(T - T_0) + \beta k(T - T_0) + \cdots$$

Write down a complete expression for the temperature dependence of D.

15. Consider a metal in which diffusion occurs via an interstitialcy mechanism. Here self-interstitials form a $\langle 100 \rangle$ split dumbbell configuration in an FCC lattice. What is the jump distance of the center of mass of the defect? If the motion is uncorrelated, write down an expression for the temperature dependence of D.

16. Consider the two dimensional lattice that follows. In the middle of the diagram, denoted by the broken lines, is a vacant site. The black circle represents a solute atom.

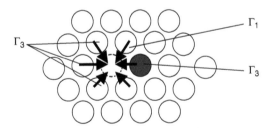

This diagram is meant to illustrate the fact that the diffusion of solute via a vacancy mechanism can be more complex than the case of self-diffusion. The vacancy often has strong interactions with the solute atom. Consequently they may remain nearest neighbor pairs. The rate at which the solute atom exchanges places with the vacancy is Γ_2. The rate at which the other atoms exchange sites with the vacancy are Γ_1 and Γ_3, as shown in the diagram. Throughout the diffusion process they remain nearest neighbor pairs. How, then, does the solute atom diffuse?

There are two possibilities, 1) $\Gamma_2 \gg \Gamma_1 \gg \Gamma_3$ or 2) $\Gamma_1 \gg \Gamma_2 \gg \Gamma_3$. Write down an expression for the diffusion coefficient of the solute atom for each of these extreme cases. Assume that the nearest neighbor jump distance is a.

17. If you measured $\Delta l/l$ and $\Delta a/a$ for a cubic dilute alloy containing vacancies, some of which interact with the solute atoms. What information, precisely, does this experiment give you?

18. Calculate the chemical potential for vacancies and for vacancy-solute pairs using Eq. 3.67. Then determine equations 3.72 and 3.74.

19. The free energy difference between a sample and another otherwise identical sample without vacancies is,

$$g = N_v G_v^f + N_I G_I^f N_{Iv} + N_{Iv} G_{Iv}^f - kT \ln \Omega_I \Omega_v \Omega_{IV}$$

We assume that all three entities, impurities, vacancies, and vacancy substitutional-impurity pairs exist. In the foregoing equation G_{Iv}^f is the free energy associated with the formation of a vacancy substutional-impurity pair, with N_{vI} such pairs, and that there are N_I impurities, and associated with each impurity is the formation energy, G_I^f. The expressions for the number of ways to arrange these entities on the lattice follow. Assuming that $N_v \ll N_L$ and that $N_I \ll N_L$, show that the chemical potentials are

$$\mu_v = \frac{\partial g}{\partial N_v} = G_I^f - kT \ln \frac{N_I}{(N_L - N_I)}$$

$$\mu_I = \frac{\partial g}{\partial N_I} = G_{Iv}^f + kT \ln z - kT \ln \frac{N_{Iv}!}{\left(\frac{z}{2}N_L - N_{Iv}\right)!}$$

$$\mu_{Iv} = \frac{\partial g}{\partial N_{Iv}} = G_v^f - kT \ln \frac{N_v}{(N_L - N_v)!}$$

and show that

$$X_{Iv} = zX_vX_Ie^{-(G_{Iv}^F - G_v^F - G_I^F)/kT}$$

State all assumptions.

20. Estimate the activation volumes in the data in Fig. 3.20. Explain any assumptions.

21. Consider the motion of a particle in a two-dimensional square lattice. The particle moves in such a manner that it makes unit displacements of distance L. These displacements occur in any direction with equal probability.

    a) What is the mean square displacement $\langle r^2 \rangle$ of the molecule after N steps (N is large)?

    b) If the diffusion coefficient is D and t is the time, calculate $\langle r^2 \rangle$ only in terms of D and t (in three dimensions).

    c) How many steps does the particle take after one hour if $L = 0.3$ nm, $D = 10^{-6}$ cm^2/s.

3.16 References and Additional Reading

Simmons, R.O., Baluffi, R.W., "Measurement of equilibrium concentration of vacancies in copper," *Physical Review* 129, 1533 (1963).

K. Nordlund and R.S. Averback, "Role of self-interstitial atoms on the high temperature properties of metals," *Phys. Rev. Lett.* 80, 4201 (1998).

Kovenski, I.I. Diffusion in BCC metals, American Society of Metals, Conference of Diffusion, 1964, Metals Park, OH.

Pandey, K.C., "Diffusion without vacancies and interstitials: A new concerted exchange mechanism," *Phys. Rev. Lett.* 57, 2287 (1986).

Sorensen, M.R. Mishin, Y. Voter, A.F., "Diffusion mechanisms in copper grain boundaries," *Phys. Rev. B.* 62, 3658 (2000).

Kwok, T. Ho, P.S. Yip, S., "Molecular dynamics studies of grain boundary diffusion," *Phys. Rev. B* 29, 5363 (1984).

Brown, A.M. and Ashby, M.F., "Correlations for diffusion constants" *Acta Metallurgica* 28, 1085 (1980).

Glicksman, M.E., *Diffusion in Solids*, Wiley and Sons 2000.

Borg R.J. and Dienes, G.J., *An Introduction to Solid State Diffusion*, Academic Press, NY, 1988.

Girifalco, L., *Statistical Physics of Materials*, John Wiley and Sons, NY, 1973.

Porter D.A. and Easterling, K.E., *Phase Transformations in Metals and Alloys*, Van Nostrand Reinhold (1981).

Flynn, C.P., *Point Defects in Diffusion*, Clarendon Press, Oxford, 1972.

Shewmon, P., *Diffusion in Solids*, 2nd ed., TMS publication (1989).

Allant, A.R. and Lidiard, A.B., *Atomic Transport in Solids*, Cambridge University Press, UK 1993.

Pathria, R.K., *Statistical Mechanics*, Pergamon Press, Oxford UK, 1980.

Crawford, J.H., Slifkin, L.M., *Point Defects in Solids*, vol. 1, Plenum Press, NY, 1980.

Warren, B.E., *X-ray Diffraction*, Addison & Wesley, NY, 1969.

Mondy, J.N., "Effect of pressure on the isotope effect in sodium diffusion," *Physical & Review B*, 3, 2431 (1971).

Le Claire, A.D. and Lidiard, A.B., "Correlation effects in diffusion in crystals," *Philosophical Magazine*, 1, 518 (1956).

Howard, R.E., Lidiard, A.B., "Matter transport in solids," *Reports on Progress in Physics* 27, 161 (1964).

4

Diffusion in Ionic Crystals: Alkali Halides

4.1 Introduction

In this chapter, cationic and anionic diffusional transport in ionic crystals is of particular interest. Technologically, the interest in ionic transport in materials is associated with applications that include batteries and various electrochemical sensors.

Atomic transport is typically more complex in ionic crystals than for elemental metals, largely due to implications associated with charged defects and charge neutrality constraints. Multivalent impurities alter the overall charge carrier distribution and conductivity in ionic crystals; at low temperatures, where the intrinsic thermal defect concentration is low, they dominate charge transport.

To understand defect formation and transport in ionic crystals, it is necessary to understand structure. Consider materials which possess a sodium chloride (NaCl) type unit cell structure, as illustrated in Fig. 4.1(a). There are a total of 8 atoms per unit cell. The lattice consists of equal numbers of alternating Na^+ and Cl^-, cations and anions, respectively. These ions are organized such that each has 6 nearest neighbors of the opposite type, and, moreover, each ion occupies a sublattice of cubic symmetry. Table 4.1 lists some compounds with the sodium chloride structure, together with their lattice spacings.

Another common type of structure is the CsCl-type structure. The unit cell is illustrated in 4.1(b). Each cell is occupied by 1 CsCl molecule, where the Cs^+ cation occupies the center of the unit cell with Cl^- anions at the corner of the cell. Each Cs^+ cation has 8 nearest neighbor Cl^- anions. Similarly, each Cl^- anion has 8 nearest neighbor Cs^+ cations. The unit cell possesses cubic symmetry. In fact, the Bravais lattice would be simple cubic. Crystals which possess this structure include CsI, CsBr, TlBr, and TlI.

The goal of this chapter is to introduce basic concepts of ionic transport in crystals. In order to achieve these goals, three examples will be discussed: one involving Frenkel defects, a second involving Schottky defects, and a third which describes the effect of multivalent defects on the conductivity.

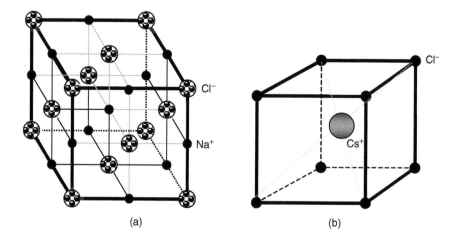

(a)   (b)

FIG. 4.1

Two types if ionic crystal structures (a) Structure of NaCl. Eight atoms comprise the NaCl unit cell; (b) Structure of CsCl.3.

TABLE 4.1

List of ionic compounds that possess the NaCl-like structure

Crystal	Lattice spacing (nm)	Crystal	Lattice spacing (nm)
LiF	0.402	KCl	0.629
LiCL	0.523	AgCl	0.555
LiBr	0.550	SrO	0.516
LiI	0.600	CsF	0.601
NaF	0.462	AgBr	0.577
NaCl	0.564	AgF	0.492

Data taken from Kittel, 1976

4.2 Defects in Ionic Crystals

Frenkel and Schottky defects are typical point defects encountered in ionic crystals. A Frenkel defect is formed when a cation vacancy (negatively charged) is created with the removal of a cation from a lattice and becomes

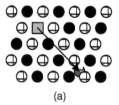

FIG. 4.2A

An ion leaves a normal site and occupies an interstitial site, Frenkel defect.

(a)

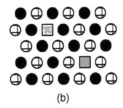

(b)

FIG. 4.2B
A cation vacancy and an anion vacancy form a Schottky pair.

located in an interstitial site (see Fig. 4.2(a)). Equal numbers of cation vacancies and cation interstitials form Frenkel pairs. While Frenkel defects also reside on the anion sublattice. These defects occur with low probability because anions are typically larger than cations. A Schottky defect is formed when an anion vacancy and a cation vacancy are created, as shown in Fig. 4.2(b). Equal numbers of cation and anion vacancies in a crystal constitute Schottky pairs (charge neutrality condition).

Although different distributions of defects might exist in ionic crystals, it is generally understood that certain types of defects are prevalent in some systems. For example, Schottky defects are believed to be dominant carriers in NaCl and it is thought that the cation vacancies move at a faster rate than anion vacancies. In AgCl and AgBr, cation Frenkel defects are most prevalent. Typically, one would need to know specifics about the defect structure to be certain of the nature of the entities that contribute to the overall conductivity.

4.3 Frenkel Defect Concentration

The free energy difference between a perfect crystal and one with Frenkel defects is

$$g = N_{iv}G_F^f - kT \ln \Omega_v \Omega_i \qquad 4.1$$

where N_{iv} is the number of vacancy-interstitial pairs, and G_F^f is the relevant free energy of formation per pair, and, accordingly,

$$\Omega_i = \frac{N_I!}{N_i!(N_I - N_i)!} \qquad 4.2$$

is the number of ways of arranging N_i interstitials on N_I interstitial sites. In addition,

$$\Omega_v = \frac{N_L!}{N_v!(N_L - N_v)!} \qquad 4.3$$

is the number of ways of arranging N_v vacancies on N_L lattice sites. Using Stirling's approximation and the fact that $N_{iv} = N_v = N_i$,

$$\frac{\partial \Delta g}{\partial N_{iv}} = \ln \frac{(N_L - N_{iv})(N_I - N_{iv})}{N_{iv}^2} + \frac{G_F^f}{kT} = 0 \qquad 4.4$$

from which it follows that

$$N_{iv} = [(N_I - N_{iv})(N_L - N_{iv})]^{1/2} e^{-G_F^f/2kT} \qquad 4.5$$

Since $N_L \gg N_{iv}$ and $N_I \gg N_{iv}$ and that $N_L = N_I$, then we obtain the following expression for the fraction of Frenkel defects,

$$X_{iv} \approx e^{-G_F^f/2kT} \qquad 4.6$$

In light of our prior discussion of defect concentrations, this is an intuitive result in that the Boltzmann factor dictates the fraction of such defects in the system at equilibrium. Why does the factor of 1/2 exist in the exponent?

4.4   Schottky Defect Concentration

Based on the similarities of all these calculations and results, it is readily surmised that, for Schottky defects, the fraction of cation vacancies $X_{V_M'}$ is equal to the number of anion vacancies, $X_{V_X^\bullet}$

$$X_{V_M'} = X_{V_X^\bullet} \approx e^{-G_S^f/2kT} \qquad 4.7$$

Here we assumed that the energy associated with the creation of a cation vacancy is equal to that used to create an anion vacancy. Throughout this chapter, we will often rely on the following notation: $N_{N_M'}$ is the number of cation vacancies, and the fraction of cation vacancies is $X_{V_M'} \equiv [V_M']$ and, similarly, the fraction of anion vacancies is $X_{V_X^\bullet} \equiv [V_X^\bullet]$.

We now show how the prediction (Eq. 4.7) arises. The free energy change associated with a cation vacancy has a contribution from the charge, q, and from the local potential, V

$$G_{V_M'} = G_{V_M'}^f - q|V| \qquad 4.8$$

A similar relation exists for anion vacancies,

$$G_{V_X^\bullet} = G_{V_X^\bullet}^f + q|V| \qquad 4.9$$

The fraction of cation vacancies would be specified by the Boltzmann factor

$$X_{V_M'} = e^{-(G_{V_M'}^f - q|V|)/kT} \qquad 4.10$$

Accordingly, the fraction of anion vacancies is

$$X_{V_M^{\bullet}} = e^{-(G^f_{V_X^{\bullet}} + q|V|)/kT}$$ 4.11

The product of the two yield

$$X_{V_M^{\bullet}} X_{V_X^{\prime}} = e^{-G^f_S/kT}$$ 4.12

where $G^f_S = G^f_{V_M} + G^f_{V_X^{\bullet}}$. Since $X_{V_M^{\bullet}} = X_{V_X^{\prime}}$ and one may reasonably assume that $G^f_{V_M} = G^f_{V_X^{\bullet}}$. Therefore the fraction of Schottky defects is $X_{V_M^{\bullet}} = X_{V_X^{\prime}} \approx e^{-G^f_S/2kT}$, as stated earlier. It should now be clear where the factor of 1/2 originates.

4.5 Diffusional Transport of Cationic and Ionic Defects

Thus far our discussion of diffusion has been limited primarily to single component systems in which dynamics proceed in the absence of external driving forces. In the presence of an external driving force, atomic migration occurs to reduce the Gibbs free energy of the system; in other words, species migrate to reduce chemical potential gradients. Essentially, the real effect of an external driving force on the system is that, in the presence of the force, species exhibit a greater than random probability to migrate in the direction of the force.

In general, for an n-component system, the flux of species k under the influence of a chemical potential driving force, $\nabla \mu_k$, is

$$\bar{J}_i = -\sum_{k=1}^{n} L_{ik} \nabla \mu_k$$ 4.13

where the coefficients L_{ik} are the phenomenological Onsager coefficients which are related to the mobilities of the species. Explicitly, for the n-component system, the fluxes are

$$J_1 = -L_{11}\nabla\mu_1 - L_{12}\nabla\mu_2 - L_{13}\nabla\mu_3 - L_{14}\nabla\mu_4 - \cdots - L_{1n}\nabla\mu_n$$
$$J_2 = -L_{21}\nabla\mu_1 - L_{22}\nabla\mu_2 - L_{23}\nabla\mu_3 - L_{24}\nabla\mu_4 - \cdots - L_{2n}\nabla\mu_n$$
$$\cdot$$
$$\cdot$$ 4.14
$$\cdot$$
$$J_n = -L_{n1}\nabla\mu_1 - L_{n2}\nabla\mu_2 - L_{n3}\nabla\mu_3 - L_{n4}\nabla\mu_4 - \cdots - L_{nn}\nabla\mu_n$$

The chemical potentials would be

$$\mu_k(\bar{r},c) = \mu_k^0 - \bar{F}_k \cdot \bar{r}_k + kT \ln \gamma_k c_k$$ 4.15

In this equation, c_k is the concentration of species k; F_k is the external force exerted on species k; and γ_k is the activity coefficient.

From Eq. 4.15,

$$\nabla \mu_k = -\vec{F}_k + \frac{kT}{c_k}\left(1 + \frac{\partial \ln \gamma_k}{\partial \ln c_k}\right)\nabla c_k \qquad 4.16$$

The more familiar expression for the flux, as discussed in Chapter 1, is

$$\vec{J}_i = -\sum_{k=1}^{n} D_{ik}\vec{\nabla}c_k \qquad 4.17$$

Equation 4.16 provides the connection between 4.13 and 4.17. If the activity coefficient is constant and if the spatial dependence of the concentration is ignored, then

$$\nabla \mu_k = -\vec{F}_k \qquad 4.18$$

With the use of Eq. 4.13,

$$\vec{J}_k = \hat{L}_k \vec{F}_k \qquad 4.19$$

where $\hat{L}_k = \Sigma_{k=1}^{n} L_{ik}$. Since the flux is the product of the concentration of particles, c_k, the mobility and the driving force, is $\hat{L}_k = c_k \hat{B}_k$, where $\hat{B}_k$ is the mobility tensor. With the use of the Einstein relation, Eq. 4.19 may be rewritten as

$$\vec{J}_k = \frac{c_k \hat{D}_k}{kT}\vec{F}_k \qquad 4.20$$

Recall that this is the Nernst-Einstein relation we derived earlier in Chapter 2 for one dimension.

In general, the external driving force could be due to a number of things. It could be due to an electric field, $\vec{E}$, where

$$\vec{F}_k = q_k \vec{E} \qquad 4.21$$

and q_k is the change on species k. The force may also be due to a temperature gradient,

$$\vec{F}_k = -\frac{Q_k^*}{T}\nabla T \qquad 4.22$$

where Q^* is associated with heat transport. If it is due to a stress field potential, U, then

$$\vec{F} = -\nabla U \qquad 4.23$$

We are interested in the effect of an external electric field on the charge carriers. The current due to these carriers is

$$\bar{I} = q_k \bar{J}_k = \frac{(n_q)c_k(\hat{D}_k \bar{F}_k)}{kT} = \frac{q_k c_k (\hat{D}_k q_k \bar{E})}{kT} = \hat{\sigma} \cdot \bar{E} \qquad 4.24$$

where $\hat{\sigma}$ is the conductivity tensor. Since $I = \hat{\sigma}\bar{E}$ it follows that the ionic conductivity is

$$\hat{\sigma}_k = \frac{q_k^2 c_k}{kT} \hat{D}_k \qquad 4.25$$

The discussion thus far has ignored the influence of interactions between the species. We will deal with the issue of spatial correlations on conductivity in Chapter 7 when we discuss ionic transport in network glasses.

In what follows, we will discuss three examples, one involving Frenkel defects, a second involving Schottky defects, and a third where the effect of multivalent impurities on ionic conductivity of ionic crystals is illustrated. We conclude with general comments regarding ionic transport in alkali halide type systems.

4.6 Diffusivity of Frenkel Defects

We now examine the diffusivity of defects in silver bromide, which possesses a NaCl-type structure. Frenkel defects are known to form in this material. The reaction is

$$Ag_{Ag} \Leftrightarrow Ag_i^{\bullet} + V'_{Ag} \qquad 4.26$$

where Ag_{Ag} refers to Ag in a normal Ag site, $Ag_i^{\bullet}$ is an Ag cation in an interstitial site and V'_{Ag} is a vacant Ag site of the opposite charge. Recall that, for a reaction in which the reactants and products are identified as A_k and in which v_i is the stochiometric coefficient ($v_k < 0$ corresponds to reactants and $v_k > 0$ correspond to products),

$$\sum_k v_k A_k = 0 \qquad 4.27$$

In terms of the Gibbs free energy in the standard state, the equilibrium constant, K_{eq}, for the reaction is

$$\Delta G^0 = -RT \ln K_{eq} \qquad 4.28$$

where

$$K_{eq} = \prod a_k^{v_i} \qquad 4.29$$

and a_k is the activity.

For the Frenkel defect reaction, the equilibrium constant is

$$K = e^{-G_F^f/kT} = \frac{a_{Ag_i^\bullet} a_{V_{Ag}'}}{a_{Ag_{Ag}}} = [Ag_i^\bullet][V_{Ag}']$$

(4.30)

because the concentrations are dilute ($\gamma = 1$) and the activity of Ag in the Ag site is 1. Moreover, since the charge neutrality condition must be satisfied, $[Ag_i^\bullet] = [V_{Ag}']$, then

$$[Ag_i^\bullet] = [V_{Ag}'] = e^{-G_F^f/2kT}$$

(4.31)

This result should be familiar since it is derived earlier (Eq. 4.6). If the diffusion of the silver ion occurs via a vacancy mechanism on the silver FCC sublattice, then D_{Ag} is determined by the fraction of available vacant sites and by the correlation coefficient, (cf. Eq. 4.33)

$$D_{Ag} = a^2 f v_D e^{-G_F^m/kT}[V_{Ag}']$$

(4.32)

where $[V_{Ag}^-]$ is given by 4.31. This result reveals that both the diffusion coefficient and the product of the conductivity and the temperature, σT, exhibit Arrhenius dependencies on temperature, $\sigma T \propto D$.

  If the ionic conductivity of the Ag ions was measured in an experiment, and Eq. 4.32 used to extract D_{Ag}^σ, the actual value of D_{Ag}^σ would be larger than the value of D_{Ag} measured in a tracer diffusion experiment. This is because the conductivity is not sensitive to correlations in the same way that the diffusion coefficient is affected, as discussed in Chapter 3. Equation 4.32 would have to be modified by replacing D_{Ag}^σ with $D_{Ag} = D_{Ag}^\sigma/f$, in one dimension,

$$\sigma_k = \frac{n_k^2 q_k^2 c_k D_k}{fkT}$$

(4.33)

4.7   Diffusion of Schottky Defects

Our second example involves Schottky defects in sodium chloride. This reaction may be written as

$$Na_{Na} + Cl_{Cl} \Leftrightarrow V_{Na}' + V_{Cl}^\bullet + Na_{Na} + Cl_{Cl}$$

(4.34)

The equilibrium constant for the reaction is $K = e^{-G_S^f/kT} = [V_{Na}'][V_{Cl}^\bullet]$ and, since charge neutrality must be preserved, $[V_{Na}'] = [V_{Cl}^\bullet]$, then

$$[V_{Na}'] = [V_{Cl}^\bullet] = e^{-G_S^f/2kT}$$

(4.35)

This equation should be familiar since it is identical to Eq. 4.7. The diffusion of the anions and cations takes place via a vacancy mechanism on their respective FCC lattices, so

$$D_{Na} = a^2 f v_D e^{(S_{Na}^m + S_S^f/2)/k} e^{-(H_{Na}^m + H_S^f/2)/kT} \qquad 4.36$$

and

$$D_{Cl} = a^2 f v_D e^{(S_{Cl}^m + S_S^f/2)/k} e^{-(H_{Cl}^m + H_S^f/2)/kT} \qquad 4.37$$

These data indicate that the temperature dependencies of diffusion and σT are Arrhenius.

4.8 The Effect of Multivalent Impurities on Conductivity

The influence of small concentrations of multivalent impurities on diffusivity can be significant. In this third example, the influence of multivalent solute impurities on ionic diffusion in NaCl is discussed. We will consider, as an example, the effect of the presence of a small concentration (~0.005%) of $CdCl_2$ on the ionic diffusivity of NaCl. As illustrated in Fig. 4.3, the conductivity of "pure" NaCl is represented by the thick solid line. The ionic conductivity and diffusion data for NaCl doped with $CdCl_2$ show two Arrhenius regions, one at higher temperatures and the other, which possesses a smaller slope, at lower temperatures. An important feature of this system is that the slopes

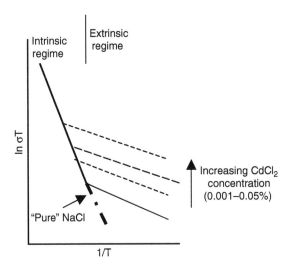

FIG. 4.3
Effect of multivalent impurities on the conductivity, $\sigma T \propto D$ is shown here. The sharpness of the "knee" is somewhat exaggerated in this figure. Pure NaCl does not show a break. The magnitude of the prefactor increases with the $CdCl_2$ concentration, but the slope remains unchanged.

in the low and high temperature regimes differ by a factor of $H_v^f/2$. Each line in the lower temperature regime is associated with a different concentration of dopant; in fact, D_0 increases with increasing dopant (CdCl$_2$) concentration. At high temperatures, the activation energy is $Q = H_v^m + H_v^f/2 = 1.8$ eV and the prefactor for the diffusivity is $D_0 = 3.1$ cm^2/s for NaCl sample containing 0.005% CdCl$_2$. In the lower temperature region, known as the extrinsic region, $H_v^m = 0.77$ eV, $H_s^f = 2.06$ eV and $D_0/D_0' = 2 \times 10^6$ cm^2/s.

The diffusion of other dopants in NaCl has also been examined. SrCl$_2$ doped ("0" to 370 ppm) NaCl exhibits similar behavior to that shown in Fig. 4.3. With respect to the conductivity on Na$_2$S doped NaCl, anion vacancies are the primary contributors to the conductivity (Hooton and Jacobs). The subsequent model is discussed to explain the temperature dependence of the diffusivity.

Two possible "reactions" might be considered in this system. In the first, Case 1, the Cd^{++} cations are accommodated substitutionally by the sodium sites,

$$CdCl_2 \Leftrightarrow Cd_{Na}^{\bullet} + V_{Na}' + 2Cl_{Cl} \qquad 4.38$$

The second possibility would be the formation of Cd^{++} interstitials

$$CdCl_2 \Leftrightarrow Cd_i^{\bullet\bullet} + 2V_{Na}' + 2Cl_{Cl} \qquad 4.39$$

In the former, which is known to be a more likely scenario, the charge neutrality condition (total number of positive charges is balanced by the total number of negative charges) dictates that

$$[Cd_{Na}^{\bullet}] + [V_{Cl}^{\bullet}] = [V_{Na}'] \qquad 4.40$$

When Schottky defects are formed, the following condition is always true,

$$[V_{Na}'][V_{Cl}^{\bullet}] = K = e^{G_s^f/kT} \qquad 4.41$$

It follows from substituting 4.40 into 4.41 that

$$[V_{Na}']\{[V_{Na}'] - [Ca_K^{\bullet}]\} = e^{-G_s^f/kT} \qquad 4.42$$

In the pure state, the concentration of vacant Na-sites is

$$[V_{Na}']_0 = e^{-G_s^f/2kT} \qquad 4.43$$

which implies that

$$[V_{Na}']\{[V_{Na}'] - [Ca_{Na}^{\bullet}]\} = [V_{Na}']_0^2 \qquad 4.44$$

This is a quadratic equation with solution

$$[V_{Na}'] = \frac{[Ca_{Na}^{\bullet}]}{2}\left\{1 \pm \left[1 + 4\left(\frac{[V_{Na}']_0}{[Ca_{Na}^{\bullet}]}\right)^2\right]^{1/2}\right\} \qquad 4.45$$

Two limiting situations can be obtained from this equation. In the first, the concentration of vacancies far exceeds the impurity concentration, $[V_{Na}'] \gg [Ca_{Na}^{\bullet}]$. This would necessarily occur at high temperatures where the thermal vacancy concentration is high. Equation 4.45, under these conditions, indicates that the vacancy concentration in the system is controlled by thermal vacancies,

$$[V'_{Na}] \approx [V'_{Na}]_0 \qquad 4.46$$

Consequently, the diffusion coefficient (assuming a vacancy mechanism is operational) is

$$D = z\gamma f r^2 e^{-G_s^m} \qquad 4.47$$
$$= D_0 e^{-(H_s^m + H_s^f/2)/kT}$$

This temperature regime is identified as the so-called intrinsic regime (see problem 1).

In the other limiting case $[Cd^\bullet_{Na}] \gg [V'_{Na}]_0$, Eq. 4.45 reveals that the conductivity in this regime is determined by the cation concentration. This is the so-called extrinsic temperature regime since the conductivity is dominated by the impurity carrier concentration and

$$D \approx \frac{a^2 f}{4}[Cd^\bullet_{Na}]e^{-S_s^m/k}e^{-H_s^m/kT} = D'_0 e^{-H_s^m/kT} \qquad 4.49$$

Equations 4.47 and 4.48 reveal that the enthalpy of migration in the low temperature range (so-called extrinsic range) differs from the high temperature intrinsic range, where vacancies control diffusion, by $H_s^f/2kT$. Second, the magnitude of the prefactor increases with increasing amounts of the impurity. Both predictions are observed experimentally (Fig. 4.3).

4.9 Comments on Transport in Alkali Halide Crystals: Transport Coefficients

In the foregoing example, ionic conductivity is determined by the impurity concentration at low temperatures, whereas, at high temperatures, where the thermal vacancy concentration was high, the conductivity is due primarily to the transport of cation vacancies.

Notwithstanding the aforementioned comments, it is reasonable to consider that, when high temperatures such as the melting temperature (T_m) are approached, there would exist a finite concentration of self interstitials ($T > T_m$ the system is disordered). The implication therefore is that Frenkel defects would contribute to the ionic conductivity at sufficiently high T. It is possible that self interstitials may exist in a split dumbbell configuration, rather than a single interstitial with the symmetry of a lattice point. Frenkel defects on both lattices have been shown to provide an important contribution to the ionic conductivity of KCl and RbCl. The larger lattice constant of these systems compared to that of NaCl would more easily accommodate the formation of self interstitials. The temperature dependence of the ionic conductivity of both doped and undoped RbCl, i.e., $\ln(\sigma T)$ versus $1/T$ ($\sigma T \propto D$) behavior is not Arrhenius, except at the very high temperature range. Instead of exhibiting a "knee" denoting the transition from intrinsic to extrinsic, the Log (σT) vs. $1/T$ exhibits considerable curvature, as illustrated in Fig. 4.4. This curvature would be indicative of the presence of more than one mechanism of transport simultaneously operational within the material.

If both Schottky and Frenkel defects are present, vacancy and interstitialcy mechanisms of transport would ensue. Moreover, vacancy-impurity inter-actions would also have to be considered in any model developed to describe transport in these systems.

In light of this, charge carrying entities in doped and undoped ionic crystals, would include anion vacancies, anion intersitials, cation vacancies, and cation interstitials. To this end, it is customary to define a transport number t_i, such that

$$\sum_{i=1}^{n} t_i = 1 \qquad\qquad 4.50$$

where i refers to a charge carrier entity and

$$\sigma = \sum_{i=1}^{n} \sigma_i \qquad\qquad 4.51$$

where $\sigma_i = \sigma t_i$ (the contribution due to electrons is also included in the fore-going equation). The transport numbers are temperature dependent because they reflect the dominance of different charge carrying entities in different temperature ranges. Figure 4.5 shows the temperature dependence for the transport number for a hypothetical cation and for an anion doped sample. In the cation doped sample, the effect of divalent impurities on the conduc-tivity is apparent at lower temperatures, whereas, at higher T, the fraction of thermal cation and anion vacancies (Schottky defects) and interstitials (Frenkel defects) increases and the cation vacancy transport number, t_{cv}, decreases. In the anion doped case, t_{cv} is approximately zero in the extrinsic region because it is dominated by the anion transport. However, with increasing temperature, t_{cv} increases due to the increase in the relative fractions of the carriers.

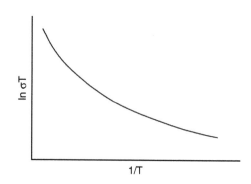

FIG. 4.4
The temperature dependence of σT for RbCl.

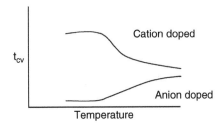

FIG. 4.5
A sketch of the temperature dependence of the cation vacancy transport coefficient, t_{cv}, in two hypothetical samples: one is anion doped, whereas the other is cation doped is shown here.

4.10 Problems for Chapter 4

1. Calculate tcv for the $CdCl_2$ doped NaCl sample.

2. The following is an expression for vacancy diffusion of silver ions in AgCl, $D_{Ag} = a^2 f v_D e^{-(G_{F(Ag)}^f + G_i^m)/kT}$, where $f = 0.83$ and $a = 0.555$ nm and $G_{F(Ag)}^f$ is the free energy of formation of the defect. If $G_F^{m(i)}$ is the free energy of migration of an interstitial and $v_D = 10^{14}$ s^{-1} is the Debye frequency, write down the complete expression for the diffusion coefficient for silver ions in AgCl if diffusion occurred via (i) interstitial mechanism and (ii) interstitialcy mechanism. How would one be able to determine whether diffusion occurred by an interstitialcy mechanism instead of a vacancy mechanism in this system? Discuss any assumptions.

3. Determine the temperature at which the diffusion changes from intrinsic to extrinsic in the $CdCl_2$ doped NaCl system. Neglect the migration enthalpies and assume that the jump distance is 3.98×10^{-10} m, $v_D = 2 \times 10^{14}$ sec^{-1}.

4. Assuming that the reaction $CdCl_2 \Leftrightarrow Cd_i^{\bullet\bullet} + 2V_{Na}' + 2Cl_{Cl}$ is feasible in the NaCl system, calculate an expression for the diffusion of sodium, assuming a vacancy mechanism.

5. The formation enthalpy and entropy of Schottky defects in NaCl are $H_s^f = 2.4$ eV, $S_s^f/k = 8.99$.

   a) Calculate the fraction of cation vacancies at 500 K and at 1000 K in this material.

   b) The enthalpy and entropy of cation migration are $H_s^m = 0.626$ and $S_s^m/k = 1.065$. The enthalpy and entropy of anion migration are $H_s^m = 0.744$ and $S_s^m/k = 2.27$. Determine the magnitude of the ratio of the diffusion of cation vacancies to that of anion vacancies (state and justify any assumptions).

   c) The binding, or association, enthalpies and entropies of cation vacancy-impurity pairs are $H_{IS}^b = -0.64$ and $S_{IS}^b/k = -2.33$ and the binding, or association, enthalpies and entropies of anion vacancy-impurity pairs are $H_{IS}^b = -0.75$ and $S_{IS}^b/k = -1.42$. Using the Lidiard model (i) calculate the fraction vacancy-impurity pairs, p, and (ii) calculate the mobility of vacancy-interstitial pairs (state and discuss any assumptions).

d) Discuss the relative contributions of the pairs to the overall conductivity at low temperatures.

6. The tables below list defect formation energies and migration energies in RbCl and RbCl + Sr^{2+}. The data in the tables we extracted from (Jacobs et al. 1997) (From P.W. M. Jacobs, M.L. Vernon, 1007 (1997).)

a) Comment on the formation and migration thermodynamic parameters in relation to that of simple elemental metals.

b) Compare the fraction of anion and cation vacancies and interstitials of these systems (pure and doped) as a function of temperature.

c) Estimate the temperature dependencies of the transport numbers for anion and cation vacancies in RbCl and RbCl + Sr^+.

TABLE 4.2

Defect formation energies in RbCl systems

Formation energies (eV)	RbCl	RbCl + Sr^+
H_s^f	2.5	2.5
S_s^f	8.7	9.3
H_{FC}^f	3.5	3.5
S_{FC}^f (FC-Frenkel-Cation)	7.	7.4
H_{FA}^f	3.5	3.5
S_{FA}^f (Frenkel Anion)	19	9.1
H_{CD}^f (cation-defect assoc.)	-9.9	-0.6
S_{CD}^f	-2.2	-2.9

TABLE 4.3

Defect migration energies in RbCl systems

Migration energies (eV)	RbCl	RbCl + Sr^+
H_{cv}^m	0.66	0.66
S_{cv}^m (cation vacancies)	2.1	1.9
H_{av}^m (anion vacancies)	0.73	0.72
S_{av}^m	3	3.3
H_{ci}^m (cation interstitial)	.21	.21
S_{ci}^m	5.6	5.6
H_{ai}^m (anion interstitial)	0.19	019
S_{ai}^m	7	6.5

4.11 References and Additional Reading

Hooton, I.E., and Jacobs, P.W.M., "Ionic Transport in Crystals of Pure and Doped Sodium Chloride," *J. Phys. Chem. Solids* 51, 1207 (1990).

C.P. Flynn, *Point Defects in Diffusion*, Clarendon Press, Oxford, 1972.

P. Shewmon, *Diffusion in Solids*, 2nd ed., TMS publication (1989).

Atomic Transport in Solids, A.R. Allantt and A.B. Lidiard, Cambridge University Press, UK, 1993.

C. Kittel, *Introduction to Solid State Physics*, 5th ed., Wiley, NY (1976).

Jacobs, P.W.M., Vernon, M.L., "Ionic Transport in Rubidium Chloride," *J. Phys. Chem. Solids* 58, 1007 (1997).

5

Diffusion in Semiconductors

5.1 Introduction

Atomic migration processes in semiconductors are discussed in this chapter. A primary impetus for studying atomic migration in semiconductors is associated with reliable processing and fabrication of materials for optoelectronic and microelectronic (silicon-based) technologies (see for example Haynes 2000). Processing and fabrication of semiconducting materials may include molecular beam epitaxy, plasma etching, and chemical vapor deposition. Atomic migration processes are critical throughout different stages of the fabrication of devices. One critical stage during device fabrication involves the introduction of dopants, often accomplished using ion implantation. Ion implantation and other forms of irradiation create defects, particularly vacancies and interstitials, in the host material. Subsequent annealing of the sample induces redistribution of atomic species. Here the atomic migration process is characterized by large initial transients and relatively immobile populations of atomic species residing in the near-surface region of the sample, all of which are largely due to the influence of the nonequilibrium-point defect population distributions throughout the sample. The defect distribution is not well understood and is potentially problematic, particularly for thin films and the related implications associated with the fabrication of small devices. A second example in which atomic transport is critical involves the fabrication of metal-on-insulator (MOS) devices, wherein an important step in the process involves oxidation. Atomic diffusion in the presence of oxidation is enhanced, often in ways that are often not well understood. Third, semiconductor lasers which operate in the 1.3- to 1.5-mm wavelength range are well suited for optical fiber telecommunication. InGaNAs-based heterostructures are promising materials. The diffusion of N and In between the layers needs to be understood and controlled during processing at elevated temperatures because the quantity and spatial distribution of In and N are critical for device performance (emission intensity, band gaps, photoluminescence, etc.).

Although the situation involving diffusion in semiconductors would appear straightforward at first glance, particularly in light of the wealth of information regarding diffusion in metals, the analysis of semiconductors from a diffusion viewpoint has proven to be quite troublesome. This is largely because the experimental data often are not subject to unambiguous interpretation and point defects can exist in various charge states. The development of sophisticated atomistic simulation techniques has made it possible, only very recently, to develop further fundamental insight into the nature of various mechanisms of atomic transport that are possible. The goals of this chapter are to discuss self-diffusion and the diffusion of dopants in semiconductors.

5.2   Structure and Point Defects in Silicon

Figure 5.1 shows the structure of pure silicon. The silicon unit cell possesses the so-called diamond structure where the bonding in this system is characterized by tetrahedral symmetry. There are 8 atoms per unit cell. This is a relatively open structure with an atomic packing fraction of 0.34 and a lattice constant of 0.543 nm. Both diamond and germanium possess similar structures; however, the lattice constant is 0.356 nm for diamond and 0.565 nm for germanium. The open and directional bonding structure of these materials has a profound influence on atomic transport.

In silicon, there are two predominant point defects: vacancies and self-interstitials. The vacancies can exist in different charge states (positive, neutral, or negative), and the charge state dictates the local distortion of the lattice. By extension, the migration enthalpies are also a function of the charge state of the defect (Watkins). Self-interstitials constitute an important type of point defect in pure silicon. In pure silicon, the self-interstitial can possess different configurations. Shown in the aforementioned Fig. 5.1(a) and 5.1(b) are a hexagonal and a $\langle 110 \rangle$ split interstitial, where two silicon atoms are oriented along a $\langle 110 \rangle$ direction, respectively. Other interstitial configurations, such as $\langle 100 \rangle$ split and tetragonal interstitials, are also possible in silicon but are less stable in the neutral state. The hexagonal and the $\langle 110 \rangle$ split state energies are comparable.

5.3   Self-Diffusion in Silicon and Germanium

The situation involving self-diffusion in silicon has been a matter of concern for some time and appears not to be completely resolved. During the mid 1980s, it was suggested that three diffusion mechanisms are operational in pure silicon under normal conditions: an interstitial mechanism, a vacancy mechanism, and an exchange mechanism. The latter is a so-called concerted

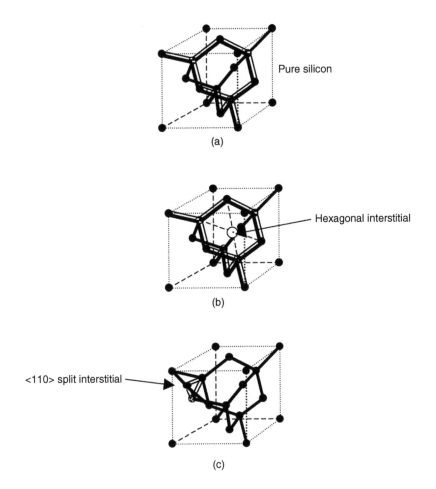

(a)

Pure silicon

(b)

Hexagonal interstitial

(c)

<110> split interstitial

FIG. 5.1
The structure of Si (Ge and C(diamond)) is shown here in part (a). Hexagonal and $\langle 110 \rangle$ split interstitials are shown in (b) and (c), respectively. Other interstitial configurations are possible.

exchange mechanism. It involves rotation of an Si-Si bond so that two Si atoms can exchange places. The rotation occurs such that the number of bonds that are broken to facilitate the exchange process is minimized (Pandey, 1986).

Since the three mechanisms of diffusion are independent, the total diffusion coefficient is a sum of three terms each representing the contribution from one mechanism,

$$D = D_v + D_i + D_X \qquad 5.1$$

where D_X is the contribution due to the exchange mechanism. There have been disagreements for some time about the dominant diffusion mechanism in this system. First-principles calculations capable of making an accurate assessment of lattice rearrangements and monitoring the diffusive motions in real time provide important insight into the dominant mechanism responsible for diffusion in this system (Jääskelainen et al. 2001). Based on these calculations, the formation entropy for a self-interstitial has been determined to be 11.2 k. The formation entropy contains both vibrational and a configurational components. In the case of a vacancy, the formation entropy is 8.8 k. The enthalpy of formation is 3.80 eV for an interstitial and 3.92 eV for a vacancy. The migration energies are 1.37 eV for an interstitial and 0.1 for a vacancy. These predictions are in good agreement with the data of Bracht et al. These calculations follow early predictions by Blochl et al., who predicted similar trends in the relative contributions of the vacancy and interstitial mechanisms. A plot of the self-diffusion data for silicon is shown in Fig. 5.2 (data of Bracht et al). For the interstitial mechanism and for the vacancy mechanism

$$D_i = 2980e^{-4.95/kT} \text{ cm}^2/\text{s} \qquad\qquad 5.1(a)$$

$$D_v = 0.92e^{-4.14/kT} \text{ cm}^2/\text{s}. \qquad\qquad 5.1(b)$$

It is clear from these equations that the activation energy (units of eV) for diffusion is large compared to that for metals, which is associated with the fact that the formation energies of the defects are much larger. The prefactor for the interstitial mechanism is orders of magnitude larger than the prefactor for self-diffusion via vacancies, which is related to the differences between the entropies of formation and migration. In addition, the activation energy for the interstitial mechanism is larger than that associated with the vacancy

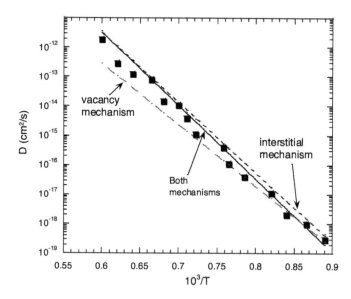

FIG. 5.2
The temperature dependence of the self diffusion of silicon (data extracted from H. Bracht, E. E. Bracht, E. E. Haller, R. Clark-Phelps, Physical Review Letters, 81, 393 (1998))

mechanism. The implication is that, although the vacancy and interstitial mechanisms are always operational, the interstitial mechanism is the dominant mechanism at high temperatures. The contribution due to the exchange mechanism is evidently small (Needs, 1999). The exchange mechanism has been ruled out as a dominant mechanism by simulations which indicate that the entropic contribution, which determines the prefactor, is too small (Ural et al.). Specifically, although the activation energy for the exchange mechanism was calculated to be 4.3 eV, the entropy was estimated to be 3.3 k, which might be considered a lower bound.

Although the three aforementioned mechanisms have received extensive attention in the literature, there is evidence of a divacancy diffusion mechanism at high temperatures. Recent *ab initio* calculations by Hwang and Goddard indicate that the divacancy moves as one entity, as opposed to a successive dissociation-recombination process. The activation energy for migration is calculated to be 1.35 eV, in agreement with an experimental value of 1.3 eV (G.S. Hwang and W.A. Goddard III 2002).

With regard to self-diffusion of germanium, the situation is much more clear cut. It is reasonably well established that self-diffusion in germanium occurs predominantly via a vacancy mechanism. The prefactor for self-diffusion in Ge is $D_0 = 0.12$ cm^2/s (other estimates range between 0.078 and 0.44 cm^2/s); the entropy associated with self-diffusion is 9 k (H.D. Fuchs et al. 1995).

In summary, the prefactors that control self-diffusion in semiconductors are large compared to that of metals. This is because the entropic component is larger, $S/k \sim 9$–11, compared to $S/k = 2$–4 for metals. The activation energies are also large $Q = H^f + H^m$ (~4 eV) compared to metals.

5.4 Diffusion of Dopants

The structural and electronic characteristics of the dopant have a profound impact on diffusional transport in semiconductors. Group V donor atoms (P, As, and Sb) form substitutional impurities in silicon. When a group V atom forms four covalent bonds with silicon, a single electron is left over in the valance band (p-type donor impurity). Group III atoms (B, Ga, In, and Al) also typically reside in a vacant lattice site, forming substitutional (n-type) acceptor impurities.

Defects in semiconductors may exist in different charge states. The free energy associated with the formation of a singly negatively charged vacancy, for example, is larger than that of a neutral vacancy. The Gibbs free energy associated with the formation of a charged vacancy is a combination of G_V, the free energy associated with the formation of a neutral vacancy, and the difference between E_F, the energy of the Fermi level, and E_{-v}, that of the energy level of the negative defect within the band gap, $E(-v) = E_{-v} - E_F$. $E(-v)$ possesses both entropic and enthalpic contributions. The entropic contributions are associated with local distortions associated with bond formation; this distortion is a function of the charge state. A natural consequence of the different charge state is that, because the concentration of charged defects is dictated by this energy through the Boltzmann factor, the fraction of singly negatively charged vacancies,

$$X_{v-} = X_v e^{-E(-v)/kT}$$
5.2

is lower than that of neutral vacancies, X_v. In general, for a singly negatively charged defect, denoted by y, where y could be a vacancy or an interstitial,

$$X_{y-} = X_y e^{-E(-y)/kT}$$
5.3

For a positively charged defect,

$$X_{y+} = X_y e^{-E(+y)/kT}$$
5.4

where $E(+y) = E_F - E_{+y}$.

The activation energies for the diffusion of substitutional dopants are slightly higher for donors (P, As, Sb, Bi) than for acceptors (B, Al, Ga, In). In these cases, one must account for a binding energy between the impurity and the defect, which has the overall effect of reducing the free energy.

Interactions between dopants and defects are strong, and the defects typically play an especially prominent role in the diffusion of dopants in semiconductors compared to the transport of tracer species in metal hosts. As mentioned earlier, dopants typically form substitutional impurities. If a dopant resides next to a vacancy, then it may form a dopant-vacancy pair (*DV*). If it resides next to an interstitial, then it may form a dopant-interstitial pair (*DI*). The interactions between the dopants and the defects are such that, when the dopant concentration is low (below the solubility limit), the dopants are ionized and will interact with the vacancies. Especially strong associations exist between Group V elements (e.g., phosphorous) and vacancies, compared to Group III elements. Interactions between interstitials and Group III elements (e.g., boron) are more probable than interactions between interstitials and Group V elements. The lower activation energies for impurity transport are believed to be associated with interactions between the impurity and the defects (vacancies or interstitials). Generally, the dopants diffuse at a faster rate than self-diffusion, in part because of this lower activation energy for transport. With this said, it should be emphasized that slight variations exist in values quoted for activation energies and that some debate continues.

5.4.1 Mechanisms of Atomic Transport

Standard mechanisms of transport that occur in impurity-doped silicon are now discussed. In one mechanism, the substitutional impurity, A_S, interacts with a self-interstitial (I) to create an intermediate impurity-interstitial pair (AI) that can undergo associative and dissociative interactions

$$A_S + I \underset{k_{AI}}{\overset{k_{AI}}{\Leftrightarrow}} AI \qquad 5.5$$

The AI coupling may be due to a Coulombic attraction or an attraction that minimizes the local lattice distortion or it may be due to a bonding interaction. The association/dissociation constants, k_{AI} and k'_{AI}, are different. Related mechanisms occur such that the substitutional impurity forms an interstitial, A_i, and (1) diffuses as an interstitial or (2) $A_s - A_i$ exchanges occur.

The substitutional impurity may also form an impurity-vacancy pair,

$$A_S + V \underset{k_{AV}}{\overset{k_{AV}}{\Leftrightarrow}} AV \qquad 5.6$$

where they associate and dissociate as a pair during diffusion.

Other mechanisms of diffusion are also possible in semiconductors. The "kick out" mechanism is possible wherein an *interstitial impurity* displaces a lattice atom, which then becomes a self-interstitial,

$$A_i \Leftrightarrow A_S + I \qquad 5.7$$

The self-interstitial subsequently diffuses. Alternatively, an interstitial impurity could interact with a vacancy

$$A_i + V \Leftrightarrow A_S \qquad 5.8$$

5.4.2 Examples

In this section, examples involving the diffusion of two different dopants are discussed. These examples provide insight into the diverse diffusion mechanisms that may occur in different systems. The first involves the diffusion of iridium, Ir, into silicon. Atypical Ir concentration profiles suggest that more than one mechanism may control Ir diffusion in silicon. Ir may behave as an interstitial impurity as well as a substitutional impurity. Two mechanisms believed to be simultaneously operational, (1) a "kick-out" mechanism where an Ir interstitial "kicks out" a silicon atom from a lattice site and the *Si* atom subsequently diffuses as a self-interstitial,

$$Ir_i \Leftrightarrow Ir_S + I \qquad 5.9$$

and (2) an associative mechanism involving vacancies,

$$V + Ir_i \Leftrightarrow Ir_s \qquad 5.10$$

wherein the *Ir* interstitial (Ir_i) interacts with a vacancy and forms substitutional Ir impurity, (Ir_s). The relative contribution of each mechanism is temperature dependent, with the "kick out" mechanism more dominant at higher temperatures.

Our second example concerns effects due to ion implantation of high concentrations of a dopant. The most common method used for doping semiconductors is ion implantation followed by thermal annealing. A well-known phenomenon, transient enhanced diffusion, occurs when the *near-surface* layer is implanted with a dopant and, upon subsequent annealing the profile broadens by creating a large tail covering many times the equilibrium diffusion distance. The initially large diffusion transient that created this tail is generally attributed to the large number of interstitials and vacancies created by the implantation process and the related formation of mobile impurity-interstitial and impurity-vacancy complexes. Consider, for example, the effect of interstitials. The enhancement of the diffusion coefficient is determined by the ratio of the nonequilibrium concentration of interstitials $[A_i]$ compared to the concentration of interstitials at thermal equilibrium $[A_i^{eq}]$

$$D = D_i^{eq} \frac{\left[A_i^{eq} \right]}{[A_i]}$$

                                                                                5.11

Although this is a reasonable approximation, theory and simulations provide additional insight into this process in some systems. We might choose as an example the implantation of silicon with a large concentration of boron ($>10^{18}/cm^3$), a p-type dopant. Substitutional boron, B_s, and silicon interstitial pairs, $B_s - Si_i$, were believed for some time to be primarily responsible for diffusion via a "kick out" mechanism. Atomistic simulations by Hwang and Goddard show that, when boron concentration is high, multiboron complexes are also formed. Substitutional and interstitial boron-boron pairs ($B_s - B_i$) form, and they are mobile. The activation energy for diffusion is 1.81 eV. The simulations show that, for high concentrations of boron at high temperatures, the concentration of $B_s - B_i$ pairs exceeds the number of $B_s - Si_i$ pairs. If the concentration of such pairs is sufficiently high, then their influence on diffusion and hence the shape of the concentration profiles is not trivial. In summary, the overall concentration profile is determined by the diffusion of various defect complexes, fermi level, controlled by boron concentration and by the temperature.

5.5   Concluding Remarks

Mechanisms of transport in doped and undoped semiconductors are diverse. The dominant mechanism of transport will depend on the concentration of defects and complexes within the system and on the migration and the formation energies of these complexes. Implantation produces a large number of point defects that interact with dopants, and the transport mechanisms necessarily become more complex. Different defect configurations are possible, and transitions between defect configurations occur. This is a very active area of research, and theoretical and atomistic simulations will play a critical role toward understanding various mechanisms of transport in different materials.

5.6 Problems for Chapter 5

1. If it is assumed that self-diffusion is determined exclusively by vacancy and interstitial mechanisms,

    a) write down complete expressions for diffusion coefficients.

    b) estimate the prefactors based on equivalent jump distances, nearest neighbor jump distances, etc. The values quoted in the chapter for the formation and migration entropies may be used.

    c) Do the same for Ge.

    d) If the three mechanisms (vacancy, interstitial, and exchange) are operational, estimate the temperatures at which each would become dominant.

2. If a divacancy mechanism is operational, along with the single vacancy and interstitial mechanism, and at 5/6 Tm, the diffusivity is enhanced by 20% due to the divacancy.

    a) Write down an expression for the total diffusivity

    b) Estimate the cross-over temperature at which the divacancy mechanism becomes dominant (state any assumptions, if necessary).

3. If a donor impurity interacts strongly with a vacancy silicon site, determine an expression for the diffusivity of the impurity. (justify any assumptions).

4. If a donor forms an associated pair, $As - As$ in silicon,

    a) indicate the possible mechanisms by which this pair could diffuse.

    b) write down a possible expression that would describe this process.

5. Imagine that P and Al are impurities added to silicon. How might diffusional transport occur in these systems (P-doped Si and Al-doped Si). Second, using estimates for activation energies for migration, estimate the relative diffusion rates of these impurities.

5.7 References

P.E. Blochl, E. Smargiassi, R. Carr, D.B. Laks, W. Andreoni and S.T. Pantelides, "First-principles calculations of self-diffusion constants in silicon," *Phys. Rev. Lett.*, 70, 2435 (1993).

L.J. Munro, "Defect migration in crystalline silicon," *Phys. Rev.* B. 59, 3969 (1999).

Pandy, K.C. "Diffusion without vacancies and interstitials: A new concerted exchange mechanism," *Phys. Rev. Lett.*, 57, 2287 (1986).

A. Jääskelainen, L. Colombo, and R. Nieminen, *Physical Review* B, 64, 233203 (2001).

R.J. Needs, "First principles calculations of self-interstitial defect structures and diffusion paths," *J. Phys. Condensed Matter* 11, 10437 (1999).

P.A. Stolk, H.-J. Grossmann, D.J. Eaglesham, D.C. Jaconson, C.S. Rafferty, G.H. Gilmer, M. Jaraiz, J.M. Poate, H.S. Luftman, and T.E. Hanes, "Physical mechanisms of transient enhanced dopant diffusion in ion-implanted silicon," *J. Appl. Phys.*, 81, 6031 (1997).

P.M. Fahey, P.B. Griffin, and J.D. Plummer, "Point defects and dopant diffusion in silicon," *Reviews of Modern Physics*, 61, 289 (1989).

G.D. Watkins, "Intrinsic defects in silicon," *Materials Science in semiconductor processing*, 3, 227 (2000).

S. List and H. Ryssel, "Atomistic analysis of the vacancy mechanism of impurity diffusion in silicon," *Journal of Applied Physics* 83, 7585 (1998).

P.A. Stolk, H.-J. Grossman, D.J. Eaglesham, D.C. Jacobson, C.S. Rafferty, H.S. Luftman, and T.E. Haynes, "Physical mechanisms of transient enhanced dopant diffusion in ion-implanted silicon," *J. Appl. Phys.*, 81, 6031 (1997).

G.S. Hwang and W.A. Goddard III, "Diffusion and dissociation of neutral divacancies in crystalline solids" *Phys. Rev.* B, 65, 233205 (2002).

G.S. Hwang and W.A. Goddard III, "Diffusion of the diboron pair in silicon," *Phys. Rev. Lett.*, 80, 055901 (2002).

A. Ural, P.B. Griffin and J.D. Plummer, "Self-diffusion in silicon: Similarity between properties of native point defects," *Phys. Rev. Lett.*, 83, 3454 (1999).

H.D. Fuchs, W. Walukiewicz, E.E. Haller, W. Dondl, R. Schorer, G. Abstreiter and A.I. Rudnev, "Germanium $^{70}Ge/^{74}Ge$ heterostructures: An approach to self-diffusion studies," *Phys. Rev.* B. 51, 16817 (1995).

S. Obeidi and N.A. Stolwijk, "Diffusion of irridium in silicon: Change over from a foreign-atom-limited to a native-defect-controlled transport mode," *Phys. Rev.* B. 64, 113201 (2001).

Haynes, T.E. Editor, Defects and Diffusion in Silicon Technology Materials Research Society Bulletin, June 2000.

Part III

Diffusional Transport in Systems That Lack Long-Range Structural Order

If the arrangement of atoms, or molecules, that compose a solid material lacks long-range order, then the material is identified as a glass

Packing irregularities (free volume) and excess configurational entropy (in relation to the crystal state) are factors that control the temperature dependent long-range dynamics of glass forming systems. In the supercooled state, well known phenomena like the Stokes-Einstein relation, which connects translational diffusion to the viscosity, are not obeyed in some systems. Transport in disordered media is a very diverse topic and our discussion is necessarily limited. In Chapter 6, the topic of the dynamics and vicsoelasticity of polymer melts is introduced. Chapter 7 addresses the structure and transport in inorganic network glasses. Part III is concluded with general comments on the dynamics of systems in the supercooled state.

6

Transport and Viscoelasticity of Large Macromolecules

6.1 Introduction and Context

The goals of this chapter are ultimately to provide a microscopic picture that explains the time-dependent behavior of long-chain polymeric liquids. It will be shown that these long-chain molecules form a dense and complex entangled mesh, and that the existence of this mesh has far-reaching consequences for the dynamics of the chains.

General comments regarding the dynamics of simple and complex liquids are now presented to provide a context for the subsequent discussion of polymer dynamics. In a simple homogeneous liquid, diffusional transport is isotropic and the diffusion coefficient, D, is related to the viscosity, η, of the liquid and to the temperature, T, in a manner dictated by the Stokes-Einstein equation $\eta D = kT/6\pi r$. The mechanical properties of this simple, homogeneous, liquid are specified entirely in terms of its viscosity. For a simple Newtonian fluid, the stress, $\sigma(t)$, is proportional to the shear strain rate, $d\gamma/dt$,

$$\sigma = \eta \frac{d\gamma}{dt} \qquad 6.1$$

where the constant of proportionality is the viscosity. A simple liquid below its melting temperature is crystalline and its response to a mechanical force is elastic, wherein the stress is proportional to the strain (provided the deformation is sufficiently small) and the constant of proportionality is the elastic (Young's) modulus. Generally the elastic modulus is highly anisotropic, specified mathematically by a tensor.

By virtue of their molecular architecture and organization, the behavior of a range of more complex liquids or *soft materials* can be quite unexpected. For the purposes of illustration we compare the behavior of three "soft materials," mayonnaise, mustard, and honey. Honey will flow under the influence of gravity whereas mayonnaise and mustard do not. Interestingly, mayonnaise and mustard each possess a low yield stress and are easily sheared by a knife. Honey, which does not possess a yield stress, is generally not as easily sheared. Mayonnaise is an emulsion (liquid dispersed in a liquid: droplets of vegetable oil in vinegar stabilized by a surfactant) and mustard is a suspension (paste: ground mustard seeds in water, vinegar, and salt). The properties of these materials reflect differences in the structural organization and composition of their constituents. Indeed, constitutive relations more complex than Eq. 6.1 are required to understand the relationships between the stresses and strains and their rate dependencies for such materials.

In this chapter we are interested in the diffusion and viscoelastic properties of a different class of "soft" materials, long chain polymeric molecules. Polymeric materials are especially interesting because they are capable of responding to external stresses, behaving like elastic solids at sufficiently short time scales and like viscous liquids over long time scales. This is the essence of the *viscoelastic* response of these materials. Silly putty, which is well known to most of us, exhibits these characteristics under fairly ordinary circumstances. Specifically, if rapidly deformed (e.g., thrown into a wall at high velocity) its response is similar to that of a rubber (elastic behavior (immediate response)), whereas if left alone on a table under the influence of gravity for a few hours it will appear to have flowed like a viscous, Newtonian liquid (time-dependent response).

Other aspects of the flow properties of polymers or other viscoelastic liquids are equally intriguing. An experiment that might be performed to distinguish between the flow properties of a simple liquid and a viscoelastic liquid involves stirring different liquids in a beaker. Newtonian liquids like water or simple oils will form a vortex around the stirrer whereas polymer solutions, and cake batter, "climb up" the stirrer, as if there existed an attraction. With increasing shear rates, it becomes easier to shear viscoelastic liquids. This is the phenomenon of *shear thinning*, which is further illustrated in Fig. 6.1.

In Fig. 6.1(a) it is shown that the viscosity of the liquid decreases with increasing shear rate, a natural consequence of the shear thinning phenomenon. Indeed, a related, classic everyday experience with which kids become intimately familiar, is the shearing motions of chewing gum between their fingers. Kids are well aware that it is easier to shear chewing gum with increasing shear rates, whereas pulling, in contrast, requires more effort. The opposite, less common, behavior, shear thickening (Fig. 6.1(b)), associated with the increasing viscosity with increasing shear rate, is exhibited by some suspensions. Indeed, the shear rate dependence of the viscosity is an important property that distinguishes polymeric liquids from simple liquids.

Our discussion of the topic of polymer dynamics (diffusion and viscoelasticity) necessarily begins by introducing a description of the basic properties of the polymer chain. This discussion is followed by a more detailed, yet phenomenological, description of the time-dependent viscoelastic behavior of polymers. Our discussion of these two topics will establish the context for the subsequent discussion of the microscopic model for polymer dynamics.

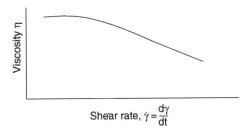

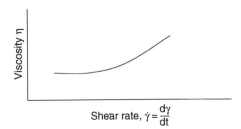

FIG. 6.1
a) Shear thinning of a complex liquid such as a polymeric melt.
b) Shear thickening of a complex liquid. Some concentrated suspensions exhibit this behavior.

6.2   Classification of Polymers

A simple *linear* homopolymer chain is composed of a sequence of covalently bonded monomers, as shown in Fig. 6.2. Monomers possess a diverse range of chemical structures and they can be synthetic or natural. Common examples of synthetic monomers include: 1) ethylene, the basic building block of polyethylene (PE), used for applications that include films, electrical insulation, and tubing. The structure for polyethylene is denoted $-[CH_2\text{-}CH_2]_n-$, where n is the number of monomers. 2) Styrene is the monomer from which

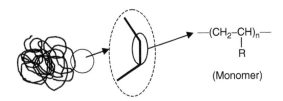

(Monomer)

FIG. 6.2
A polymer chain is composed of repeat units (monomers, R) that are covalently connected. The polymer chain assumes the configuration of a random coil in the melt. The monomer identified here is a vinyl monomer and when R is a hydrogen atom, the polymer is polyethylene. If R is a phenyl ring (C_6H_5) the polymer is polystyrene. Polyvinyl chloride is the polymer if the repeat unit is chlorine.

polystyrene (PS) is synthesized, and cups (Styrofoam) and packaging materials are common uses. Its structure is $-[CH_2-CH-(C_6H_5)]_n-$. 3) The monomer
vinyl chloride, the basic building block of polyvinyl chloride (PVC), material
for pipes used for lawn irrigation systems, possesses the following chemical
structure, $-[CH_2-CH-Cl]_n-$. These three polymers belong to a class of polymers
known as vinyl polymers and have the generic structure $-[CH_2-CHR]_n-$.
Another common type of monomer is tetrafluoroethylene $-[CF_2-CF_2]_n-$, the
basic building block of polytetrafluoroethylene (Teflon), the coating on non-
stick frying surfaces. Our final example of a synthetic polymer is polymeth-
ylmethacrylate (Plexiglass, Lucite, airplane windows, transparent sheets)
and the structure of this monomer is $-[CH_2-C-(CH_3)-(COOCH_3)]_n-$.

There exist many examples of polymeric molecules that are not linear
chains. Some molecules are branched, as illustrated in Fig. 6.3(a). In fact, the
architecture of some PE molecules, the so-called low density PE (LDPE), is
branched and the extent of branching determines the nature the applications
of this polymer. Its high density analog, HDPE, which possesses a lower
degree of branching, and associated higher crystallinity, is used for applications that require comparatively higher strength. LDPE is typically used for
packaging applications (e.g., trash bags) whereas HDPE is often used to
make liquid containers (e.g., milk containers). In many common situations
polymeric molecules form a permanent, yet flexible, cross-linked network,
elastomers (automobile tires, elastic bands, etc.), Fig. 6.3(b). In fact one might
classify polymers in three broad areas, thermoplastics, elastomers, and
thermosets. The last of these are generally network polymers which are
structurally very rigid (intractable) due, in part, to a very high degree of
cross-linking.

We now comment further on the architecture of polymers with regard to
another class of polymers, copolymers. When the polymer is composed of

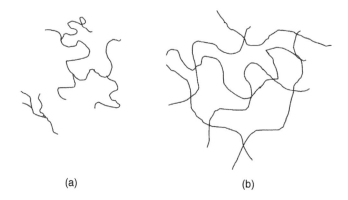

(a)   (b)

FIG. 6.3
Schematics of a) branched and b) network chains.

two types of monomer it is identified as a copolymer. If a group of monomers (-A-A-A-A-A-A-) of identical structure is bonded covalently to another group, or block, of monomers (-B-B-B-B-B-) the polymer is identified as a diblock copolymer. The category of copolymers also includes triblock copolymers (-A-A-A-A-A-A-B-B-B-B-B- ..... -B-A-A-A-A- ...), alternating copolymers (-A-B-A-B-A-B-A-B-A-B-), random copolymers, graft copolymers, and star block copolymers. There exist, of course, a diverse range of naturally occurring polymers such as DNA, cellulose, proteins (proteins are copolymers), and carbohydrates.

The foregoing discussion was meant to provide a basic, yet brief, introduction to the different types of polymers and to set the stage for the discussion of chain conformations.

6.3 Properties of a Single Polymer Chain

6.3.1 Freely Jointed Chain Model

The conformation (spatial organization of monomers) of linear polymer chains is now discussed. For the purposes of our discussion, we consider a linear, flexible polymer chain composed of a large number of monomers, typically 10^3 to 10^6 monomers. Vinyl polymers are examples of flexible polymers. Double stranded DNA, on the other hand, is considered stiff, by contrast (note, however, that over sufficiently large length scales the DNA molecule might be considered flexible). In the melt, the chain forms a random coil. A model is required to describe the structure of the chain and the simplest model is the so-called freely jointed chain model. In this model the chain is composed of n bonds (or links), each of length l_i and $n + 1$ atoms. We note that for vinyl polymers the number of backbone bonds is 1 per backbone carbon atom, or 2 per monomer. There are no restrictions on the bond angles nor bond orientations in this model; two monomers are not prohibited from occupying the same space in this analysis.

The contour length of the chain is $L = nl$ ($l = |\vec{l}_i|$) and the end-to-end vector, $\vec{R}$, defined in terms of $n + 1$ position vectors, $(\vec{R}_0, \vec{R}_1 .... \vec{R}_n)$, as illustrated in Fig. 6.4,

$$\vec{R} = \vec{R}_n - \vec{R}_0 = \sum_{i=1}^{n} \vec{l}_i \qquad 6.2$$

In this model all the bond lengths are equal. As one might anticipate, the ensemble average of $\vec{R}$, $\langle \vec{R} \rangle = 0$ because the end-to-end vectors in the large collection of chains are not correlated. Equivalently, the end-to-end vectors of a chain taken at different time intervals sufficiently far apart are not

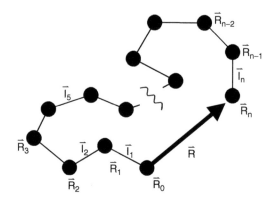

FIG. 6.4
Schematic of the freely jointed chain composed of n bonds and $n + 1$ bond vectors.

correlated. The mean-square end-to-end vector of this freely jointed chain is obtained from an ensemble average of $\vec{R} \bullet \vec{R}$,

$$\langle R^2 \rangle = \sum_{i=1}^{n} \langle l_i^2 \rangle + 2 \sum_{j=1}^{n-1} \sum_{i=1}^{n-j} \langle \vec{l_i} \bullet \vec{l}_{i+j} \rangle$$

6.3

$$= nl^2 \left\{ 1 + \frac{2}{n} \sum_{j=1}^{n-1} \sum_{i=1}^{n-j} \langle \cos \theta_{i,i+j} \rangle \right\}$$

Since all values of θ are equally probable, the rotations of the chain are unrestricted, then $\{\frac{2}{n} \sum_{j=1}^{n-1} \sum_{i=1}^{n-j} \langle \cos \theta_{i,i+j} \rangle\} = 0$. In other words, there exists no correlations between the between the backbone vectors along the chain. Therefore

$$\langle R^2 \rangle_f = nl^2$$

6.4

$(\sum_{i=1}^{n} \langle l_i^2 \rangle = n \langle l^2 \rangle = nl^2)$. The subscript f in Eq. 6.4 denotes the fact that this is the prediction of the idealized freely jointed chain model. Note that Eq. 6.4 could have easily been anticipated based on a simple random walk model, discussed earlier in Chapter 2.

6.3.2 Freely Rotating Chain Model

Realistically, the bond angle θ is subject to restrictions. Moreover an azimuthual angle ϕ is also required to characterize the conformation of the chain, as illustrated in Fig. 6.5a. A somewhat more realistic model would involve keeping θ ($90° < \theta < 180°$) fixed and allowing ϕ to be unrestricted $\theta = 109.5$ for the tetrahedral carbon-carbon bond angle. This is called the freely rotating chain model and the mean square end-to-end vector becomes

$$\langle R^2 \rangle_{fo} = C_n nl^2$$

6.5

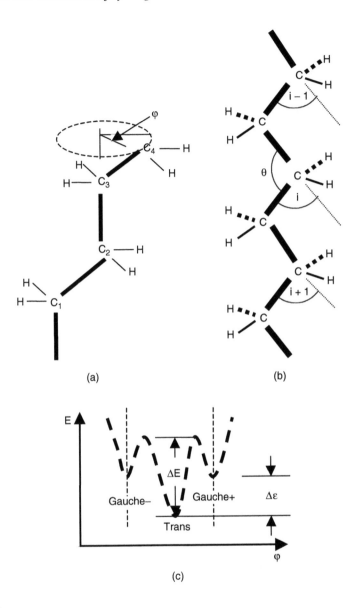

FIG. 6.5
Models of the chain: A) In the freely jointed model, θ and φ are unrestricted. B) The conformation shown in part B corresponds to $\varphi = 0$, with C-atoms lying in the plane of the page. This the fully extended planar zig-zag (all trans) conformation. The broken lines denote H-atoms below the plane of the page while the other H-atoms reside above the plane of the page. C) Energy versus φ diagram.

TABLE 6.1

Molecular characteristics (C_∞, σ, $\langle R^2 \rangle_0/M$, b) of some polymers

Polymer	$T(K)$	C_∞	b (cm)	σ	$\langle P^2 \rangle_0/M$ (Å^2 mol/g)
Polyisobutylene	298	6.7	1.54×10^{-8}	1.8	0.57
Polystyrene (atactic)	463	9.5	1.54×10^{-8}	2.18	0.434
Polymethylmethacrylate (atactic)	473	9.0	1.54×10^{-8}	2.01	0.425
Polydimethyl siloxane	298	6.8	1.62×10^{-8}	1.39	0.422
Poly(α-methyl styrene)	273	10.1	1.54×10^{-8}	2.25	0.442
polyethylene	463	7.4	1.54×10^{-8}	1.87	1.25
Polyvinyl acetate	330	9	1.54×10^{-8}	2.08	0.49

Source: L.J. Fetters, D.J. Lohse, and S.T. Milner, *Macromolecules*, 32, 6847 (1999).

L.J. Fetters, D.J. Losche, D. Richter, T.A. Witten, and A. Zirkel, *Macromolecules*, 27, 4639 (1994).

Jozef Bicerano, *Prediction of Polymer Properties*, Marcel Dekker, New York, 1996.

and

$$C_n = \left\{ 1 + \frac{2}{n} \sum_{j=1}^{n-1} \sum_{i=1}^{n-j} \langle \cos\theta_{i,i+j} \rangle \right\} \qquad 6.6$$

where C_n is called the characteristic ratio. Recall that in Chapter 3, this expression described the correlations between hops of an atom on a lattice! Here C_n represents the average of all main-chain bond angles and is always greater than 1 for real polymers and approaches a constant value for large n. It is left as an exercise to the reader to show that in the limit of large n, $\lim_{n \to \infty} C_n = C_\infty = \frac{(1+\cos\theta)}{(1-\cos\theta)}$. With this in mind, then

$$\langle R^2 \rangle_{fo} \approx C_\infty n l^2 \qquad 6.7$$

The characteristic ratio tends to be larger for chains with bulkier side groups, associated with which is a larger degree of steric hindrance. There exist a number of exceptions to this generalization. An example is PE, which has a characteristic ratio that is comparable to that of some polymers which possess more complex structures (Table 6.1) suggesting that other factors contribute to the magnitude of C_∞. This will be revisited in Section 6.7 after the topic of dynamics has been introduced.

6.3.3 Hindered Rotation Chain Model

There is an additional factor that needs to be considered. Because of steric hindrance and an intrinsic potential, φ assumes only a limited range of values. Consider, for example, Fig. 6.5b. Here the carbon atoms of the PE molecule all reside in the plane of the page. The broken lines represent bonds connecting H-atoms that reside below the page, whereas the other H-atoms reside above the page. This segment of the PE molecule (the segment might be viewed as an n-alkane) exists in a trans conformation, where $\varphi = 0$. For a very short molecule (example butane), this would be the lowest energy conformation because the steric effects are minimum. Two other minima exist in the gauche (±) positions where $\varphi = \pm120°$. This is illustrated in Fig. 6.5c. The difference between the energies, $\Delta\varepsilon$, of these conformations dictates the relative portions of trans and gauche conformations in the chain. If $\Delta\varepsilon < kT$, then the chain is flexible. On the other hand if $\Delta\varepsilon$ is large compared to kT, then the chain is considered to be stiff because the chain would reside in a predominantly trans conformation. The relative amounts of trans versus gauche conformations would be reflected in the Boltzmann factor ($e^{\Delta\varepsilon/kT}$). The dynamics of the chain is facilitated by the rate of trans-gauche transitions and the ease with these transitions occur determine the flexibility of the chain in a dynamical sense. If the energy barrier, ΔE (Fig. 6.5c) is at least comparable to kT, then the transitions would be rapid, with a relaxation time $\tau \propto e^{\Delta E/kT}$. The prefactor is probably comparable to a Debye frequency ~10^{-14} Hz. On the other hand, a larger barrier height would reflect much slower dynamics.

If we assume that φ is fixed then it may be shown, based on a so-called hindered rotation model, that

$$\langle R^2 \rangle_o = nl^2\left(\frac{1+\cos\theta}{1-\cos\theta}\right)\left(\frac{1+\langle\cos\theta\rangle}{1-\langle\cos\theta\rangle}\right) \qquad 6.8$$

In this case $C_\infty = \frac{1+\cos\theta}{1-\cos\theta}\frac{1+\langle\cos\phi\rangle}{1-\langle\cos\phi\rangle}$. Clearly, in spite of the additional restrictions, the mean square end-to-end vector continues to scale as nl^2. Often it is convenient to define a steric parameter σ, such that $C_\infty = \frac{1+\cos\theta}{1-\cos\theta}\sigma^2$. The parameter σ accounts for short-range steric repulsions.

The foregoing appears to have complicated matters, somewhat, but the situation is simplified by realizing that an equivalent freely jointed chain is

$$\langle R^2 \rangle = Nb^2 = C_\infty nl^2 \qquad 6.9$$

where $N = \frac{R_{max}^2}{C_\infty nl^2}$ and $b = C_\infty nl^2/R_{max}$ and R_{max} is the length of the fully extended chain, subject to the fixed bond angle constraint; $R_{max} = nl\cos(\theta/2)$. Note that under these conditions, the chain is in a trans conformation and R_{max} is identified as the contour length of the chain $\langle R^2 \rangle = bR_{max}$.

This final result indicates that despite the restrictions imposed on the angles, the mean square end-to-end vector still varies as the square of the step length (random walk!), of course with a new effective bond length, b. Table 4.1 shows values of the characteristic ratio and the effective bond length, often called the *Kuhn segment length*, for some common polymers. Note that C_∞ is approximately 8 for many polymers and the Kuhn segment length is approximately 1.5 nm for most polymers.

6.3.3.1 *Persistence Length*

In the above, it should have been evident that the bond vectors are correlated. The direction of one bond vector is determined by its connected neighbors and as the distance between monomers on the chain increases this correlation approaches zero. In view of this, one can think in terms of a persistence length. In principle, it reflects the tendency of the chain to reside in a trans state and in this regard, the persistence length, l_p, is proportional to $e^{\Delta\varepsilon/kT}$. In general, one can define a chain as stiff, locally. However, at much longer length scales it could be identified as flexible, in particular if l_p/L (recall that L is the contour length) is small. A more reliable estimate of the persistence length can be obtained by considering the following. We begin by considering a freely rotating (fixed θ) model and identify one bond on the chain. Now consider the average projection of the kth bond in the direction of this bond. Note that the angle between each successive bond is fixed so this average projection is $l(\cos\theta)^{k-1}$. It is left as an exercise to show that the sum of the projections of these bonds is $l_p = l/(1-\cos\theta)$, the persistence length. This model is called the Porod-Kratky worm-like model (see for example Flory 1969).

Finally, it is convenient to describe the configuration of the chain in terms of a radius of gyration, R_g, instead of a mean square end-to-end distance. The radius or gyration of a collection of particles, you should recall from freshman physics, is the root-mean-square distance of particles from their common center of mass. $\langle R_g^2 \rangle = \frac{\langle R^2 \rangle}{6}$ is a more meaningful parameterization particularly in the case of branched molecules. Moreover, for many common techniques such as light scattering, R_g is the relevant parameter that is measured.

6.3.4 Single Chain Statistics: Excluded Volume Effects

The foregoing discussion largely addresses an ideal chain. For a real chain other restrictions, aside from the bond angle restrictions, need to be considered, the so-called long-range excluded volume effects. These are associated with the fact that real monomers are of finite dimensions and that monomers remotely removed from each other along the chain cannot occupy the same space. In other words, in the random walk problem "the drunk" can retrace the earlier steps on later occasions. The prohibition of such events is what is often known as the long-range excluded volume effect. In thinking about this problem, it is important to realize that the effective bond volume, and not the actual bond volume, is the appropriate parameter to be considered. This effective volume will be determined by relative monomer-monomer, monomer-solvent, and solvent-monomer interactions. One can imagine that the probability that segments will eventually cross each other increases as the chain length increases. This would naturally imply that the effect is very important for long chains and it would have the effect of increasing the dimensions of the chain, depending on the circumstances. The root mean square end-to-end vector may be rewritten $\langle R^2 \rangle' = \alpha \langle R^2 \rangle$ where α would depend on temperature and on the solvent as well as M (on N), unless the chain is in a theta solvent.

We begin by considering the interactions associated with bringing two identical monomers, initially far apart, together in the presence of a solvent. Consider, further, the situation where the monomers exhibit a somewhat stronger affinity with each other than with the solvent molecules. Of course, if the molecular forces involved are purely dispersive, then this condition is easily met. When the molecules become close to each other, they experience an attraction (attractive potential well) and as they become even closer they experience steric (so-called hard core) repulsions. If the steric repulsive and attractive forces are equal then a special condition, known as the theta condition, is met. An important consequence of the theta condition is that the polymer chain is unperturbed ($\langle R^2 \rangle \propto N^{1/2}$) because there exists no net penalty for monomer-monomer contact. In other words, the excluded volume vanishes.

In the situation where the monomer solvent interactions are stronger than the monomer-monomer interactions (good solvent) the chain is swollen. The monomers interact via their hard-core potentials to avoid overlap. In addition there is another contribution to the potential resulting from the attraction with the solvent molecules. In this case the monomers avoid each other (so-called self-avoiding random walk) and the chain is swollen wherein $\langle R^2 \rangle \propto N^{3/5}$ (excluded volume effect). In fact, at high temperatures, and even in athermal solvent conditions (monomers like each other as much as they do solvent molecules), this behavior is expected. At high T, the monomers also avoid each other because they interact via hard-core potentials. As the temperature is reduced the system eventually approaches the theta condition where the chain dimensions become unperturbed.

In dilute solution, at temperatures below the theta temperature the chain exists in a collapsed state (more likely to find monomers closer together than with solvent molecules). Perhaps the best known example is polystyrene in cyclohexane, which is a good solvent above 35°C; the theta condition is realized when $T \approx 35°C$. For $T < 35°C$, the chain resides in a collapsed conformation. This behavior (collapsed chain) is similar to the polymer in a non-solvent where the monomers would completely exclude solvent molecules. Under these conditions complete phase separation occurs (precipitation) between the polymer and solvent. The situation involving polymer melts is interesting. In a melt the chains organize to form a densely packed mesh in which the chains interpenetrate each other. In a melt of identical chains, the monomers from on a chain cannot distinguish between themselves from monomers belonging to other chains. Here the intermolecular interactions balance the intramolecular interactions and chains in polymer melts are unperturbed, $\langle R^2 \rangle = \langle R^2 \rangle_0$. In semi-dilute solution, $R^2 \propto c^{-1/4}$, where c is the concentration. Our final comments refer to measurements of chain dimensions. Very good estimates of the unperturbed chain dimensions of many polymers have been made by performing dilute solution viscosity measurements since the intrinsic viscosity, $[\eta]$, of the polymer in a theta solvent is $[\eta]_\theta = K_\theta M^{1/2}$ where the constant $K_\theta = \Phi[\langle R^2 \rangle_0/M]^{3/2}$ is determined by the unperturbed dimensions of the chain; Φ is a universal hydrodynamic constant, $\Phi = 2.5 \times 10^{21}$ $dL/cm^3 mol$. The intrinsic viscosity is obtained by measuring the viscosity, η, of dilute solutions of various concentrations, c, of polymer and $[\eta] = \lim_{c \to 0} \frac{1}{c}(\frac{\eta - \eta_0}{\eta_0})$, where η_0 is the viscosity of the pure solvent. Small Angle Neutron Scattering (SANS) is an alternative technique for measuring chain dimensions and is in fact believed to be somewhat more reliable. Nevertheless, in the absence of SANS data, dilute solution measurements are adequate.

6.3.5 Single Chain Statistics Continued: Gaussian Statistics

It is noteworthy that since the chain possesses a random walk configuration, the probability density distribution function is Gaussian

$$P(R) = \left(\frac{3}{2\pi \langle R^2 \rangle_0} \right)^{3/2} e^{-\left(\frac{3}{2\langle R^2 \rangle_0} \right) R^2} \qquad 6.10$$

This is the probability that one end of the chain is a distance R from the origin (Note that this is also the probability that a chain segment $N = n'$ steps from the origin is located at position R, provided n' is sufficiently large). The probability that the other end of this chain is a distance R from the origin within a shell of thickness dR is given by $P(R)d\bar{R}$ and since this is a probability density function then the following condition must be satisfied

$$\int_0^\infty P(R)d\bar{R} = \int_0^\infty 4\pi R^2 \left(\frac{3}{2\pi \langle R^2 \rangle_0} \right)^{3/2} e^{-\left(\frac{3}{2\langle R^2 \rangle_0} \right) R^2} dR = 1 \qquad 6.11$$

and furthermore, the mean square end-to-end distance is

$$\langle R^2 \rangle = \int_0^\infty 4\pi R^4 P(R)dR = Nb^2 \qquad 6.12$$

as expected.

Having described the basic properties of chains the phenomenology of viscoelasticity is now described. The discussion on phenomenology will lay the foundation for a subsequent development of the molecular picture.

6.4 Phenomenology of the Viscoelastic Behavior of Polymers

The viscoelastic response of the long-chain polymeric liquids will depend on the rate of deformation and for sufficiently large deformations, the magnitude of the deformation. One of the truly significant experiments that served to reveal features unique to long-chain polymeric systems is the so called *stress relaxation* experiment. In such an experiment, a sudden strain is imposed on a sample, resulting in a net displacement of molecules in the sample thereby increasing its free energy. A Rheometer is often used to perform such an experiment and the experiment is typically performed in a shear configuration (though it can be performed in tension as well).

The response, the stress relaxation modulus, $G(t)$, is determined using the Rheometer and in such an experiment, the sudden strain, γ_0, imposed on the sample is held throughout the duration of the experiment. The material responds, with the molecules attempting to get back to equilibrium, necessarily with a time dependent shear stress $\sigma_{xy}(t)$. In this case, the stress relaxation modulus is related to the response such that

$$G(t, \gamma_0) = \sigma_{xy}(t, \gamma_0)/\gamma_0 \qquad 6.14$$

If the strain is sufficiently small, then the modulus and time-dependent response are independent of strain. When this is true the stress relaxation modulus provides information intrinsically associated with the dynamics, influenced necessarily by the interactions between the molecules. The schematics in Fig. 6.6 show the response of different types of materials to the imposed strain, γ_0. Notice that the response of an elastic material is immediate and necessarily remains constant as long as $\gamma = \gamma_0$ (Fig. 6.6b). For a truly viscous (Newtonian) material the stress is not maintained, but dissipated immediately (Fig. 6.6c).

The most interesting response is exhibited by the polymeric liquid, where the response is time-dependent, illustrated in Fig. 6.6d, and persists over a much longer time scale than the simple liquids. It will be shown later that for polymers two cases can be distinguished. 1) For short chain polymers, below what is known as the critical molecular weight for entanglements, M_c, the response, $G(t)$, is an exponential decay. 2) $G(t)$ for highly entangled, long chain polymers, exhibit a characteristic plateau at intermediate time scales before reaching zero at long times when all the energy associated with the imposed strain has dissipated.

The plateau, where $G(t)$ remains constant for a time interval τ_d, is reminiscent of the behavior of an elastomer. The value of $G(t)$ at the plateau is identified as the plateau modulus, G_N^0. The width of the plateau increases as the chains become longer. In fact, a melt composed of sufficiently long chains will exhibit some degree of elastic recovery in response to a non-linear tensile deformation. This recovery observed in these molten linear chain systems is associated with the fact that the chains are entangled and these entanglements act as temporary "cross-links." In fact, G_N^0 is associated with the average molecular weight between entanglements. We will discuss this issue in further detail later when we discuss a microscopic mechanism, reptation, by which chains diffuse throughout the entangled mesh (Section 6.8). In the meantime we introduce two well known phenomenological models.

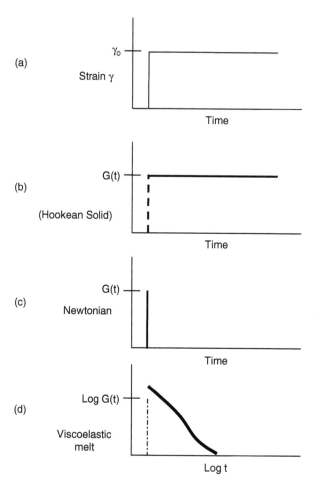

FIG. 6.6
Stress relaxation experiment: Response of various types of materials to a sudden imposed strain.

6.4.1 Maxwell and Voigt Phenomenological Models

Early theoretical attempts to understand the time dependence of $G(t)$ employed use of phenomenological models, devoid of any molecular interpretation. These models provided some insight into time-dependent material response. The models are based on the notion that the behavior of the polymer can be elastic, in which case a spring serves as model mechanical element, or viscous (dissipative) where the response is represented by a dash pot (or piston). The Maxwell model (Fig. 6.7) assumes that the rate of strain, $d\varepsilon/dt$, is related to an applied *tensile* stress, σ, such that

$$\frac{d\varepsilon}{dt} = \frac{1}{E}\frac{d\sigma}{dt} + \frac{\sigma}{\eta} \qquad\qquad 6.15$$

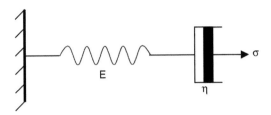

FIG. 6.7
Spring and dashpot arranged in series under the influence of an applied stress.

This equation arises from the fact that the overall strain in the assembly is the sum of the strains in each element (spring and piston connected in series),

$$\varepsilon = \varepsilon_\eta + \varepsilon_E \qquad 6.16$$

where the first term on the RHS is the strain in the dashpot and the second is that in the spring. Moreover, since the elements are in series, the stresses are equal in each of them,

$$\sigma = \sigma_E = \sigma_\eta \qquad 6.17$$

where $\sigma_E = E\varepsilon_E$ and $\sigma_\eta = \eta \frac{d\varepsilon_\eta}{dt}$. Equation 6.15 is the constitutive model that describes the relationship between the stress and strain rates of the material undergoing deformation (Note: Earlier in this chapter we used γ to denote a shear strain, whereas we now use ε to denote a tensile strain).

We might consider as an example these boundary conditions that characterize a stress relaxation experiment. In such an experiment, the boundary conditions dictate that $\varepsilon = \varepsilon_0$ and that $d\varepsilon/dt = 0$, implying that

$$\frac{1}{E}\frac{d\sigma}{dt} + \frac{\sigma}{\eta} = 0 \qquad 6.18$$

The solution to this equation is an exponential function

$$E(t) = Ee^{-\frac{t}{\tau}} \qquad 6.19$$

where $E(t) = \sigma(t)/\varepsilon_0$. The relaxation time is

$$\tau = \frac{\eta}{E} \qquad 6.20$$

This result (Eq. 6.20) indicates that the viscosity can be expressed as a product of a relaxation time and a modulus. It turns out that this is a somewhat general result and other more sophisticated models, as we see later, provide a similar prediction, $\tau \propto \eta$. In practice, an exponentially decaying function does not adequately describe the time dependence of $G(t)$ for polymeric systems. An alternate model that proves more effective involves an assembly of springs and dashpots as depicted in Fig. 6.8.

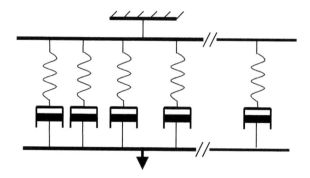

FIG. 6.8
A series of n viscoelastic components (springs and dashpots) connected in parallel.

Under such conditions, where the spring-piston assemblies are connected in parallel,

$$E(t) = \sum_{i=1}^{n} E_i e^{-t/\tau_i} \qquad\qquad 6.21$$

where i refers to the ith spring-dashpot pair. The foregoing example involves tensile stresses but the same arguments would clearly apply to shear stresses. Sometimes the stress relaxation modulus is written in terms of a stretched exponential

$$G(t) = G_0 e^{-(t/\tau)^\beta} \qquad\qquad 6.22$$

where the stretching exponent, $0 < \beta < 1$ and is a measure of the distribution of relaxation times. A value of $\beta = 1$ (single exponential) corresponds to a breadth (FWHM) of 1.144 decades, whereas $\beta = 0.5$ is about 2.5 decades wide. Equation 6.22 does a better job of describing $G(t)$ for a real system than the single exponential prediction, which should not be surprising. The behavior of the real system is characterized by a distribution of relaxation time processes.

Another common model is the Voigt model, where the spring and the dashpot are arranged in parallel, leading to the following constitutive equation

$$\frac{d\varepsilon}{dt} = \frac{\sigma}{\eta} - \frac{E\varepsilon}{\eta} \qquad\qquad 6.23$$

When the elements are arranged in parallel, the strain in each element is equal and the stress is the sum of the stresses in the elements, $\sigma = \sigma_\eta + \sigma_E$.

As another example we might consider a Creep experiment, wherein the applied stress remains constant, $\sigma = \sigma_0$. The Voigt model suggests the following differential equation,

$$\frac{d\varepsilon}{dt} + \frac{E\varepsilon}{\eta} = \frac{\sigma_0}{\eta} \qquad\qquad 6.24$$

which indicates that the strain in the material increases as

$$\varepsilon(t) = \frac{\sigma_0}{E}(1 - e^{-t/\tau})$$

6.25

The creep compliance is defined as $J(t) = \varepsilon(t)/\sigma_0$. The time-dependent increase of the creep compliance is shown in Fig. 6.9b due to a constant applied stress, σ_0.

In the melt state, the system can partially recover if the stress is released at time t_r. Hence, for $t > t_r$, the materials enters the recovery stage. If t_r is sufficiently long then the system would have had an opportunity to enter the steady state regime. In the steady state regime, the steady state compliance, J_e^0, is determined from the strain at long times, ε_∞,

$$\varepsilon_\infty = \sigma_0 J_e^0$$

6.26

During the steady state regime

$$J(t) = J_e^0 + \frac{t}{\eta}$$

6.27

where the intercept yields the steady state compliance and the slope yields the viscosity. Whereas for elastic materials $J = 1/G$, in viscoelastic materials

$$J(t) \neq \frac{1}{G(t)}$$

6.28

Instead $J(t)$ and $G(t)$ are connected through constitutive relations. This point becomes more evident later in this chapter.

Having discussed two common phenomenological models, in the next few sections (6.5 to 6.7) we now discuss further observations regarding the

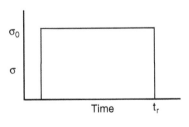

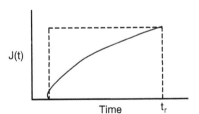

FIG. 6.9
The time-dependent response, compliance, of a viscoelastic material to a constant applied stress (Creep).

viscoelastic behavior of these materials, thereby laying the foundation for the molecular picture. We begin with a discussion of the viscosity.

6.4.2   The Viscosity: Experimental Observations

The stress relaxation function determines the shear viscosity,

$$\eta = \int_0^\infty G(t)dt \qquad\qquad 6.29$$

and using Eq. 6.21 and 6.29, an equation for the viscosity can be written as

$$\eta = \sum_{i=1}^{n} G_i \tau_i \qquad\qquad 6.30$$

which is not surprising, based on the Maxwell model.

For short *unentangled* chains; chains with molecular weights below a threshold molecular weight M_c, the viscosity changes linearly with M,

$$\eta(M) = \eta(M_c)\left(\frac{M}{M_c}\right) \qquad\qquad 6.31$$

For longer chains, the viscosity exhibits a much stronger dependence on molecular weight and

$$\eta(M) = \eta(M_c)\left(\frac{M}{M_c}\right)^{3.4} \qquad\qquad 6.32$$

M_c represents a critical molecular weight beyond which the entanglements begin to influence the flow properties of the melt.

Figure 6.10 depicts the universal plot of the molecular weight dependence of the viscosity for polymers. M_c varies from one polymer to another, as shown in Table 6.2.

In summary, the transition in the molecular weight dependence of the viscosity from M to $M^{3.4}$ reflects the influence of entanglements, topological

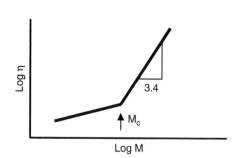

FIG. 6.10
The molecular weight dependence of the viscosity. For $M < M_c$, $\eta \propto M$ and for $M > M_c$, $\eta \propto M$ 3.4.

TABLE 6.2

Molecular characteristics: M_e, M_c, ρ, G_N^0, J_e^0, and $\langle R^2 \rangle_0 / M$ for some polymers

Polymer	$T(K)$	M_e (g/mol)	M_c (g/mol)	$*G_N^0$ (dynes/cm²)	ρ (g/cm³)	J_e^0 (cm²/dyne)
Polyisobutylene	490	10,500	17,000	2.5×10^6	0.98	
Polystyrene	490	18,100	32,000	2×10^6	0.97	1.75×10^{-6}
polymethylmethacrylate	490	13,600	29,500	4.8×10^6	1.14	
Polydimethyl siloxane	298	12,000	24,500	2.4×10^6	0.97	1×10^{-6}
Poly(α-methyl styrene)	459	13,300	28,000	3.2×10^6	1.04	1×10^{-6}
polyethylene	443	11,50	3,480	2×10^7	0.76	2.2×10^{-7}
Polyvinyl acetate	428	9,100	25,400	3.6×10^6	1.14	1.2×10^{-6}

Sources: L.J. Fetters, D.J. Losche, S.T. Milner, Macromolecules, 32, 6847 (1999).

*W.W. Graessley and S.F. Edwards, Polymer, 22, 1329 (1981).

Viscoelastic Properties of Polymers, J.D. Ferry, Wiley, NY, (1980).

constraints, on the translational dynamics of the chains. The molecular weight dependencies arise from the fact that the longest relaxation times (translational motions) associated with the dynamics of chains exhibit a characteristic dependence on chain length. This will become clear in Section 6.8 when we discuss the microscopic dynamics models.

6.4.2.1   Temperature Dependence of the Viscosity

The viscosity of a polymer melt exhibits a strong dependence on temperature and is often described by the so-called Williams-Landel-Ferry (WLF) equation,

$$\log \frac{\eta(T)}{\eta(T_0)} = \frac{-c_1^0(T - T_0)}{c_2^0 + T - T_0} \qquad 6.33$$

where c_1^0 and c_2^0 are constants characteristic of the material and T_0 is a reference temperature. This temperature dependence can be rationalized in terms of "free volume" theory.

Free Volume Theory

The free volume, v_f, is defined as

$$v_f(T) = v(T) - v_0(T) \qquad 6.34$$

where v_0 is the so-called occupied (or van der Waals) volume, and v is the equilibrium specific volume. v_f arises from packing irregularities in the structurally disordered system. Both v_0 and v_f decrease with temperature. With decreasing temperature and, more importantly decreasing specific volume, the molecules of the system move around in an increasingly restrictive (decreasing free volume) environment. In crystals, the decrease in v_0 is largely associated with the anharmonicity in the interatomic potentials (thermal

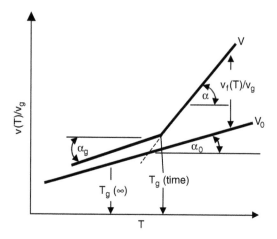

FIG. 6.11
Temperature dependencies of the free volume and specific volumes are plotted here. V_g is the free volume of T_g.

expansion). Below T_g they possess similar temperature dependencies; their expansion coefficients are similar $\alpha_g = \alpha_0$. Above T_g, v_f possesses a larger temperature dependence, $\alpha > \alpha_0$. The temperature dependencies of v and v_0 are illustrated in Fig. 6.11.

The temperature dependence of the fractional free volume is often approximated

$$f = f_g + \alpha_f(T - T_g) \qquad 6.35$$

where $f = v_f/v$ and α_f is the difference between the thermal expansion of the liquid and the glass. It has been empirically shown that the viscosity of a large number of glass forming liquids could be written as

$$\ln \eta = const + \frac{B(v - v_f)}{v_f} \sim B\left(\frac{1}{f} - 1\right) \qquad 6.36$$

This result indicates that in essence the longest relaxation time, and hence the viscosity, depends exponentially on the available free volume, $\tau \propto e^{Bv_f/v}$. The WLF equation (Eq. 6.33) follows by considering the quantity, $\ln\eta(T) - \ln\eta(T_g)$, with

$$c_1^g = B/2.303 f_g$$

$$c_2^g = f_g/\alpha_f$$

$$f_g = B/2.303 c_1^g \qquad 6.37$$

$$\alpha_f = B/2.303 c_1^g c_2^g$$

where in the above the reference temperature, T_0, is now taken to be T_g.

In the aforementioned, we have shown that the temperature dependence of the viscosity, particularly its increase with decreasing T can be rationalized in terms of free volume. This discussion highlights the kinetic aspect of the vitrification process. Later the vitrification process will be discussed in terms of the configurational entropy, thereby highlighting the thermodynamic aspect of the glass transition.

6.4.3 Time-Temperature-Superposition and Shift Factors

Having introduced the molecular weight and temperature dependencies of the viscosity we now return to the stress relaxation modulus. $G(t)$ is determined by a mechanical experiment and in typical experiments $G(t)$ is measured over a limited time range, say 10 seconds to 1000 seconds, at a particular temperature which we denote as T_1. During this time interval, $G(t)$ will decrease over a only a limited range of values, and the extent of the decrease depends on temperature. Changes in $G(t)$ at different temperatures will reflect the underlying behavior of $\eta(T)$, as suggested by Eq. 6.29.

Figure 6.12 shows the temperature dependence of $G(t)$. To the left of this figure, $G(t)$ is shown at different temperatures. Collectively, these curves show how the relaxation modulus varies at different temperatures during the same

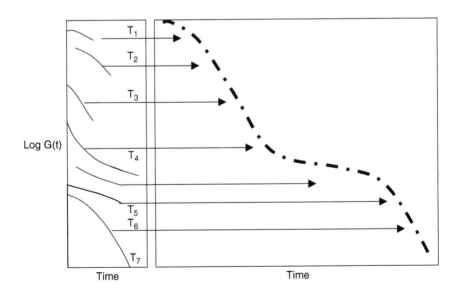

FIG. 6.12

Log $G(t)$ is shown at the left for different temperatures. Each curve can be shifted horizontally along time-axis until it overlaps the curves at adjacent temperatures. The curve to the right is the *master curve*. A slight vertical shift may be required to account for changes in density and elastic modulus associated with temperature.

time interval. One can begin by choosing a reference temperature arbitrarily, say T_1. The curve at T_2, $G(t,T_2)$ can be shifted horizontally along the time axis and, only slightly in the vertical direction, to coincide with the longer time-scale values of $G(t,T_1)$. This can be expressed using the more formal notation,

$$\frac{G(T_1,t)}{\rho(T_1)T_1} = \frac{G(T_2,t/a_T)}{\rho(T_2)T_2} \qquad 6.38$$

The time-scale is normalized by the shift factor a_T, reflecting the horizontal shift along the time axis between temperatures, T_1 and T_2. It is important to note, parenthetically, that the elastic modulus is proportional to temperature and to the density. Therefore, the stress relaxation modulus must be normalized by the density at the appropriate temperature, which is reflected in the slight vertical shift of the curve.

The shift factor, a_T, used to quantify the magnitude of the shifts required to superimpose $G(t)$ at different temperatures, is dependent on temperature. The temperature dependence of a_T is specified by the WLF equation with the constants c_1 and c_2, characteristic of the polymer,

$$\log a_T = \frac{-c_1^0(T-T_0)}{c_2^0 + T - T_0} \qquad 6.39$$

The temperature dependence of the shift factor may be constructed for any polymer by measuring appropriate segments of $G(t)$ at different temperatures and shifting them to an appropriate reference temperature, T_0. This procedure would yield appropriate WLF constants for that polymer at the appropriate reference temperature. It follows that if we choose T_0 as the reference temperature, then

$$a_T = \frac{\eta(T)}{\eta(T_0)} \frac{T_0 \rho(T_0)}{T\rho} \qquad 6.40$$

hereby providing a direct connection between the temperature dependencies of $G(t)$, the viscosity and the shift factor. Note that $T_0\rho(T_0)/T\rho \ll \eta(T)/\eta(T_0)$ because typically the viscosity changes many orders of magnitude over the same temperature range where the density and temperature would only change by less than a factor of 2. This implies that $a_T \cong \eta(T)/\eta(T_0)$, as is often quoted.

The aforementioned is the time-temperature-superposition principle. The implications are that if the viscosity is measured at a given temperature, then with the use of T_0, c_1 and c_2, the viscosity can be calculated at any other temperature. Table 6.3 shows WLF parameters for some common polymers.

The shift factor, and therefore the viscosity, have a strong dependence on temperature and varies from one polymer to another. Figure 6.13 illustrates the temperature dependence for four different polymers. Note that close to T_g the viscosity decreases with T more rapidly than it does at higher

TABLE 6.3

WLF parameters and glass transition temperatures for some polymers

Polymer	T_0	Tg (K)	T_∞	C_1	C_2
Polyisobutylene	298	205	101	8.61	200.4
Polystyrene (atactic)	373	373 (373)	325	12.7 (13.7)	49.8 (50)
Polymethylmethacrylate (atactic)	381	381	301	34	80
Polydimethyl siloxane	303	150	81	1.9	222
Poly(α-methyl styrene)	445	445	396	13.7	49.3
Poly(methyl acrylate)	324	276	222	8.86	101.6
Polyvinyl acetate	349	305	258	8.86	101.6

Source: *Viscoelastic properties of polymers*, J.D. Ferry, Wiley, NY, 1980.

temperature. Note that we could have performed a similar analysis of the compliance, $J(t)$ instead of $G(t)$.

6.4.4 Oscillatory Shear Measurements

In many situations it is more convenient to perform experiments in the frequency domain, instead of the time domain, wherein the material is subject to oscillatory shear deformations, as depicted in Fig. 6.14. In this section

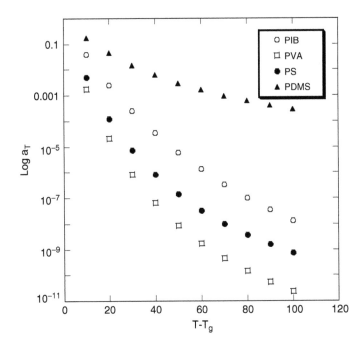

FIG. 6.13

The temperature dependencies of the shift factors for four polymers are shown here. The points were calculated with the WLF equation using the constants in Table 6.3.

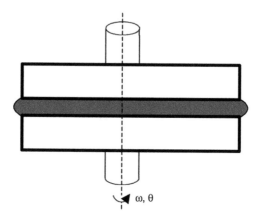

FIG. 6.14
Schematic of a parallel plate construct employed in some rheological experiments to measure viscoelastic moduli.

we derive equations appropriate for the oscillatory shear deformations. The response of many materials that experience deformation is not immediate (in the case of elastic materials the response is immediate). Such materials exhibit a time-dependent response that may be described by a complex modulus, or complex compliance, that accounts for the in-phase (elastic) response and the out of phase (loss) response.

This notion can be explored in further detail by considering a creep experiment, where the sample is subjected to a stress, σ. The in-phase component of compliance modulus is $J' = \gamma'/\sigma$ and the out of phase compliance is $J'' = \gamma''/\sigma$. It is common to describe the connection between J' and J'' in terms of a phase lag, δ, where tan (δ) is defined

$$\text{Tan}(\delta) = \frac{J''}{J'} = \frac{G''}{G'} \qquad 6.41$$

Where $\text{Tan}(\delta) = G''/G'$ would be obtained from an oscillatory shear experiment. Physically, $\text{Tan}(\delta)$ represents the ratio of the energy dissipated to that stored, per deformation cycle. The complex modulus is

$$G^* = G' + iG'' \qquad 6.42$$

and the complex compliance is

$$J^* = J' - iJ'' \qquad 6.43$$

The complex number i is used only as a matter of convenience since in the complex plane it represents a straightforward way to denote vectors that are orthogonal.

In the next section connections between the stress relaxation modulus, $G(t)$, and the frequency dependent moduli, $G'(\omega)$ and $G''(\omega)$, and between $J(t)$ and

$J'(\omega)$ and $J''(\omega)$ are discussed. In order to do this a constitutive relation must be developed.

6.4.5 Connections between $G(t)$ and Frequency Domain Experiments

We proceed by recognizing that the behavior of the material at time, t, is related to its deformation history. For a creep experiment, the time-dependent imposed strain

$$\gamma(t) = \sigma_0 J(t) \qquad 6.44$$

where $J(t)$ is the compliance. In a stress relaxation experiment the time-dependent response is $\sigma(t) = G(t)\gamma_0$. In the case of the creep experiment a stress σ_1 is applied at an arbitrary earlier time u_1, so

$$\gamma(t) = \sigma_1 J(t - u_1) \qquad 6.45$$

If the Boltzmann superposition principle applies, assuming the strains are sufficiently small (the response of the material to a consecutive series of perturbations is the sum of the responses to each of these perturbations), then the total strain that arises from n such stress increments is

$$\gamma(t) = \sum_{i=0}^{n} \sigma_i J(t - u_i) \qquad 6.46$$

Should the time intervals between stresses be infinitesimally small, then the sum can be converted to an integral and the constitutive relation now becomes

$$\gamma(t) = \int_{-\infty}^{t} \frac{\partial \sigma(u)}{\partial u} J(t - u) du \qquad 6.47$$

This result indicates that the total strain is based on the previously applied stresses, a very important message.

For a stress relaxation experiment, it is similarly shown that the relevant constitutive equation is

$$\sigma(t) = \int_{-\infty}^{t} \frac{\partial \gamma(u)}{\partial u} G(t - u) du \qquad 6.48$$

We now consider an experiment in which the sample is subjected to a periodic deformation, which we represent mathematically

$$\gamma(t) = \gamma_0 \sin \omega t \qquad 6.49$$

The related stress is out of phase by a phase angle δ

$$\sigma(t) = \sigma_0 \sin(\omega t + \delta) = \sigma_0[\cos\delta\sin\omega t + \sin\delta\cos\omega t] \qquad 6.50$$

An expression for the stress may also be obtained with the use of the constitutive relation, Eq. 6.48. Upon making the transformation $s = t - u$,

$$\sigma(t) = \int_0^\infty G(s)\omega\gamma_0\cos(\omega t - s)\,ds \qquad 6.51$$

leading to

$$\sigma(t) = \gamma_0\left[\int_0^\infty \omega G(s)\sin\omega s\,ds\right]\sin\omega t + \gamma_0\left[\int_0^\infty \omega G(s)\cos\omega s\,ds\right]\cos\omega t \qquad 6.52$$

The terms within brackets are only functions of frequency and so

$$\sigma(t) = \gamma_0 G'(\omega)\sin\omega t + \gamma_0 G''(\omega)\cos\omega t \qquad 6.53$$

where

$$G'(\omega) = \int_0^\infty G(s)\omega\sin(\omega s)\,ds \qquad 6.54$$

is the elastic modulus, and the loss modulus is

$$G''(\omega) = \int_0^\infty G(s)\omega\cos(\omega s)\,ds \qquad 6.55$$

If the variable s is replaced with t then it is evident that these equations identify the connection between $G(t)$ and the frequency dependent moduli, $G'(\omega)$ and $G''(\omega)$. Note that they are Fourier transforms of the stress relaxation modulus (response function). Moreover, a plot of $G'(\omega)$ versus ω should show opposite trends to a plot of $G(t)$ versus t.

The expression for $\tan(\delta)$ can realized if Eq. 6.50 and 6.53 are compared where

$$G' = \frac{\sigma_0}{\gamma_0}\cos\delta \quad\text{and}\quad G'' = \frac{\sigma_0}{\gamma_0}\sin\delta \qquad 6.56$$

Before discussing an example involving $G(t)$, expression for the magnitude of G^* and of $\tan(\delta)$ are discussed. If the strain is written as $\gamma(t) = \gamma_0 e^{i\omega t}$ then $\sigma(t) = \sigma_0 e^{i(\omega t + \delta)}$ and the magnitude of the complex modulus G^* is the ratio $\gamma(t)/\sigma(t)$ and

$$|G^*| = \frac{\sigma_0}{\gamma_0} \qquad 6.57$$

Example

A specific example involving a relaxation function that decays with a single exponential is now considered. The stress relaxation modulus is

$$G(t) = Ge^{-t/\tau} \qquad \text{6.57}$$

and using Eq. 6.55 to 6.57 yields

$$G'(\omega) = G\frac{(\omega\tau)^2}{(\omega\tau)^2 + 1} \qquad \text{6.58}$$

and

$$G''(\omega) = G\frac{\omega\tau}{(\omega\tau)^2 + 1} \qquad \text{6.59}$$

The Maxwell model, incidentally, would have yielded the same predictions, not surprising. It is clear from these equations that G'' exhibits a maximum at $\omega\tau = 1$ and that it decreases as $\omega\tau$ at low frequencies. G', on the other hand, decreases as $(\omega\tau)^2$ and reaches a plateau at large ωt. (Problem 8) These equations, however, do not provide an adequate description of the real situation, depicted in Fig. 6.15. They only capture the low frequency behavior and the maximum in $G''(\omega)$. In a later section we will provide a microscopic model, Reptation, which does a remarkable job at describing the actual observations.

In the meantime, we note that the connection between these moduli and the viscosity can be established in a straightforward manner. Since $\eta = \frac{\sigma(t)}{d\gamma/dt}$, then

$$\eta' = \frac{G''(\omega)}{\omega} \qquad \text{6.60}$$

and

$$\eta'' = \frac{G'(\omega)}{\omega} \qquad \text{6.61}$$

where

$$\eta^* = \eta'' - i\eta' \qquad \text{6.62}$$

The viscosity of interest in many practical situations is η', which tells us that in the limit of zero frequency one recovers the viscosity in the limit of zero shear,

$$\eta = \lim_{\omega \to 0} \frac{G(\omega)}{\omega} \qquad \text{6.63}$$

Before concluding this section, one final point should be mentioned regarding the temperature dependence of the elastic modulus. At low temperatures,

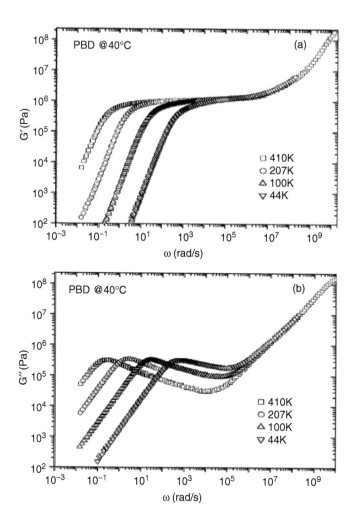

FIG. 6.15
(Reproduced with permission) Master curves of $G'(\omega)$ and $G''(\omega)$ are shown here for monodis-
perse polymers(linear 1,6-polybutadiene) of varying molecular weights. (S. Wang, Shi-Qing
Wang, A. Halasa, and W.-L Hsu, *Macromolecules*, 2003, 36, 5356–5371.). Note that as the molecular
weight increases, the breadth of the plateau in G' extends to lower frequencies. The "knee" of
the curve is sometimes called the terminal relaxation time, τ_{REP}, and the height of the plateau is
identified as a the rubbery plateau modulus G_N^0. Both parameters will be discussed in due course.

far below the glass transition temperature, side groups along the backbone
of the chain undergo oscillatory motions at characteristic frequencies. Just
beyond these temperatures, the range of motion increases where eventually
groups of monomers undergo short-range rotational segmental motions at
characteristic frequencies. Throughout this temperature range, the elastic
modulus decreases slightly with temperature. With increasing temperature,
in the vicinity of T_g, the elastic modulus exhibits a significant decrease. For
$T > T_g$ the chains can undergo long-range, center of mass, excursions. The

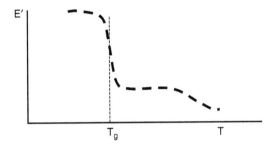

FIG. 6.16
The typical temperature dependence of the elastic modulus of thermoplastic polymer is shown here.

behavior in the temperature range just above T_g is often described as rubbery. At yet higher temperatures the behavior of the sample is liquid-like and the elastic modulus decreases to zero. The typical temperature dependence of the elastic modulus of polymers is illustrated in Fig. 6.16. The shape of this curve, $E(T)$ versus T is very similar to $E(t)$ versus t (or equivalently $G(t)$ versus t). The underlying behavior that gives rise to the $G(t)$ versus t behavior will be discussed later during our discussion of the molecular picture.

6.5 Microscopic Model for Diffusion and Viscoelasticity in Polymer Melts

A microscopic model that accounts for the role of the topological constraints imposed on the dynamics of a chain is now discussed. The model, based on the notion that a chain executes translational motion along its own contour, within the constraints of a virtual tube, created by its neighbors, has enjoyed enormous success at describing many of the dynamical features of polymer melts. In 1971 deGennes calculated the translational diffusion coefficient of a long chain as it moved within the confines of fixed obstacles. He predicted that the translational dynamics of a chain of N links would be characterized by a relaxation time $\tau_d \propto N^3$ and that a natural consequence of this would be that its translational diffusion coefficient would be $D \propto N^{-2}$.

His work was later extended by Doi and Edwards, in 1978, to describe other aspects of the viscoelastic behavior of polymers. Edwards had earlier introduced the concept of the "tube" in his theoretical developments in the field of rubber elasticity in order to qualitatively illustrate the confinements imposed by neighboring chains. This strategy has had a profound impact on further developments toward the understanding of the dynamics of long-chain polymer melts. These theoretical developments nevertheless suffered from shortcomings. The molecular weight dependence of the viscosity was predicted to be M^3 (or equivalently N^3), because $\tau_d \propto N^3$, while experimentally the viscosity was observed to be $M^{3.4}$. This may at first glance appear to be minor but, as we show later, resolving this discrepancy to the satisfaction of most researchers proved to be more vexing than one might have anticipated. Discrepancies associated with the frequency dependence of $G''(\omega)$ and failure to account for the effect of the dynamics of the surrounding environment on the long-range dynamics of the probe chains in a host of varying molecular weights provided further ammunition for detractors. In the intervening years a number of researchers have made modifications (contour length fluctuations, constraint release) that appear to have adequately accounted for most of the shortcomings. Today models based on the notion of the "tube" are widely accepted. In the next section we discuss the Rouse model which provides a good description of the dynamics of unentangled chains in a melt.

6.5.1 Rouse Model: Unentangled Chains

Rouse (1953) originally developed this model to describe polymer solutions, but it later turned out to provide a better description of the dynamics of unentangled melts in which long-ranged hydrodynamic effects are absent. In this model the chain is divided into a series of submolecules, each of which obeys Gaussian statistics. This is not an unreasonable assumption since it can be shown that the spatial organization of monomers that compose a sufficiently long chain segment also obeys Gaussian statistics. The submolecules are connected by beads and the locations of the beads are identified by the position vectors ($\vec{R}_1, \vec{R}_2 ... , \vec{R}_N$).

For reasons we are about to describe, the submolecules behave as springs, The Helmholtz free energy, A, can be expressed in terms of the number of accessible states, Ω, $A = -kT\ln\Omega$, as discussed earlier in Chapter 1. At constant temperature, T, the force associated with extension is

$$f = \nabla A(R) \qquad\qquad 6.64$$

which follows from $dA = dU - TdS$ ($A = U - TS$) and $fdr = dU - TdS$. Since $\Omega \propto P(R)$, where $P(R)$ is specified by Eq. 6.10, then the force is

$$f = KR \qquad\qquad 6.65$$

where the spring constant is $K = 3kT/Nb^2$. Note that the "spring constant" is proportional to T/N. In fact it is for this reason that the modulus of an elastomer is proportional to temperature. To this end, the Langevin equation, if used as the primary dynamical equation, can be written as

$$\zeta_0 \frac{dR_n}{dt} = F_n + f_n(t) \qquad\qquad 6.66$$

where R_n is the location of a point (bead) along the chain at time t and $f_n(t)$ is the random force on the beads. The frictional drag is assumed to be distributed uniformly throughout the chain so each monomer experiences an average frictional drag of ζ_0. In the Rouse model, the dynamics are not determined by long-ranged interactions but determined by localized interactions along the chain. The function F_n is associated with the elastic forces, $-K(\vec{R}_n - \vec{R}_{n+1} + \vec{R}_n - \vec{R}_{n-1})$. The solution to this equation will not be treated in detail here and the reader is referred to the book by Doi and Edwards. It nevertheless suffices to mention that the problem amounts to solving a system of equations describing a series of coupled oscillators. The normal strategy for solving such a problem is to identify the normal mode coordinates for the system. In the normal mode coordinate system, the vibrational modes are independent, thereby simplifying the analysis considerably. The solution to this equation, in terms of the normal mode coordinates $\vec{X}_p$, is

$$\vec{R}_n = \vec{X}_0 + 2 \sum_{p=1}^{\infty} \vec{X}_p \cos\left(\frac{p\pi n}{N}\right) \qquad 6.67$$

The normal coordinate $\vec{X}_0$ ($p = 0$) is the location of the center of mass of the chain, $\vec{R}_{CM}$. The mean square displacement of the center of mass of the chain can be shown to be

$$\langle (\vec{R}_{CM}(t) - \vec{R}_{CM}(0))^2 = 6 \frac{kT}{N\zeta} t \qquad 6.68$$

Knowledge of the center of mass of the chain enables calculation of the center of mass, or translational, diffusion coefficient, D_{Ro}, and by definition

$$D_{Ro} = \lim_{t \to \infty} \frac{1}{6t} \langle (\vec{R}_{CM}(t) - \vec{R}_{CM}(0)) \rangle \qquad 6.69$$

which yields

$$D_{Ro} = \frac{kT}{N\zeta_0} \qquad 6.70$$

This result is actually quite intuitive. You might recall that if the frictional drag is assumed to be distributed along the chain, then using the Stokes-Einstein relationship $D = kT/\xi$, with $\zeta = N\zeta_0$ Eq. 6.70 follows! ζ_0 is the friction coefficient of a monomer.

The other property of interest is the correlation function that describes the displacements of the Rouse segments (Doi and Edwards, McLeish)

$$\langle (\vec{R}_n(t) - \vec{R}_n(0))^2 \rangle = 6D_{Ro}t + \frac{2Nb^2}{3\pi^2} \sum_{p=1}^{\infty} \frac{1}{p^2} \cos^2\left(\frac{n\pi p}{N}\right)[1 - e^{-p^2 t/\tau_{Ro}}] \qquad 6.71$$

The first term in this equation describes the center of mass displacement of the Rouse chain and the second describes the internal relaxation modes of the chain. An important result of this analysis is the relaxation times of modes, p, associated with submolecules, N/p segments long are

$$\tau_p = \zeta_0 \frac{Nb^2}{3kTp^2\pi^2} \propto \zeta_0 \frac{N^2}{p^2} \qquad 6.72$$

This result indicates that the relaxation time of the modes increase as N^2 with the longest relaxation mode ($p = 1$)

$$\tau_{Ro} = \frac{N^2 b^2 \zeta_0}{3\pi^2 kT} \qquad 6.73$$

This equation describes the relaxation time of the center of mass of the chain and, moreover, what emerges from this result is that this time scale is determined by two parameters, N and the monomer friction factor, ζ_0.

We can examine the dynamics of the chain in further detail. If $t \ll \tau_{Ro}$, then, by replacing the sum with an integral, it can be shown that

$$\langle (\bar{R}_n(t) - \bar{R}_n(0))^2 \rangle \approx 6R_{Ro}t + \frac{Nb^2}{3\pi^2}\left(\frac{t}{\tau_{Ro}}\right)^{1/2} \qquad 6.74$$

This equation suggests that it is only at longer times, $t > \tau_{Ro}$, that the Fickian diffusion process (root mean square displacement $\sim t^{1/2}$) is active. At earlier times the displacement has a weaker, sub-Fickian, the root mean square displacement $\sim t^{1/4}$. It is left as an exercise for the reader to derive the above Eq. 6.74 and plot $\langle (\bar{R}_n(t) - \bar{R}_n(0))^2 \rangle$ as a function of $\log t$ to illustrate this important point.

6.5.2   Reptation: Dynamics of Entangled Chains

In the melt, chains are organized into a dense, entangled, mesh, as illustrated in Fig. 6.17, where an arbitrary (probe) chain is clearly identified. The interactions of the chain with its neighbors are such that any excursions it attempts to execute beyond a certain distance, a, normal to its contour, are prohibited.

Such interactions with neighbors is tantamount to motion restricted within a virtual tube of cross sectional dimensions of order a. Therefore the chain is destined to undergo translational motions only along to its own contour. One can think of the conformation of the "tube" as a random walk of Z submolecules and associated with each submolecule is a step length of a. The number of chain segments per submolecule is N_e.

As first suggested by deGennes, the mechanism by which the chain moves is through the formation and propagation of "kinks" along its contour (Fig. 6.18).

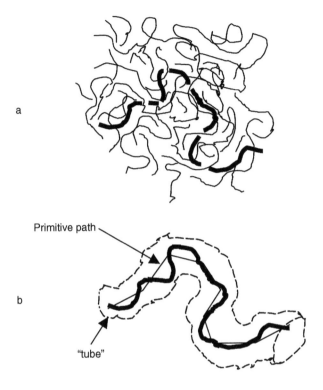

FIG. 6.17
a) Schematic of a dense melt in which the probe chain is constrained to move along its own contour. b) The effect of the neighboring chains is tantamount to forming a virtual tube.

The chain undergoes Brownian motion along a so-called primitive path (introduced by Doi and Edwards), the average trajectory of the chain, identified in Fig. 6.18. The length of the primitive path is

$$L = Za \qquad\qquad 6.75$$

where $Z = N/N_e$ is the number of submolecules, or equivalently the number of primitive steps, each of length a, along the path. This equation

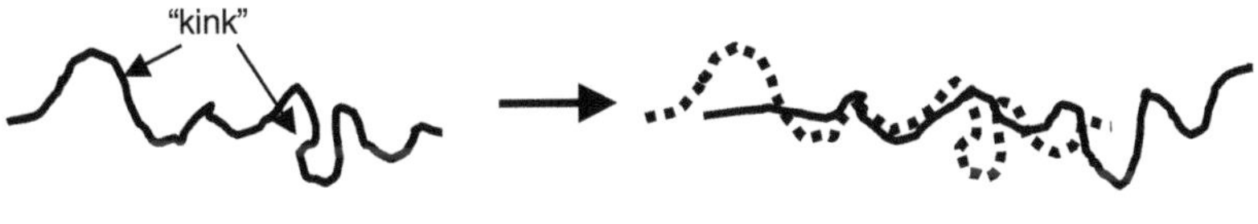

FIG. 6.18
Schematic of the mechanism of translational diffusion of a chain *segment*. Motion occurs through the formation and propagation of "kinks."

implies that $Z^{1/2}a$ should also be the mean end-to-end distance of the chain, $N^{1/2}b$. The dynamics of the chain along its contour within the "tube" is assumed to be determined by Rouse dynamics. As the chain diffuses its ends move in random directions. These displacements create new tube segments ahead while abandoning segments (Fig 6.19). Over the relaxation time interval, τ_R, the chain will have formed a completely new "tube" and will have lost complete memory of the old. In this regard the translational dynamics of the chain are not correlated over the time scale $t > \tau_R$. In what follows, we calculate the stress relaxation modulus, $G(t)$, the longest relaxation time, τ_R, and the Reptation (center of mass) diffusion coefficient. The solution to this complex multiple-body dynamics problem initially appeared to be a somewhat terrifying prospect, but the notion that this problem could be reduced to translation along a contorted tube has simplified things considerably, as shown by deGennes. The time scales of the dynamics are determined by two parameters, the friction coefficient per monomer and the chain length. It will become clear that the temperature dependence of the Reptation diffusion coefficient and of the viscosity are determined by ζ_0. In the original formulation of the Reptation model, the tube length or, equivalently, the primitive path length is fixed.

We now describe consequences of the model subject to the assumptions mentioned heretofore. It should be apparent that knowledge of the correlation function of the end-to-end vector of the chain will enable calculation of the dynamical properties. The end-to-end vector of the chain is

$$\bar{P}(t) = \bar{R}(L,t) - \bar{R}(0,t) \qquad\qquad 6.76$$

At an arbitrary time, $t = 0$, the chain is confined within its tube. At a later time t, only a fraction of the original tube remains occupied. The end-to-end vector of the chain at time $t = 0$ is $\bar{P}(0)$ and this end-to-end vector, defined

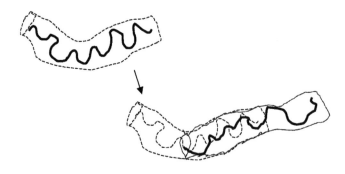

FIG. 6.19

As the chain diffuses, its ends choose random directions. And during this process it created new "tube" segments and abandons old segments. The above is a schematic of a chain the translated to the right. The motion of the chin within the tube is Brownian.

in terms of a sum of vectors, $\bar{P}(0) = \bar{p}_1(0) + \bar{p}_2(0) + \cdots + \bar{p}_N(0)$ (similar to the discussion in Section 6.2). The chain ends subsequently choose random directions in which to move and at a later time $t > 0$, the new vector becomes $\bar{P}(t) = \bar{p}_1(t) + \bar{p}_2(t) + \cdots + \bar{p}_N(t)$. For a moment consider the schematic illustrated in Fig. 6.20. It shows a chain at time $t = 0$ the chain $(A_0 - D_0)$ resides in its original tube and again at time t $(A_t - D_t)$ after it has undergone Brownian motion for time t. At time t, only the segments between N' and N'' $(B_0 - C_0)$ remain in the original tube. The vectors that connect points A_0 to B_0 and A_t to B_0 are uncorrelated and so are the vectors that connect C_0 to D_0 and C_0 to D_t. Only the vectors $\bar{p}_{N'}$ to $\bar{p}_{N''}$ (between B and C) are correlated. The quantity that we need to calculate is $\langle m(t) \rangle$, the fraction of the original "tube" that remains occupied at t (B – C). $\langle m(t) \rangle$ is related to the correlation function for the new end-to-end vector

$$\langle \bar{P}(t) \bullet \bar{P}(0) \rangle = \langle (\bar{p}_{N'} + \cdots + \bar{p}_{N''}) \bullet (\bar{p}_{N'} + \cdots + \bar{p}_{N''}) \rangle = a < m(t) \rangle \qquad 6.77$$

where

$$\langle m(t) \rangle = \int_0^{L_t} \mu(s,t)ds \qquad 6.78$$

In order to calculate $\langle m(t) \rangle$ we first need to enquire about the probability, $\Phi(s,t)$, that the tube remains occupied after time t has elapsed. For the moment, consider a coordinate system along the contour of the primitive path and use ξ to denote the distance along this path such that the point on the chain at location A is the origin, $\xi = 0$ and the end of the chain is $\xi = L$. The fraction of the tube that would be occupied in the region between ξ and $\xi + \delta\xi$ after time t is $\Phi(\xi,s,t)d\xi$. The boundaries of the variable ξ lie between

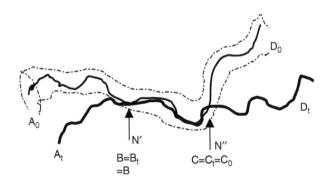

FIG. 6.20
The chain occupies a "tube" at time $t = 0$ and its end-to-end vector is characterized by $P(0)$. At a later time t, after the chain has undergone Brownian motion, only a fraction of the tube remains occupied by segments between N' (location B) and N'' (location C). The vector between N' and N'' is specified as $P'(t)$ and has a contour length of $m(t)$.

$\xi = s$ and $\tilde{\xi} = s - L$, where the variable s denotes a segment on the tube. For a concrete example that illustrates this point, consider Fig. 6.20 again. At the left of the figure, the chain end crossed the initial tube at point B_0 and the other chain end crossed the tube boundary at point C_0. Point B would correspond to $\xi = s_B$ and point C would be $\xi = s_C$. When the chain ends reach these points on the tube $\Phi(\xi,s,t) = 0$, indicating abandonment. Integration of $\Phi(\xi,s,t)$ between the boundaries $\xi = s - L$ and $\tilde{\xi} = s$ provides the probability that the tube remains occupied at time t,

$$\mu(s,t) = \int_{s-L}^{s} \Phi(\xi,s,t)d\xi \qquad 6.79$$

The probability function, $\Phi(\xi,s,t)$ satisfies the diffusion equation (Fick's 2nd law),

$$\frac{\partial \Phi}{\partial t} = D_{Ro}\frac{\partial^2 \Phi}{\partial \xi^2} \qquad 6.80$$

with boundary conditions: $\Phi(\xi,s,0) = \delta(\xi)$ and $\Phi(\xi,s,t) = 0$ for $\xi = s$ and $\xi = s - L$, respectively. The solution to this equation, using separation of variables, is

$$\Phi(\xi,s,t) = \sum_{p=1}^{\infty} \frac{2}{L_t}\sin\left(\frac{ps\pi}{L_t}\right)\sin\left(\frac{p\pi(s-\xi)}{L_t}\right)e^{-p^2t/\tau_R} \qquad 6.81$$

where τ_R is the Reptation relaxation time

$$\tau_{Rep} = \frac{L_t^2}{D_{Ro}\pi^2} = \frac{b^4\varsigma_0}{a^2\pi^2kT}N^3 \qquad 6.82$$

(recall: $L = Za = Nb$). This relaxation time is proportional to N^3 in contrast to N^2, determined for the Rouse chain, revealing the influence of the topological constraints on the translational dynamics of entangled chains. The longest relaxation time could also have been obtained from a scaling argument, as follows. The diffusion coefficient along the primitive path is the Rouse diffusion coefficient ($D_{Ro} = kT/N\varsigma_0$) and the length of the primitive path is $L = (N/N_e)a$. The time that the chain takes to traverse the entire tube is $\tau_{Rep} \sim L_t^2/D_{Ro}$, leading to

$$\tau_{Rep} \approx \frac{a^2\varsigma_0}{N_e^2kT}N^3 = \frac{b^4\varsigma_0}{a^2kT}N^3 \qquad 6.83$$

With the exception of π^2, Eq. 6.83 is identical to Eq. 6.82.

We now discuss the end-to-end vector correlation function. It follows from Eq. 6.77 to 6.81, that

$$\langle \bar{P}(t) \bullet \bar{P}(0)\rangle = Nb^2\mu(t) \qquad 6.84$$

and

$$\mu(t) = \frac{1}{L_t} \int_0^{L_t} \mu(s,t)ds \qquad\qquad 6.85$$

where

$$\mu(s,t) = \sum_{p;odd} \frac{4}{p\pi} \sin\left(\frac{ps\pi}{L_t}\right) e^{-p^2 t/\tau_{Rep}} \qquad\qquad 6.86$$

A solution to Eq. 6.85 leads to

$$\mu(t) = \sum_{p;odd} \frac{8}{p^2\pi^2} e^{-p^2 t/\tau_{Rep}} \qquad\qquad 6.87$$

This equation specifies the fraction of the original tube that remains occupied after time t. It is also known as the tube survival probability.

It is worthwhile to make an additional comment regarding the dynamics of the chain within the tube. In the foregoing analysis, the contour length remained fixed, not allowed to fluctuate. Segments near the end of the chain escape the tube readily and their new locations no longer remain correlated with their previous positions, loss of memory. The segments near the middle of the chain are more likely to remain confined within the tube for longer times than the ends. It is left as an exercise to show this by considering $\mu(s,t)$ the probability that a segment, s, on the chain remains trapped in the tube at time t.

6.5.3 The Stress Relaxation Modulus, the Viscosity, and the Steady State Compliance

In this section an expression for stress relaxation modulus, $G(t)$, and the viscosity are derived based on the Reptation model. These predictions are subsequently compared with experiment and the limitations of the approach, at least in relation to these properties, discussed.

In the stress relaxation experiment, the sample is deformed and the strain maintained while the response of the sample is monitored. Microscopically, the tube is deformed and the chain must diffuse to relieve the associated increase in the free energy. At very short times, $t < \tau_e$, segments can undergo rapid and unobstructed relaxations, because they do not know that they are trapped in a tube, Rouse dynamics. It is important to note that $\tau_e < \tau_{Ro}$, since it occurs on length scales less than a, the tube diameter. For $t > \tau_e$, motion is restricted to the confines of the tube. The other relevant time scale is τ_{Ro}, the Rouse relaxation time along the primitive path. Note that in the regime between τ_{Ro} and τ_{Rep}, the relaxation time varies as N^2 whereas for $t > \tau_{Rep}$ the relaxation time is determined by Reptation and $\tau_{Rep} \propto N^3$. For time, $t > \tau_{Rep}$, segments of the chain that escape the tube will lose memory of the deformation. In other words, the stress becomes relaxed as a result of escape of the chain from its original tube.

We now determine $G(t)$ for $t > \tau_e$. The segments in the center of the tube will remain in the tube for a longer time, as discussed earlier, as the ends rapidly move around. Necessarily, these segments of the chain that remain in the tube will maintain memory of this deformation for a longer period of time than the ends.

The stress relaxation will therefore be proportional to the fraction of the tube that remains occupied at time t, where the constant of proportionality would be the so-called rubbery plateau modulus, G_N^0,

$$G(t) = G_N^0 \mu(t) \tag{6.88}$$

$G(t)$ versus t is plotted in Fig. 6.21, where it is shown that for $t > \tau_{Rep}$ the stress is completely relaxed. The plateau modulus is given by

$$G_N^0 = \frac{\rho RT}{M_e} \tag{6.89}$$

where ρ is the density of the polymer, R is the universal gas constant and M_e is the average molecular weight between entanglements. There are two ways to think about the plateau modulus. The elastic modulus for a polymer based on the theory of rubber elasticity, is $E = \rho RT/M_x$, where M_x is the molecular weight between cross-links. The rationale is that since the entanglements act as temporary cross-links, it is appropriate to use the same form and replace M_x with M_e. In this theory, the deformation is

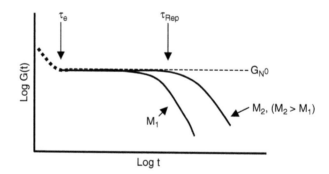

FIG. 6.21
The theoretical prediction for the stress relaxation modulus is shown here. For $t > \tau_e$, the Reptation prediction is represented by the solid line. Since the Reptation relaxation time (or equivalently the tube disengagement time) depends on M, the width of the plateau region increases with M. The time scale $t < \tau_e$ is due to the rapid time-scale segmental relaxations that are not influenced by the tube and occur on length-scales less than the tube diameter a.

assumed to be affine, wherein each junction experiences the same displace-ment as the sample (melt).

In a melt, it is necessarily true that the vast majority of entanglements would not behave as permanent cross-links, since they would fluctuate in position. It has been suggested by Doi that the actual equation for the modulus should be

$$G_N^0 = \frac{4}{5} \frac{\rho RT}{M_e}$$

6.90

where the factor of 4/5 reflects the somewhat weaker contributions of the entanglements compared to that of permanent crosslinks to the modulus.

We have not mentioned the relaxation modulus for the Rouse chain, but it is important for our understanding of stress relaxation during the initial stage ($t < \tau_e$) when the stress begins to decay. In this regime the segments are not aware of a tube and, moreover, the chains are not entangled. The contributions to the stress relaxation modulus in this case are, simply, unre-laxed Rouse modes. The contribution of each unrelaxed Rouse mode is proportional to the thermal energy kT. The number of submolecules per unit volume is c/N. The time dependence of the stress relaxation modulus is calculated explicitly by Doi and Edwards to yield

$$G_{Ro}(t) = \frac{c}{N} kT \sum_{p=1}^{\infty} e^{-2tp^2/\tau_{RO}}$$

6.91

The time-dependent quantity ($\sum_{p=1}^{\infty} e^{-2tp^2/\tau_{RO}}$) represents the decay of the end-to-end chain vector. In the time-scale $t \ll \tau_e$ the sum can be approximated by an integral to yield (see Problem 11),

$$G(t) = \frac{c}{2\sqrt{2}N} kT \left(\frac{\tau_{Ro}}{t} \right)^{1/2}$$

6.92

It follows that in the sub-Fickian regime, the stress relaxation modulus decays as $t^{-1/2}$ and over longer times ($t > \tau_{Ro}$), when the Fickian diffusion mechanism takes over, for unentanged chains, the decay is exponential. For entangled chains the plateau region develops after the initial Rouse regime. The connection between the regimes $\tau < \tau_e$ and $\tau > \tau_e$ when the chain diffuses along the primitive path, is determined by noting that at time $t = \tau_e$, the plateau modulus $G_N^0 = G(\tau_e) \cong \frac{c}{N} kT (\frac{\tau_{Ro}}{\tau_e})^{1/2}$ (from the equation above). There-fore in summary,

$$G(t) = \begin{cases} G_N^0 \mu(t) & t > \tau_e \\ G_N^0 \left(\frac{\tau_e}{t} \right)^{1/2} & t < \tau_e \end{cases}$$

6.93

It is left as an exercise to show that the relaxation time scale τ_e can be calculated from Eq. 6.73 in terms of the tube diameter a.

$$\tau_e = \frac{N_e^2 b^2 \zeta_0}{3\pi^2 kT} = \frac{a^4 \zeta_0}{3\pi^2 b^2 kT} \qquad\qquad 6.94$$

The shape of the predicted $G(t)$ corresponds well with experiment. One shortcoming nevertheless is than the terminal time τ_{Rep} scales experimentally as $N^{3.4}$, not N^3. In this regard, the tube length fluctuations, described later should reconcile differences with experiment.

6.5.3.1 Summary of Chain Segmental Dynamics

At early times ($t < \tau_e$) the chain segments within the tube relax rapidly and are unrestricted (i.e., as free Rouse chains), until their excursions reach the tube boundaries at time τ_e. Another regime occurs between τ_e and τ_R, where the behavior is sub-Fickian. The chain exhibits center of mass diffusion (Rouse dynamics) for $\tau > \tau_{Ro}$ along the primitive path ($\tau_R < t < \tau_{Rep}$). After a sufficiently long time, $t > \tau_{Rep}$, the tube disengagement time (Reptation relaxation time) the chains are free of their original tube ($\tau_{Rep} \propto N^3$) and the dynamics are controlled by Reptation.

6.5.4 The Entanglement, the Molecular Weight, and the Critical Molecular Weight

With regard to the dynamics of polymers, the molecular weight between entanglements, M_e, and the critical molecular weight, M_c, are fundamentally important "material" parameters. M_e determines the plateau modulus, $G_N^0 \propto M_e^{-1}$, and is characteristic of the polymer. M_c is the molecular weight beyond which entanglement effects influence viscous flow, where the molecular weight dependence of the viscosity increases from M^1 to $M^{3.4}$. Historically, it was believed that $M_c \approx 2M_e$; however as information regarding a larger collection of polymeric systems became available it became apparent that this relationship was no longer true.

Both M_e and M_c are determined by the "packing" of chains in the system. To understand the connection between M_e and M_c, it is necessary to understand the origins of M_e (Graessley and Edwards 1981; Fetters et al 1994; and Fetters et al 1999). Recall that in the melt the chain is a coil that pervades a volume specified by M/ρ, where M is the molecular weight and ρ is the density of the melt. A coil is interpenetrated by other chains. The unperturbed dimensions of this chain are given by Eq. 6.9 and since $\langle R^2 \rangle_0 = 6 \langle R_g^2 \rangle$,

$$\langle R_g^2 \rangle = C_\infty \frac{M}{m_0} \frac{l^2}{6} \qquad\qquad 6.95$$

where m_0 is the monomer molecular weight.

If the volume of the smallest sphere that encloses this molecule is specified by $V = A\langle R_g^2 \rangle^{3/2}$, where A is a constant (A ~ 1), then the number of chains of molecular weight M that would completely fill this volume, V, is

$$\hat{N} = V \frac{\rho N_A}{M} \qquad 6.96$$

where N_A is Avagadro's number. It follows that the number chains, which pervades this volume and would therefore be entangled, is $(\hat{N} - 1)$. The molecular weight at which just one other chain pervades the volume is M_e and this, of course, would correspond to the condition $\hat{N} = 2$. Consequently, an expression for M_e is

$$M_e = \frac{1}{B^2(\langle R^2 \rangle_0 / M)^3 (\rho N_A)^2} \qquad 6.97$$

It is left as an exercise to the reader to determine B, a numerical constant. This equation may be modified by introducing an additional parameter, identified as the packing length, p. The packing length is determined by the average size of the coil and by the volume that the coil pervades,

$$p = \frac{M}{\langle R^2 \rangle_0 \rho N_A} \qquad 6.98$$

If the average volume of a chain per bond is $v_0 = m_0 / [\rho N_A]$, then the packing length may be rewritten

$$p = \frac{v_0}{[C_\infty l^2]} \qquad 6.99$$

Further insight into the significance of p might be gleaned by approximating the volume of a step of length b (Kuhn segment length $b = C_\infty l$) by a cylinder of volume $v_0 \sim S_A b$, where S_A is the cross-sectional area of the cylinder. The packing length is therefore $p \sim S_A / b$, the ratio of the cross sectional area of the cylindrical region swept out by the segment of length b [see Fetters et al. *Journal of Polymer Science: Polym. Phys.* 37, 1023 (1999)]. This may in fact be a better indicator of stiffness that C_∞, per se.

These equations indicate that the average molecular weight between entanglements is determined by the packing length, p, and in fact increases as the third power of p,

$$M_e = \frac{\rho N_A}{B^2} p^3 \qquad 6.100$$

An analysis of an extensive range of flexible polymers by Fetters indicates that

$$M_e = (21.3 \pm 7.5\%)^2 \rho N_A p^3 \qquad 6.101$$

This prediction indicates that if the plateau modulus is

$$G_N^0 = \frac{\rho k T N_A}{M_e} \qquad 6.102$$

(the factor 4/5 is omitted), then the packing length also determines the plateau modulus

$$G_N^0 \propto \frac{kT}{p^3} \qquad 6.103$$

The connection between M_c and M_e is now discussed. Based on an analysis of a wide range of polymers, Fetters et al have shown that the critical molecular weight may be written as,

$$M_c = 1918 N_A \rho p^{2.35 \pm 0.15} \qquad 6.104$$

This result indicates that while M_c also depends strongly on the packing length its dependence is weaker than that of M_e. This is a significant result because it indicates that not only is $M_c/M_e \neq 2$, but the ratio is not constant and in fact depends on the packing length,

6.5.5   The Viscosity of Polymers

We now discuss the viscosity. The viscosity of unentangled chains can be calculated using Eqs. 6.29 and 6.88 for unentangled chains to indicate that is scales as N

$$\eta \propto \zeta_0 N \qquad 6.105$$

and for entangled chains,

$$\eta_0 = \frac{\pi^2}{12} G_N^0 \tau_R \qquad 6.106$$

which indicates that the viscosity increases as N^3.

We now examine the steady state compliance which is specified by

$$J_e^0 = \frac{1}{\eta_0^2} \int G(t) t \, dt = \frac{6}{5 G_N^0} \qquad 6.107$$

This result indicates that the steady state compliance has no molecular weight dependence. Physically, the steady state compliance is a measure of the elastic deformation during the steady-state flow process. The product $G_N^0 J_e^0$ is a universal constant which is consistent with experiment. However, experimentally, the magnitude of the constant is approximately twice as large as predicted. While this might appear to be a minor issue, there exists a

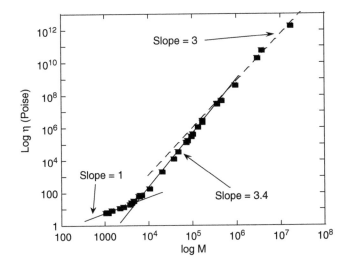

FIG. 6.22
The viscosity molecular weight dependence of polybutadiene, measured by Colby et al, is shown here. The data suggest that the molecular weight dependence of the viscosity approaches an asymptotic limit of M^3 at sufficiently large M.

larger underlying concern. While in the case of unentangled chains the predictions and experiments for the viscosity are in accord, $\eta \propto N$, the actual experimental data indicates that the chain length dependence of the viscosity should be $N^{3.4}$, not N^3, for entangled chains, as predicted.

The current construct of the Reptation model assumes that the tube length is fixed, fluctuations are not allowed In principle the model should be valid at asymptotically large values of (M/M_e), suggesting that the data on the molecular weight dependence of the viscosity should eventually scale as M^3, provided the molecular weights are sufficiently large. Extensive rheological measurements by Colby et al (Fig. 6.22 and 6.23) provide some preliminary evidence that the viscosity at very large M/M_e values would eventually reach an asymptotic limit of M^3. To account for the larger M-dependence, Doi suggested that there would have to exist another mechanism that would relax the stresses in the system at a faster rate than Reptation and that this mechanism would have to become unimportant at sufficiently large values of M/M_e. This will be addressed in section that follows our discussion of diffusion.

6.5.6 The Diffusion Coefficient of Entangled Chains

We imagine that the (one-dimensional) translation of the chain along the primitive path occurs such that its center of mass undergoes incremental displacements of distance λ at a rate of Γ, where $\Gamma = 2D_{Ro}/\lambda^2$. λ/L is therefore the fraction of the length of the primitive path along which the center of

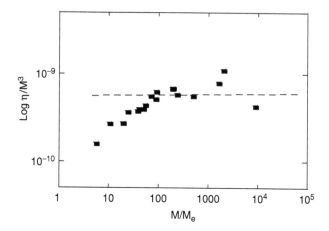

FIG. 6.23

η/M^3 is plotted as a function of M/M_e, illustrating that the "turn-over" occurs approximately beyond $M/M_e \sim 100$.

mass traverses during the time interval $1/\Gamma$. D_{Rep}, the center of mass diffusion coefficient of the entire chain moves in real space is

$$D_{Rep} = \frac{\langle \Lambda^2 \rangle \Gamma}{6} \qquad\qquad 6.108$$

where $\langle \Lambda^2 \rangle = (\frac{\lambda}{L})^2 \langle R^2 \rangle$, and $\langle R^2 \rangle$ is the end-to-end vector. It follows from the aforementioned that since $\langle R^2 \rangle = Na^2$ and $L = Na$, then

$$D_{Rep} = \frac{D_{Ro}}{3N} \propto \left(\frac{N}{N_e} \right)^{-2} \zeta_0 \qquad\qquad 6.109$$

This result can be derived by an alternate and very simple scaling argument by recognizing that during the time interval τ_{Rep}, the center of mass of the chain moves a distance on the order of its radius of gyration, leading to

$$D_R = \frac{R_g^2}{\tau_R} \propto \left(\frac{N}{N_e} \right)^{-2} \zeta_0 \qquad\qquad 6.110$$

This result illustrates an important point: Due to the topological constraints the center of mass diffusion coefficient of the chain decreases as N^{-2}, whereas in the case of unentangled chains it scales as N^{-1}.

Experiments on polymer-polymer diffusion may be divided into two classes. The first involves *self-diffusion* (D_s) experiments where the diffusion of a polymer of type A of molecular weight M diffuses into an identical host of polymer A also of molecular weight M. The second series are *tracer diffusion* (D_t) measurements wherein trace quantities of probe chains of polymer A of molecular weight M diffuse into a host of A-chains of molecular weight P ($P \neq M$). Tracer diffusion data for a large body of tracer experiments in a chain of molecular weight M diffuses into a host of molecular weight P, where $P \gg M$, are adequately be described by

$$D_s = D_0 \left(\frac{M_e}{M} \right)^{-\upsilon} \qquad\qquad 6.111$$

where $\upsilon = 2$ in accordance with predictions, Eq. 6.108. Tracer diffusion data DM^2 versus M/M_e are plotted in Fig. 6.24 for two different polymers. In self

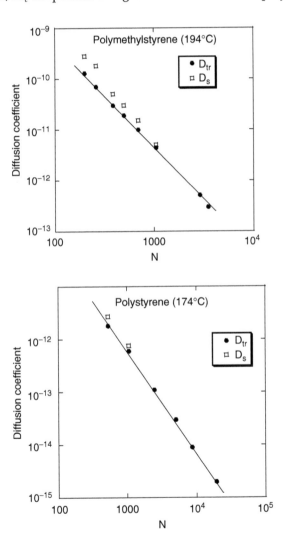

FIG. 6.24
Diffusion coefficient (cm²/s) versus N (degree of polymerization) for two different polymers are shown here. 1) polymethylstyrene (M. Antonietti, J. Coutandin, H. Sillescu, *Macromolecules*, 19, 793 (1986)), $D_{tr} \sim M^{-2}$ whereas Ds has a slightly stronger dependence on M. 2) polystyrene (Green et al, *Phys. Rev. Lett.* 53, 2145 (1984)).

diffusion coefficients, $M = P$, it is shown that for sufficiently large values of M/M_e, the self diffusion coefficient is also adequately described with the exponent $v = 2.0$. However, when $M/M_e < 10$, the exponent increases to approximately 2.3 ± 0.1. This is also illustrated in Fig. 6.24.

With this in mind, it is important to attempt to reconcile some issues regarding the exponents that govern the M-dependence of diffusion and viscosity (T.P. Lodge, *Phys. Rev. Lett.* 16, 3218 (1999)). The prediction that the translational diffusion coefficient scales as M^{-2} is based on the notion that $\tau_d \propto M^3$ and by extension $\eta \propto M^3$. In the asymptotic limit, large M/M_e, these molecular weight dependencies should be valid ($\eta \propto M^3$ and $D \propto M^{-2}$) and indeed the experiments reveal this to be true. However, as M/M_e decreases, finite size corrections need to be made and the original picture based on Reptation alone is not capable of handling this complication. Specifically, as the chain length decreases, relaxation processes that occur at time-scales (tube length fluctuations and constraint release) faster than Reptation are believed to be responsible for the change in the self-diffusion and exponents. We will address this issue in due course, but in the meantime, in the next section, we first discuss the temperature dependence of diffusion as well as estimates of the magnitude of the diffusion coefficient.

6.5.7 Temperature Dependence of Diffusion

The Reptation model indicates that the relaxation time, τ_R and by extension *the monomer friction coefficient*, ζ_0, determines the time scale of both the viscosity and the diffusion coefficient. This would suggest that D_R/T and η^{-1} should have similar temperature dependencies,

$$\log \eta = -\log \frac{D_R}{T} + const \qquad\qquad 6.112$$

Earlier in Section 6.4.2.1 the WLF equation (Eq. 6.33) was introduced and it was shown how it described the temperature dependence of the viscosity. Another equation which describes the temperature dependence of the viscosity of a variety of glass-forming liquids, and which predates the WLF equation, is the Vogel-Tammann-Fulcher (VTF) equation,

$$\ln \eta = A + \frac{B}{T - T_\infty} \qquad\qquad 6.113$$

which indicates that the viscosity increases appreciably with decreasing temperature. Both T_∞ and B are parameters characteristic of the material. T_∞ is the temperature at which the viscosity, or equivalently the longest relaxation time, would in principle, diverge. The equivalence between WLF and VTF equations becomes apparent upon recognizing that $c_1^0 = B/(T_0 - T_\infty)$ and $c_0^2 = T_0 - T_\infty$. In Fig. 6.24 we have plotted the temperature dependence of the diffusion coefficient of polystyrene of molecular weight $M = 430$ kg/mol,

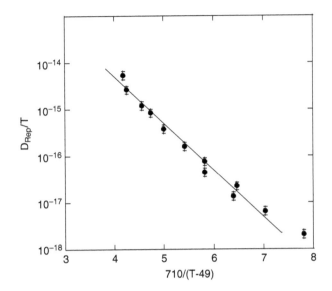

FIG. 6.25
The temperature dependence of the tracer diffusion coefficient of polystyrene of molecular weight 430 kg/mol. Specifically, D/T (Kelvin) is plotted as a function of $B/(T - T_\infty)$, where $B = 710$ and $T = 49°C$, the Vogel constants for polystyrene, obtained from the viscosity. (Note that the data at the lower temperatures were measured for chains of $M = 110$ kg/mol and normalized by a factor $\{(110/430)^2\}$.

which diffused into a PS host of molecular weight 20,000 kg/mol. D_{Rep}/T is plotted as a function of $\frac{B}{T-T_\infty}$, where $B = 710$ and $T_\infty = 49°C$, the Vogel constants for polystyrene. The line drawn through the data is of slope unity, indicating that knowledge of the temperature dependence of the viscosity provides a reasonable assessment of the temperature dependence of the Reptation diffusion coefficient.

Graessley (W.W. Graessley, J. Polym. Sci. 18, 27 (1980)) pointed out that one could in principle calculate the diffusion coefficient with knowledge of the viscoelastic parameters,

$$D_0 = \frac{1}{135}\left(\frac{R_g^2}{M}\right)G_n^0 \frac{M_c M_e^2}{\eta(M_c)} \qquad 6.114$$

Estimates of the diffusion coefficient based on viscoelastic parameters reveal reasonable estimates of the magnitude of D_0. The data in Table 6.4 show some representative comparisons.

We have, heretofore, discussed experimental observations that illustrate a connection between diffusion and viscosity, however, the dependence of the exponents on M/M_e require further discussion. Specifically, the influence of different relaxation modes that may be operational, in addition to Reptation, are discussed.

TABLE 6.4

A comparison of theoretical and experimental values of D_0 for some common polymers is shown here

Polymer	$T°C$	D_0 (theory)	D_0 (exp)
PS	174	0.06	0.075
PMMA	185	0.00009	0.00011
PBD	176		0.7
PE	176	0.32	0.43

Data taken from P. F. Green, 1996

6.5.8 Tube Length Fluctuations

It should be recalled that a critical assumption in the original Reptation theory is that the primitive path is assumed to be fixed. A significant consequence of this assumption is that predicts $\eta \propto M^3$. Doi argued that an additional mechanism would be responsible for relaxing the stress at a faster rate than τ_d. These are possibly fluctuations in the tube length. In principle the chain would contract and expand as it Reptates. The relaxations (tube contractions and expansions) at the chain ends and the orientations at the chain ends would be forgotten at a faster rate than the Reptation mechanism would suggest. In other words, a fraction of the tube a distance s_d from the ends $(0 < s_d < 1)$ would relax at a faster rate than it would due to Reptation. Toward the center of the chain the effect would be diminished and the relaxations of those segments would be dominated by Reptation. In the asymptotic limit, when M/M_e is sufficiently large the influence of the fluctuating chain ends would become unimportant, as observed experimentally.

The fluctuation in the contour length, according to Doi and Edwards is

$$\frac{\langle \Delta L^2 \rangle}{\langle L \rangle} \approx \frac{1}{\sqrt{Z}} \qquad 6.115$$

which indicates that the fluctuations can be neglected for $Z \gg 1$. If contour length fluctuations are important then the effective relaxation time is

$$\tau^f_{Rep} = (\langle L \rangle - \langle \Delta L \rangle)^2 / D_{Ro} \qquad 6.116$$

Essentially, the relaxation time, τ_{Rep} is modified (Problem 19)

$$\tau^f_{Rep} = \tau_{Rep} \left(1 - \frac{C}{\sqrt{N/N_e}} \right)^2 \qquad 6.117$$

where C is a constant of order unity and the second term is less than unity. In principle, the distance that a segment needs to diffuse via Reptation is reduced by a factor of $(1 - \frac{C}{\sqrt{N/N_e}})$.

The contour length fluctuations appear to reconcile the discrepancy between the theoretical and experimental $\eta - M$ dependencies.

The prediction of the self-diffusion coefficient, assuming that the contour length fluctuations are present, is,

$$D_{CLR} \approx D_0 \left(\frac{M}{M_e} \right)^{-2} \left(1 - C \sqrt{\frac{M_e}{M}} \right)^2 \qquad 6.118$$

This equation does a better job at describing the finite size effects associated with diffusion as M/M_e gets smaller than ~20. Similarly, the viscosity becomes,

$$\eta = \eta_{Rep} \left(1 - C \sqrt{\frac{M_e}{M}} \right)^2 \qquad 6.119$$

6.5.9 Constraint Release Mechanism

We have discussed an important consequence of the fixed tube length assumption and showed that certain important relaxation modes may have been omitted. By allowing the tube length to fluctuate the finite chain length corrections to the M-dependence of the viscosity and the diffusion coefficient could be reconciled. A second assumption which has proven to have a critical influence on the dynamics of the probe chain is that the tube is assumed to remain fixed in space. In other words, the chain would diffuse along its primitive path, which is fixed in space.

There is another important mechanism that has the effect of increasing the effective diffusion coefficient. This is the so-called constraint release mechanism. In principle the fixed tube assumption is reasonable if the rate at which the chains that compose the tube relax is slow compared to the relaxation time of the probe chain. In high molecular weight melts the condition is easily satisfied. However, as M/M_e becomes smaller the tube constraints can relax while the probe chain diffuses. This has the effect of increasing the effective diffusion coefficient of the chain. This mechanism can be responsible for affecting the exponents in self-diffusion experiments.

It turns out that this assumption that the chain moves through a stationary tube is only true for a probe chain diffusing in a host of chains that are sufficiently long that the relaxation time of the tube is appreciably slower than that of the chain. In the mechanism of constraint release, the surrounding chains of molecular weight P are able to diffuse away from the probe chain and relax the constraints, as illustrated in Fig. 6.26.

At rates that are fast compared to τ_{Rep}. In Fig. 6.25 chain C_1 and C_2 are two constraints in the tube and when they diffuse away and release the constraint, the center of mass of the probe chain can change location. A number of authors have examined this problem.

A semi-quantitative picture of the situation, as suggested by Klein is now described. Consider the tube as a Rouse chain, composed $Z = N/N_e$

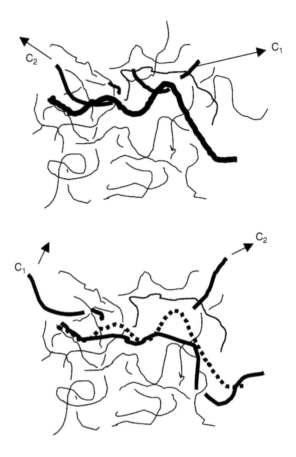

FIG. 6.26
Illustration of the constraint release mechanism. When the chains C_1 and C_2 diffuse, releasing the constraint on the probe chain, the probe chain relaxes. The release of the constraints that compose the tube is tantamount to motion of the tube.

submolecules. The relaxation time of the (Rouse) tube would be

$$\tau_{tube}(N) = \left(\frac{N}{N_e}\right)^2 \tau'$$
6.120

where the basic relaxation rate would be determined by the relaxation times of the host molecules, each of P segments. Consequently, τ' is the Reptation relaxation time, $\tau' = \tau_{Rep}(P)$. The diffusion coefficient of the tube is now

$$D_{tube}(N,P) = \frac{R_g^2(N)}{\tau_{tube}(N)} = \kappa \frac{NN_e^2 D_{Rep}(N)}{P^3}$$
6.121

where κ is a constant. The displacement of the center of mass of the probe chain occurs by two independent processes, Reptation and by displacements of the tube due to this constraint release mechanisms. In this regard the diffusion coefficient of the chain is due to the sum of two contributions

$$D^* = D_{Rep}(N) + D_{tube}(N,P) = D_{Rep}(N)\left\{1 + \kappa \frac{N_e^2 N}{P^3}\right\} \qquad 6.122$$

A number of authors have examined this process, including Graessley, Daoud, deGennes, and others. The CR mechanism is unimportant for $P \gg N$ and for very large N/N_e in the self-diffusion experiments mentioned heretofore. For finite molecular weights, it is responsible for an increase in the effective exponent due to self-diffusion. The increase, however, is not sufficient to account for the larger exponents. Tracer diffusion experiments were performed using deuterated polystyrene of molecular weight M into different hosts of PS chains of varying molecular weight P, and Eq. 6.120 is found to describe the data extremely well (Fig. 6.27).

The combined effect of tube length fluctuations and constraint release could also be included and this would reduce the effective relaxation time of the constraints. The end result would be somewhat more of an enhancement on the scaling exponent for self diffusion, as suggested by Milner and McLeish.

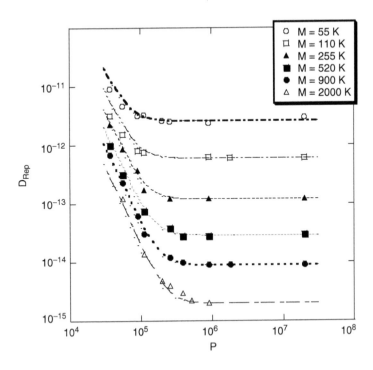

FIG. 6.27

The effect of constraint release of the P-host chains on the translational diffusion of the M-chains is demonstrated here. Equation 6.122 was used to fit the data. The constant $k = 10.9$. These data are replotted (Green et al. 1984, Green and Kramer, 1986).

In the next section we continue, briefly, our discussion of dynamics by examining the predictions of the dynamic moduli, G' and G''.

6.5.10 Dynamic Moduli $G'(\omega)$ and $G''(\omega)$

Thus far we have discussed the stress relaxation modulus as predicted by Reptation. With the use of the stress relaxation modulus and Eq. 6.56 and 6.57, $G'(\omega)$ and $G''(\omega)$, may be calculated, respectively,

$$G'(\omega) = G_N^0 \sum \frac{8}{\pi^2 p^2} \frac{(\omega \tau_{Rep}/p^2)^2}{1+(\omega \tau_{Rep}/p^2)^2} \qquad 6.123$$

and

$$G''(\omega) = G_N^0 \sum \frac{8}{\pi^2 p^2} \frac{(\omega \tau_{Rep}/p^2)}{1+(\omega \tau_{Rep}/p^2)^2} \qquad 6.124$$

Sketches of $G'(\omega)$ and $G''(\omega)$ equations are shown in Fig. 6.28. The predictions are represented by the solid lines for $\omega \tau_e < 1$. The range $\omega \tau_e > 1$ corresponds to the Rouse regime and is represented by the broken lines. The low-frequency slopes describing these graphs are in accordance with predictions. It is noteworthy that the intermediate frequency range for $G''(\omega)$ is not in accord with experimental observations. Actual experimental data are shown earlier in Fig. 6.14. The most noteworthy discrepancy appears in the middle of the frequency range where $G''(\omega)$ deceases at a much faster rate than observed experimentally. The discrepancy is reconciled by including contour length fluctuations in the $G(t)$ used to predict $G''(\omega)$. While other corrections that involve constraint release alone fail, it is likely that constraint release makes a minor contribution to the process.

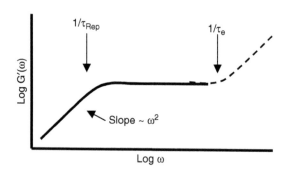

FIG. 6.28
Theoretical prediction for $G'(\omega)$ based on Reptation.

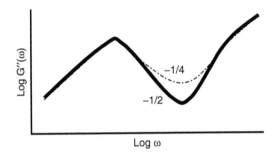

FIG. 6.29
Theoretical predictions of $G''(\omega)$ using the Doi-Edwards Reptation model (solid line) and the model modified by considering contour length fluctuations. The pure Reptation model predicts a steeper decrease in G'' in the middle of the frequency range whereas experimentally (Fig. 6.10) the decrease is much shallower. In the intermediate frequency range, experiments observe $-1/3$.

6.6   Concluding Remarks

The situation thus far involving diffusion and viscosity can be summarized as follows. The Reptation model provides an excellent description of the universal molecular weight dependence of the viscosity, the self-diffusion coefficient and the tracer diffusion coefficient when M/M_e is large, typically on the order of 10 to 20 for diffusion and at somewhat larger values for the viscosity. The situation becomes somewhat different at the lower molecular weights where the molecular weight dependence of the viscosity increases from M^3 to $M^{3.4}$ and that of self-diffusion increases from $M^{-2.0}$ to approximately $M^{-2.3}$ in most systems. The notion of contour length fluctuations, which accounts for a faster relaxation of the stress than would Reptation, reconciles the changing exponents describing the viscosity and the shape of $G''(\omega)$. It also appears to reconcile the changing exponents for self-diffusion experiments. The mechanism, however, should still be active in tracer diffusion experiments, but there is no experimental support for it, with the absence of an increasing exponent. In tracer diffusion experiments, when a probe chain diffuses into host environments of varying molecular weights, the constraint release mechanism is active and adequately explains the experiments. Other theoretical approaches such as polymer mode coupling theory developed by Schweizer and coworkers have examined the influence of non-Reptative processes on the dynamics in an attempt to reconcile the finite size effects. One of their findings was that the diffusion coefficient converged to the Reptation prediction at lower M/M_e than did the viscosity. After nearly three decades this continues to be an active topic of research. The interested reader is referred to some very recent reviews on the topic. The intent of this chapter was to introduce the reader to basic concepts regarding diffusion and viscoelasticity in polymers.

6.7 Problems for Chapter 6

1.

a) Calculate the ratio of the rms-end-to-end distance to the contour length for a macromolecule in molten polypropylene. Take the molecular weight to be $M = 10^5$ g/mol, and the C—C bond length $= 1.54 \times 10^{-8}$ cm. Assume freely flexible bond model.

b) Repeat the calculation by including the fixed bond angle requirement (tetrahedral angle 109.5°)

c) Calculate the characteristic ratio for this polymer.

d) Would you expect the characteristic ratio for polystyrene to be larger? Explain.

2. The following is a radial distribution function $P(r)$ for a Gaussian chain,

$$P(r) = \left(\frac{\beta}{\pi^{1/2}}\right)^3 e^{-\beta^2 r^2}$$

where $\beta^2 = \frac{3}{2nl^2}$. $P(r)d\bar{r}$ is the probability of finding one end of a chain within a spherical shell of thickness dr located a distance r from the other end. The other end is assumed to be fixed at the origin.

a) Based on $P(r)$ how does ratio $\langle r^2 \rangle^{1/2}/L$, depend on n? L is the contour length of the chain.

b) Show that

$$\int_0^\infty P(r)d\bar{r} = 1$$

c) Show that $P(r)$ has a maximum at $r = 1/\beta$. What is the significance of this value of r?

d) Determine $\langle r^2 \rangle$.

Integrals you will need: $I(k) = \int_0^\infty r^k e^{-kr^2} dr$

when $k = 0$, $I(0) = \frac{1}{2}(\frac{\pi}{a})^{1/2}$; for $k = 1$, $I(1) = \frac{1}{2a}$; for $k = 2$, $I(2) = \frac{1}{4}\frac{\pi^{1/2}}{a^{3/2}}$ and for $k = 3$ $I(3) = \frac{1}{2a^2}$

3. Starting with Eq. 6.2, derive Eq. 6.5

4. Starting with Eq. 6.6 show that in the limit that $\lim_{n\to\infty} C_n = C_\infty = \frac{(1+\cos\theta)}{(1-\cos\theta)}$. (Hint: Note that $\cos\theta_{i,i+j} = \cos\theta_j$ for all value of j. Also, note that because all the values of θ are equal, $\langle\cos\theta_j\rangle = (\langle\cos\theta\rangle)^j$. This will enable you to rely on a geometric series to arrive at your final answer).

5. The average sum of the projection of m bonds in the direction of the first bond is (Flory 1969)

$$l_p = l\sum_{k=0}^{m-1}(\cos\theta)^k$$

Imagine the situation in which the chain is stiff (i.e. $\theta \ll 1$), show that under these circumstances, the Kuhn segment length $b = 2l_p$.

6. Derive Eq. 6.54 and 6.55 from Eq. 6.51.

7. Derive Eq. 6.58 and 6.59 and sketch the dependencies of $G'(\omega)$ and $G''(\omega)$ on ω.

   a) Consider further the periodic deformation of a Newtonian fluid, $\gamma(t) = \gamma_0 \text{Re}(e^{i\omega t})$. Calculate and sketch $G'(\omega)$ and $G''(\omega)$ as a function of frequency.

   b) Perform the same calculation for an elastic solid.

8. The following equation describes the center of mass displacement of a Rouse chain and the dynamics of the internal modes.

$$\langle(\bar{R}_n(t) - \bar{R}_n(0))^2\rangle = 6D_{Ro}t + \frac{2Nb^2}{3\pi^2}\sum_{p=1}^{\infty}\frac{1}{p^2}\cos^2\left(\frac{n\pi p}{N}\right)[1-e^{-p^2t/\tau_{Ro}}]$$

When $t \ll \tau_{Ro}$, the sum can be replaced by an integral. If we choose the average value $(1/2)$ for the cosine term then show that

$$\langle(\bar{R}_n(t) - \bar{R}_n(0))^2\rangle \approx 6R_{Ro}t + \frac{Nb^2}{3\pi^2}\left(\frac{t}{\tau_{Ro}}\right)^{1/2}$$

Sketch $\langle(\bar{R}_n(t) - \bar{R}_n(0))^2\rangle$ as a function of time for a monomer and identify the transition between the sub-Fickian and Fickian regimes.

9. Show that, with the exception of π^2, Eq. 6.83 is identical to Eq. 6.82.

10.
   a) Plot $\mu(s,t) = \sum_{p;odd}\frac{4}{p\pi}\sin(\frac{ps\pi}{L_t})e^{-p^2t/\tau_{Rep}}$ as a function of s for $t/\tau_{Rep} = 1, 0.5$ and 0.2 and discuss a physical interpretation.

   b) Derive $\mu(t) = \sum_{p;odd}\frac{8}{p^2\pi^2}e^{-p^2t/\tau_{Rep}}$ using Eq. 6.85.

11. The modulus of the melt is given by $G(t) = \frac{c}{N}kT\int_0^\infty dp\, e^{-2tp^2/\tau_{Ro}}$. Show that for the time scale $t \ll \tau_e$,

$$G(t) = \frac{c}{2\sqrt{2}N}kT\left(\frac{\tau_{Ro}}{t}\right)^{1/2}$$

12. Using Eq. 6.71, show that the relaxation time τ_e is specified by

$$\tau_e = \frac{\pi^3 a \tau_{Ro}}{(Nb^2)^2} = \frac{a^4 \zeta_0}{kTb^2}$$

13. Show that for Rouse chains the viscosity is proportional to N.

14. Using the expression for $G(t)$ show that for entangled chains,

$$\eta_0 = \frac{\pi^2}{12}G_N^0 \tau_R$$

15. Demonstrate the equivalence between the WLF equation and the VTF equation.

16. Identify and sketch the various regimes that characterize the time dependences of the mean square displacement of monomers on a polymer chain undergoing Reptation.

17. Show that that D_R/T and η^{-1} should have similar temperature dependencies.

18. Calculate the diffusion pre factor D_0 for polybutadiene.

19. If the contour length fluctuations are specified by

$$\frac{\langle \Delta L^2 \rangle}{\langle L \rangle} \approx \frac{1}{\sqrt{Z}}$$

show that the relaxation time is given by

$$\tau_{Rep}^f = \tau_{Rep}\left(1 - \frac{C}{\sqrt{N/N_e}}\right)^2$$

Plot the chain length dependence of the viscosity using the above equation and compare the results with two power law plots, one that scales an $N^{3.4}$ and the other as N^3. Comment on your answer.

20. Derive the following equation for the self-diffusion coefficient that includes effects of contour length fluctuations,

$$D_{CLR} \approx D_0\left(\frac{M}{M_e}\right)^{-2}\left(1 - C\sqrt{\frac{M_e}{M}}\right)$$

Compare with the power-law dependence of M^{-2} and comment on the results.

21. The Maxwell model predicts that the rate of strain, $d\gamma/dt$, can be expressed as

$$\frac{d\gamma}{dt} = \frac{1}{E}\frac{d\sigma}{dt} + \frac{\sigma}{\eta}$$

where σ is the stress and η is the viscosity.

a) Starting with

$$\sigma(t) = \sigma_0 e^{i\omega t}$$

write down an expression for $d\gamma/dt$.

b) Now integrate this expression between time t_1 and time t_2 and show that one can write the complex modulus

$$\frac{1}{E^*} = \frac{1}{E} + \frac{1}{i\omega\tau E}$$

where the viscosity $\eta = E\tau$.

c) Now show that $E^* = E' + iE''$. What are E' and E''?

d) Show that

$$\tan\delta = \frac{1}{\omega\tau}$$

e) Sketch E' and E'' as a function of frequency, ω, on a log-log plot.

22. Determine the constant B in Eq. 6.97.

23. The diffusion coefficient, D, of a polystyrene chain of degree of polymerization $N = 1000$ is 10^{-13} cm^2/s at 180°C. The degree of polymerization is $N = M/M_0$, where M is the molecular weight of the chain and M_0 is the molecular weight of the monomer. For polystyrene, $M_0 = 104$ g/mol. (Assume the following: $G_N^0 = 2 \times 10^6$ dynes/cm^2; the radius of gyration is $R_g = 100$ Angstroms, $M_e = 18000$ g/mol and that the density is 1 g/cm^3.)

a) Determine the viscosity at this temperature, 180°C.

b) If the WLF constants are $c_1 = 12$ and $c_2 = 50$ at a reference temperature of 100°C, determine the viscosity at 120°C.

c) Determine D for a chain of $M = 8{,}000$ g/mol at 120°C.

d) What are the Vogel-Fulcher parameters of this material?

24. The following expression was derived for the constitutive relation between the strain, $\gamma(t)$, and the rate of stress $d\sigma(t)/du$ for a material that exhibits a time-dependent response

$$\gamma(t) = \int_{-\infty}^{t} \frac{d\sigma(u)}{du} J(t-u)du$$

This result was obtained by starting with $\gamma(t) = \sigma_0 J(t)$ and relying on the Boltzmann superposition principle.

   a) What is the Boltzmann superposition principle?

   b) Beginning with $\sigma(t) = G(t)\gamma_0$, derive the following relationship

$$\sigma(t) = \int_{-\infty}^{t} \frac{\partial \gamma(u)}{\partial u} G(t-u)du$$

   c) If one assumes that the strain can be expressed as $\gamma(t) = \gamma_0 e^{i\omega t}$ then it can be shown that

$$G^*(\omega) = G'(\omega) + iG''(\omega) = \int_{0}^{\infty} i\omega e^{-i\omega t} G(t)dt$$

from which it follows that

$$G'(\omega) = \int_{0}^{\infty} \omega \sin \omega t G(t)dt$$

and

$$G''(\omega) = \int_{0}^{\infty} \omega \cos \omega t G(t)dt$$

Using these expressions together with the expression for $G(t)$, obtained from the Reptation model, calculate expressions for $G'(\omega)$ and $G''(\omega)$.

   d) Plot the functions $G(t)$, $G'(\omega)$ and $G''(\omega)$. Use a realistic value for G_N^0.

   e) Why does the plateau exist in $G(t)$ and $G'(\omega)$?

   f) The simple Maxwell model, discussed in class, enabled us to develop a constitutive relation. Starting with $\gamma(t) = \gamma_0 e^{i\omega t}$, determine $G'(\omega)$ and $G''(\omega)$ and compare these result with the results of part d.

6.8   References

P.J. Flory, *Statistical Mechanics of Chain Molecules*, Wiley, NY (1969).
J.D. Ferry, *Viscoelastic Properties of Polymers*, Wiley, NY (1980).
J. Bicerano, *Prediction of Polymer Properties*, 2nd Ed. Marcel Dekker, Inc. N.Y. (1996).

M. Rubenstein and R.H. Colby, *Polymer Physics*, Oxford University Press, UK 2003.

S-Q. Wang, "Chain Dynamics in Entangled Polymers: Diffusion and Viscoelasticity." *Journal of Polymer Science: Polym. Physics* 41, 1589 (2003).

P-G. deGennes, *Scaling Concepts in Polymer Physics*, Cornell University Press, Ithaca, NY (1979).

P.F. Green, "Translational Dynamics of Macromolecules in Melts," in *Diffusion in Polymers*, ed. P. Neogi, Marcel Dekker, NY 1996.

T.P. Lodge, N.A. Rotstein, S. Prager, *Adv. Chem. Phys.* 79, 1 (1990).

M. Tirrell, *Rubber Chemistry and Technology*, 57, 552 (1987).

T.C.B. McLeish, "Tube Theory of Entangled Polymer dynamics," *Advances in Physics*, 51, 1379 (2002).

M. Doi and S.F. Edwards, *Theory of Polymer Dynamics*, Oxford University Press, (1986).

C.W. Macosko, *Rgeology, Principles, Measurements and Applications*, VCH Publishers., 1994.

P-G. deGennes, Simple Views on Condensed Matter, Series in *Modern Condensed Matter Physics*-Vol. 12, World Scientific Publishing, (2003).

W.W. Graessley, "Entangled Linear Branched and Network Polymer Systems: Molecular theories," *Advances in Polymer Science*, 47(67) (1982).

J. Klein "Dynamics of Entangled Polymer Systems," *Macromolecules*, 19, 105 (1986).

H. Wantanabe, "Viscoelasticity and Dynamics of Entangled Polymers," *Prog. Polym. Sci.* 24, 1253 (1999).

L.J. Fetters, D.J. Lohse and S.T. Milner, "Packing length influence in Polymer Melts on the Entanglement Critical and Reptation Molecular Weights" *Macromolecules*, 32, 6847 (1999).

L.J. Fetters, D.J. Losche, D. Richter, T.A. Witten, and A. Zirkel, "Connection between Polymer molecular weight density, chain dimensions and melt viscoelastic properties" *Macromolecules*, 27, 4639 (1994).

7

Transport Processes in Inorganic Network Glasses

7.1 Introduction

The primary goal of this chapter is to discuss transport processes, specifically viscous flow and ionic diffusional transport, in inorganic network oxide glasses. In order to accomplish this, the discussion in the first few sections of this chapter, Sections 7.2 through 7.6, are reasonably general and are not only applicable to inorganic network glasses. These sections provide a basis for the subsequent discussions of transport in network glasses. In Chapter 8 an attempt is made to discuss issues more broadly, related to the dynamics of glass forming systems.

If the arrangement of atoms, or molecules, that constitute a solid lacks long-range order, then the material is often identified as a *glass*, or is said to be in a vitreous, or noncrystalline, state. In this regard the use of the term glass does not refer to a specific material. Various materials produced using sol-gel and vapor deposition techniques possess structures that lack long-range order. On the other hand, any material, if cooled at a sufficiently rapid rate from the liquid state, would not crystallize, but would instead solidify below the glass transition temperature to form a structure which lacks long-range order. The disorder is a natural consequence of the fact that the structural entities (atoms or molecules) do not have an opportunity to organize themselves and to form an ordered, periodic structure during cooling. The rate at which a sample should be cooled before it will solidify to form this disordered structure depends on the material. Some materials, such as atactic polymers and statistically random copolymers, by virtue of their structure, exhibit a tendency to form amorphous structures upon solidifying, regardless of the cooling rate. In contrast, the formation of single component metallic glasses is accomplished only by employing extremely rapid quench rates, 10^3–10^6 Kelvin/sec. More recently, multi-elemental bulk metallic glasses have been fabricated by employing more controllable quench rates, many orders of magnitude slower, ~1 Kelvin/sec. Such bulk metallic glasses are composed of alloys of three to five metallic components of varying atomic sizes.

An example of such an alloy would be $Zr_{41.2}Ti_{13.8}Cu_{12.5}Ni_{10}Be_{22.5}$. (Busch, 2000) The atomic size mismatch frustrates the ordering process, thereby accommodating the use of more reasonable quench rates for solidification. Metallic glasses are of practical importance because they possess favorable magnetic and mechanical properties and corrosion resistance. The list of materials with reasonable, or strong, glass-forming tendencies is enormous, and it includes a diverse range of systems, such as polymers (natural and synthetic), metals,

TABLE 7.1

Table of glass formers: fragility index (m), glass transition temperature (T_g), Kuzmann temperature (T_K), and the ratio of the heat capacity in the liquid to that of the solid. (Data adopted from Huang and McKenna (2001) (T) data from Tanaka 2003 and (B) data from Böhmer at al. (1991))

Material	m	T_g	T_K	T	$C_p^l/C_p^{cryst,glass}$
GeO_2	24	818	418(T)	199	1.073
SiO_2	20	1,446	876(T)	529	1.005
B_2O_3	32	521			1.449
Soda lime	40	536(B)			
$0.25Na_2O$-$0.75SiO_2$	37	739			1.216
	30	764(B)			
$K_3Ca(NO_3)_7$	93	332			1.568
	93	343(B)			
$2BiCl_3$-KCl	85	306			1.647
$ZnCl_2$	30	370			1.279
		250	180–236(T)		
BF_2	24	590			1.000
Se	87	307			1.498
Salol	63	281			1.714
a-phenyl-o-cresol					
ethylene glycol	325	181			
glycerol	53	190			1.847
		135	127 (T)		
sorbitol	93	274			1.886
		236	224 (T)		
m-cresol	57	198.5			1.837
o-terphenyl	76	241			1.472
		200	184(T)		
toluene	96		106(T)		
	59	110			1.512
α-phenyl-o-cresol	83	220			1.457
sucrose		283	290(T)		
dibutylphthalate		179			1.527
polystyrene	116	373			1.188
polypropylene	137	260			1.268
PDMS	79	149.5			1.387
Polycarbonate	132	423			1.113
PVC	191	354			1.278
Polysulfone	141	459			1.142
PMMA	103	367			1.231
PEMA	81	344			1.189
PBMA	56	305			1.187

ionic liquids, organic small molecule liquids, and network glasses. Generally, systems that possess large entropies of fusion and large surface tensions exhibit large glass-forming tendencies (Debenedetti 1996). Table 7.1 includes a necessarily limited number of examples of materials that exhibit glass forming tendencies. The information in this table will be discussed in due course. In the next section the structure of network glasses is discussed.

7.2 The Structure of Inorganic Network Glass Formers: An Introduction

The earliest known man-made glazes date as far back as 12000 B.C. and the first production of network glass is believed to have been accomplished around 7000 B.C. Early recipes for glass production are believed to consist of sand from the sea, marine shells, and seaweed to provide essential ingredients (silica, lime, and sodium oxide). Glass blowing was invented approximately 100 B.C. Apart from vases, the first technological application of glass was for architectural windows in churches during the Middle Ages [Bunde et al 1998]. Lenses and mirrors were produced after the 1300s. The most significant manufacturing advance came with the development of the float glass process in the middle of the last century. The process involves floating a molten layer of glass on the surface of molten tin. The resulting glass is smooth, uniform in thickness, and of high quality. The process is simple and reliable and is used for the mass production of glass. (Doremus 1993; Paul 1990) The goal of this section is to provide an introduction to the structure of network forming glasses in order to establish the foundation for a discussion of primary (viscous flow) and secondary (ion hopping phenomena) relaxation processes in these systems.

The most widely used glasses for technological applications are silicates, whose structures are based on the network former Si. For this material the basic structural unit is a tetrahedron. Tetrahedra connect to form a three-dimensional covalent network that lacks long-range order. Illustrated in Fig. 7.1(a) is a schematic of two tetrahedra that are connected. A larger group of interconnected tetrahedra are shown in Fig. 7.1(b). The oxides of *network forming* elements—Si, Ge, P, and B—form single-component glasses which possess tetrahedral short-range order. The short-range structure for TeO_2 is a trigonal bipyramid, characterized by two short bonds and two long bonds, in contrast to the tetrahedral structure, which possesses four bonds of equal length.

Inorganic network glasses are used in a range of applications (Doremus 1973, Houde-Walter and Green 1998, Brow 2000, Uhlmann and Kreidl 1986). Glasses based on the network former P_2O_5 have applications that include nuclear waste storage and biomedical applications. Other recent technological applications of glass include telecommunications, flat panel displays, and

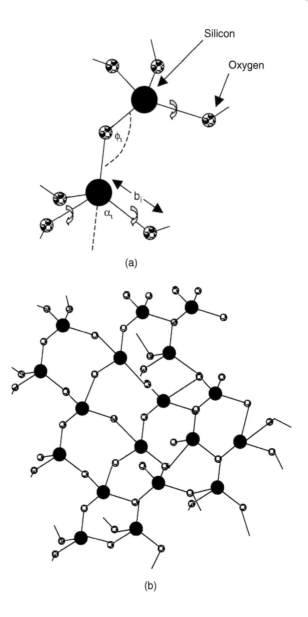

FIG. 7.1

a) Two connected SiO_2 tetrahedra are shown here. The same configuration would be true for structures representing GeO_2 and P_2O_5. The angles a_i and ϕ_i and the bond lengths are not well defined as they would in a crystal. b) The short-range structure is shown.

electrochemical sensors, including batteries. Tellurium based glasses are excellent candidates for broad band rare earth doped fiber amplifiers for wave guides and ultrafast switching applications.

The molecular structure, and hence the properties, of network glasses can be modified with the addition of alkali and alkaline earth oxides, so-called

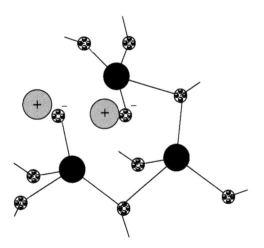

FIG. 7.2
Single charges alkali cations are shown in the vicinity of NBOs after the addition of M_2O.

network modifiers, which "depolymerize" the covalent bonding character, rendering it more ionic. Specifically, oxygen anions which formally created bridges, bridging oxygens (BOs), between the network formers become non-bridging (NBOs), now associated with a positively charged cations, as illustrated in Fig. 7.2. Here the Si ion associated with the Si-O group now becomes slightly more positive and the bond distance dilates. A nomenclature that identifies the bonding configuration is typically employed. For example, in pure silica, each Si atom is connected to four bridging oxygen atoms and the bonding configuration of the site is identified as a Q_n site where $n = 4$ is the number of bridging oxygen anions. In the case where there are three bridging oxygens, the site becomes a Q_3 site, etc.

In addition to the network modifiers, the oxides of network intermediates (e.g., Al, Be, and La) also modify the structure of the glass. Network modifiers have the opposite effect on the glass structure that network intermediates do. Network intermediates increase the covalent character of the network. It is the combination of *network formers*, *network intermediates*, and *network modifiers* that allow Glass Chemists to prepare glasses of a range of compositions and structures that exhibit varying properties for different applications.

With the level of disorder that the long-range structure network glasses possess, diffraction techniques provide limited information about the structure. A number of spectroscopic tools, including nuclear magnetic resonance (NMR), Raman spectroscopy, infrared spectroscopy, and X-ray photoelectron spectroscopy (XPS) have been used to examine the structure of network glasses (Brow 2000). In addition to these tools, extended X-ray absorption fine structure (EXAFS) has proven to be a powerful tool for elucidating short-range structure (Huang 1996; Greaves and Ngai 1995). EXAFS provides information about the identity of atomic species, average distance between species, mean square variations of distances between species as well coordination numbers. EXAFS studies indicate that depending on the alkali cation, the average alkali-oxygen (M–O) bond distances are distinct, which is not surprising because of the differing ionic diameters and polarizabilities. Moreover, there exists a distribution of M–O bond lengths around each alkali cation and the distribution is relatively distinct from one type of cation to another. The basic message is that the local environment (bond angles, bond lengths, and packing) around a given type of cation in the glass network is distinct. In K and Cs mixed alkali glasses (i.e., two distinct cations are present in the glass network) the environment is more ordered around the K (smaller of the two) than the Cs sites. Therefore, apart from the differing bond lengths (in part associated with the size of the alkali ions), the degree of order around dissimilar cations characterize the structure of mixed alkali glasses. This information revealed by EXAFS has played a central role in elucidating ion transport mechanisms in glasses, as we will see later in Section 7.11.

7.3   Bulk Transport Processes Inorganic Network Glass Formers

Inorganic network glass forming liquids, like all glass forming liquids, exhibit increasingly slow dynamics with decreasing temperature in the absence of the formation of long-range structural order. Flow/dynamics of the inorganic melt above T_g is accommodated via the breaking and reforming of bonds (Inagaki et al., 1993; Webb 1997, Stebbins et al 1995; Brow 2000). Specifically, transitions between bond character from bridging to nonbridging, via a bond interchange mechanism (transitions between Q_3 and Q_2 and Q_4 structures) are believed occur throughout the system at characteristic rates necessary to accommodate flow in the network. In phosphates it is the P-O-P exchange rates that control the viscosity. The same processes (S_i-O-S_i exchanges) occur in silicates. These exchange rates are measured using NMR and compared with the relaxation times obtained from viscosity measurements and found to be comparable (Stebbins et al 1995; Green et al 1999; Inagaki 1993). Structural changes may also contribute to changes in the viscosity with temperature. Specifically, at high temperatures, alkali tellurites have been shown to undergo a structural change from the TeO_4 trigonal bipyramid structure to a TeO_3 triganol pyramid structure (Kieffer et al 1998). The decrease of the network connectivity is associated with appreciable changes of the viscosity and elastic moduli.

It is evident from the foregoing discussion that relaxations associated with different structural entities occur within network glasses. Below the glass transition temperature, long-range ionic hopping processes dominate. These are the so-called secondary relaxations responsible for ionic conductivity and internal friction (Tanδ > 0); the primary relaxation processes are virtually frozen in this temperature range (Angell 1989, 1990, 1991). At temperatures above T_g, the main (primary) network relaxations which control the viscosity become dominant. These issues regarding the secondary and primary relaxations are revisited in Sections 7.7 through 7.11. In the meantime a phenomenological description of the glass transition is presented in the next section.

7.3.1 Temperature Dependence of the Viscosity: The VTF Equation

Angell has pointed out that the dynamics of glass forming liquids or melts (metallic glasses, polymers, network glasses, organic liquids, etc.) could be characterized as "strong" or "fragile" depending on the dependence of the viscosity, η, on temperature, T. If the log η vs. $1/T$ dependence is linear (Arrhenius), then the behavior is "strong." Conversely, the dynamics are identified as "fragile" if the log η versus $1/T$ dependence is nonlinear. Typically, the purely tetrahedral, covalent bonded, network glasses, SiO_2 and GeO_2, exhibit behavior that is strong, whereas systems in which the bonding is characterized as dispersive, or nondirectional, exhibit fragile behavior. Moreover, as shown later (Section 7.5), if the network structure is depolymerized with the addition of alkali oxides, rendering the bonding more ionic in character, the $\eta - T$ relation becomes fragile.

The temperature dependence of the viscosity and basic features of the glass transition are now discussed. The temperature dependencies of the viscosities of many glass forming liquids is often described by the Vogel-Tammann-Fulcher (VTF) equation,

$$\ln \eta = A + \frac{B}{T - T_\infty} \qquad 7.1$$

which indicates that the viscosity increases appreciably with decreasing temperature. For the case of polymers the viscosity increases many orders of magnitude as T_g/T approaches unity. Both T_∞ and B are parameters characteristic of the material, as mentioned earlier, in Chapter 6. T_∞ is the temperature at which the viscosity, or (equivalently) the longest relaxation time, would in principle diverge. T_∞ is not measured directly, but is determined by extrapolation of the viscosity data from above T_g (Debenedetti, 1996, Scheer 1990). If $T_\infty = 0$ Kelvin, then the viscosity-temperature dependence behavior is classified as strong. In this case the activation energy for flow remains constant throughout the entire temperature range.

In order to get a quantitative measure of fragility the VTF equation may be rewritten as follows,

$$\log \frac{\eta(T)}{\eta(T_g)} = -(1 - r)m \left[\frac{(1 - T_\infty/T_g)}{(1 - \alpha T_\infty/T_g)} \right] \qquad 7.2$$

where $r = T_g/T$. The parameter m is of particular significance, it is the so-called fragility index (Böhmer et al 1991) and

$$m = \frac{BT_g}{[T_g(T_g/T_\infty - 1)]^2} \qquad 7.3$$

The fragility index is a measure of the degree to which the $\eta - T$ dependence deviates from Arrhenius behavior. Larger values of m denote increasing fragility. The fragility index, in essence indicates how rapidly the viscosity changes in the vicinity of T_g (see Problem 6). In this regard m is sometimes called the steepness index. In some industrial settings, network glasses with fragile viscosity-temperature behavior are described as "short" because of the comparatively rapid rate at which they solidify over a narrow temperature range in the vicinity of T_g.

The data in Table 7.1 indicate values of the fragility indices and glass transition temperatures for various classes of materials. These data reveal that the viscosities of polymers and small molecule liquids exhibit fragile behavior. Network glasses are particularly interesting because their fragilities may be manipulated. For example, the fragility of SiO_2 can be changed from strong to fragile with the addition of alkali oxides. The effect of alkali oxides on melt fragility is explored further in Section 7.5. In the meantime the phenomenology of the glass transition is discussed.

7.3.1.1 Comments Regarding the Glass Transition

At high temperatures the translational dynamics of the molecules in the system are rapid, enabling the system to readily equilibrate. A reduction of specific volume accompanies the decrease in temperature. If the cooling rate is sufficiently rapid, then a nucleus is unable to form in order to initiate crystallization. Eventually as the temperature is lowered further, below T_g, the molecules do not have an opportunity to completely sample configurational space and the system becomes nonergodic; it falls out of equilibrium and forms a glass.

The temperature dependencies of various thermodynamic parameters reflect the transition into the vitreous state, as illustrated in Fig. 7.3. A schematic of the temperature dependence of the specific volume, v, of a sample capable of forming a glass as well as a crystal, is shown in Fig. 7.3(a). If the sample formed a crystal, its specific volume would exhibit an abrupt drop, discontinuity (water is of course a notable exception), at T_m. This is characteristic of a first order phase transition. Figure 7.3c shows a schematic of the temperature dependence of the enthalpy, h. Both v and h exhibit changes in slope in the vicinity of T_g.

Clearly, the glass transition is not a first order transition. Generally, first order phase transitions are characterized by discontinuities in the first derivative of the Gibbs free energy at the transition temperature. Specifically, the

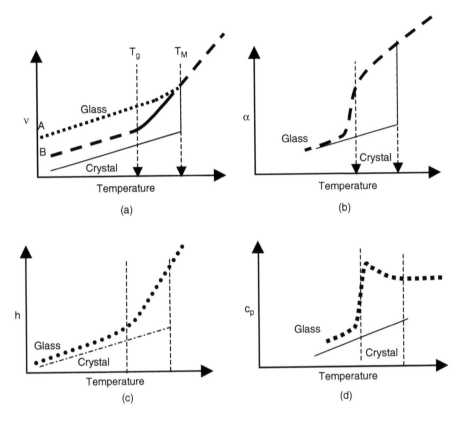

FIG. 7.3
Typical temperature dependencies of thermodynamic parameters of glass forming systems are illustrated here. a) The specific volume is plotted here for a material that is capable of forming a glass as well as a crystal. Glass transition temperatures that result from two cooling rates are shown here. T_g^A is the higher glass transition that is realized for the faster cooling rate and T_g^B is the glass transition temperature obtained from a slower cooling rate. The temperature range between T_m and the glass transition is the supercooled region; b) the thermal expansion; c) the specific enthalpy; d) the specific heat.

entropy, the specific volume and the enthalpy are related to the Gibbs free energy such that

$$s = -\left(\frac{\partial g}{\partial T}\right)_P$$

$$v = \left(\frac{\partial g}{\partial P}\right)_T \qquad 7.4$$

$$h = \frac{\partial(g/T)}{\partial(1/T)}$$

respectively.

Appreciable changes in the specific heat,

$$c_p = \left(\frac{\partial h}{\partial T}\right)_P = T\left(\frac{\partial h}{\partial T}\right)_P \qquad 7.5$$

and the thermal expansion,

$$\alpha = \frac{1}{V}\left(\frac{\partial V}{\partial T}\right)_P \qquad 7.6$$

accompany the transition of the material into the vitreous state, as illustrated in Figs. 7.3b and 7.3d. The transitions exhibited by the enthalpy and thermal expansion are smooth, not abrupt, as would be encountered in a true second order phase transition. Hysteresis effects also accompany these transitions during heating and cooling cycles. The glass transition is not well defined and occurs over a certain temperature range. The magnitude of the range depends on the system. The question of whether T_g is a true second order transition is not completely resolved.

In the temperature range $T_g < T < T_m$, the liquid resides in a super cooled state. If the sample exhibits a tendency to form a glass within a range of cooling rates below T_m, then cooling at a slower rate within this range provides the structural entities more time to sample configurational space and the glass transition occurs at a lower T; this point is illustrated in Fig. 7.3a The fact that T_g depends on the cooling rate reflects the kinetic dimension associated with the glass transition. The implications of this rate dependence are not as serious as one might initially imagine because T_g is generally a weak function of cooling rate, nearly logarithmic. In most laboratory experiments the cooling rates may vary from 0.2 to 200 K/min. Often a cooling/heating protocol is established, particularly with the use of certain techniques such as differential scanning calorimetry (DSC). With this commonly used technique, a cooling rate of 10 degrees Kelvin per minute is often used.

In addition to the signatures (changes in the temperature dependencies of specific volume, thermal expansion etc.), that denote the transition, other methods have been used to determine T_g. For example, as discussed later in Section 7.4, T_g is defined at the temperature where the viscosity of the glass forming liquid is 10^{13} poise (Moynihan 1995, Angell et al 2000). These varying methods together with the breadth of the transition can lead to ambiguities in the actual values quoted in the literature for some systems. While T_g is not a true material constant (because of its rate dependence), it is often tabulated in references for different materials because it is an important engineering parameter.

7.3.2 Temperature Dependence of the Viscosity: Adam-Gibbs Model

In Chapter 6, the Vogel-Tammann-Fulcher equation was derived on the basis of free volume theory. The basic notion is that the viscosity, or equivalently the longest relaxation time, decreases rapidly with decreasing temperature due to a reduction of the available free volume. Adam and Gibbs subsequently suggested that the glass transition, and the associated temperature dependence of the viscosity of a glass forming melt, could alternatively be described in terms of a dynamic cooperative process (Adam and Gibbs 1965). Specifically, the decrease of the relaxation time with decreasing temperature is associated with the loss of configurations available to the system as the temperature is reduced.

In their model, spatial domains (or clusters) in the sample, each composed of n molecules, are considered. Each domain is assumed to relax independently of the other. With decreasing temperature the liquid becomes more dense and the barrier to rearrangement within these clusters increases. In effect, the displacements of a small number of particles influence an increasing number of surrounding particles within the cluster with decreasing temperature. This barrier for rearrangement should scale as the size of the cluster, so $\Delta E = n\Delta e$, where $\Delta \varepsilon$ is the energy barrier per particle. Since the relaxation time of the system should depend exponentially on the activation barrier,

$$\tau = \tau_0 e^{(n\Delta\varepsilon S_c^*/kTS_c)} \qquad 7.7$$

where τ_0 is an intrinsic relaxation time associated with the dynamics of the system. S_c^* is the entropy associated with the smallest relaxing region and S_c is the configurational entropy of the system; $S_c = k \log \Omega_c$, where Ω_c is the number of configurations available to the system. Because the S_c is proportional to the logarithm of the number of available configurations, the reduction in the viscosity—or, equivalently, the longest relaxation time—with decreasing temperature is due to the reduction of the number of configurations available to the system. The equation for the relaxation time is typically written as

$$\tau = \tau_0 e^{C/TS_c} \qquad 7.8$$

where $C = n\Delta\varepsilon S_c^*/k$. This form of the equation will be used throughout the remainder of this chapter.

Since the entropy of the liquid exceeds that of the crystal, then, as the temperature decreases (at constant pressure), the difference between the entropy of the liquid and that of the crystal (the excess entropy), S_{ex}, diminishes. At a sufficiently low temperature the excess entropy vanishes. The sketch in Fig. 7.4 illustrates a typical temperature dependence of the reduction of the excess entropy in the super cooled state.

T_K is believed to be a limiting value of $T_g (T_K < T_g)$. Below T_K, the entropy of the liquid would be lower than that of the crystal (Richert and Angell, 1998, Debenedetti, 1996, Tanaka, 2003). It was originally believed that this would not be possible because the third law of thermodynamics would be violated. Some authors have referred to this as the "entropy crisis" or Kauzmann's paradox. However, Stillinger, Debenedetti, and Truskett (2001)

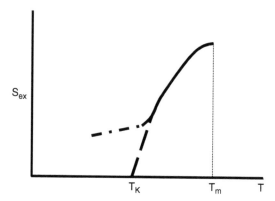

FIG. 7.4
The temperature dependence of the excess entropy is shown here for the supercooled liquid. The excess entropy is the difference between the entropy between the liquid and that of the crystal.

showed that in fact there are systems in which the entropy of the liquid falls below that of the crystal for $T < T_K$ and this does not violate the third law. The critical requirement is that the entropy of the liquid and of the crystal should be zero at absolute zero. The interested reader should consult the original references on this topic for further details.

7.4   Connection between Kinetic and Thermodynamic Fragility

Is there a possible connection between T_∞ and T_K? To answer this question, we begin by examining the excess entropy. If it is assumed that the configurational entropy accounts entirely for this difference then one can write down the configurational entropy change that a material experiences as it goes through T_g in terms of the change of the configurational heat capacity, (Angell, 1990, Hodge, 1996)

$$S_c = \int_{T_K}^{T_g} \frac{\Delta C_p(T)}{T'} dT' \qquad\qquad 7.9$$

where $\Delta C_p = C_p^{melt} - C_p^{cryst}$. This equation implicitly assumes that there is no difference between the vibrational entropies in the glass and the crystal phases. Second, if we assume that

$$\Delta C_p(T) = \frac{T_g}{T} \Delta C_p(T_g), \qquad\qquad 7.10$$

then

$$S_c = \frac{\Delta C_p(T_g)}{T_K}(T_g/T_K - 1) \qquad\qquad 7.11$$

Third, if it is assumed that $T_K = T_\infty$, then, together with Eq. 7.8, an equivalence between the VTF and the Adam Gibbs equation is established. In fact, experiments in some systems illustrate this nicely (Richert and Angell 2003).

It turns out that the equivalence between the T_K and T_∞ is often taken as proof of the connection between the kinetic and thermodynamic aspects of the glass transition. However the equivalence only works in some systems, though T_K and T_∞ are generally not far apart in magnitude in materials where the violation is noted. The excess entropy, in reality, has a vibrational component, $S_{ex} = S_c + S_{vib}$, and the contribution of the vibrational component of the glass is in fact not equal to that of the crystal. This is believed to be the source of the violation (Tanaka 1999; Johari 2000). Both VTF and Adam-Gibbs equations have been valuable with regard to our understanding of glass forming liquids. Despite considerable progress during the last decade, this continues to be a very active topic of research.

7.5   "Strong" versus "Fragile" Network Glass Melts, a Structural Connection

A structural connection between the strong and fragile temperature dependent behavior of the viscosities of glass melts is now discussed. The log η vs. $1/T$ dependence of SiO_2 is strong ($\log \eta \propto 1/T$). With the addition of alkali oxides the temperature dependence of η becomes fragile due to the depolymerization (reduction in network connectivity) of the covalent network. The fragility of glass melts is important because it influences the processing of melts and fibers and their final properties. Figure 7.5 shows the effect of adding alkali oxides to network formers on the temperature dependence of the viscosity.

In this Fig., the viscosity is plotted as function of T_g/T, where T_g is identified as the temperature at which the viscosity is $\eta = 10^{13}$ *Poise*. This is the temperature regime wherein the time scales of the enthalpy relaxations begin to deviate appreciably from the shear relaxation time (~100 sec) associated with the viscosity. While $\eta = 10^{13} P$ is meant to be an operational definition of T_g it does not coincide with the calorimetric glass transition temperature for some systems, most notably polymers. Recall that for polymers the longest relaxation time, which controls the viscosity, is a function of molecular weight ~$M^{3.4}$. For the same polymer samples with widely varying chain lengths, and associated differences in viscosites of many orders of magnitude, the T_g determined by differential scanning calorimetry (DSC) remains the same. To this

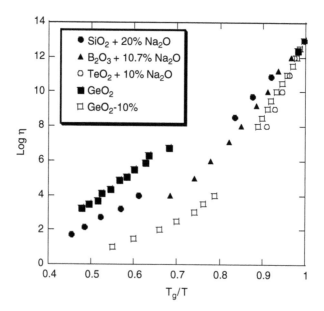

FIG. 7.5

Plot of viscosity (Poise) versus temperature illustrating the effect of alkali oxides on the $\eta - T$ behavior (*Handbook of Glass Data*, Elsevier, Oxford, New York, 1983). Physical Science Data 15, Silica Glass and Binary Silicate Glasses O.V. Mazurin et al, Single Component and Binary Non-Silicate Oxide Glasses O.V. Mazurin et al (1985).

end, it is important to note that for polymers it is the local segmental dynamics that are relevant to the glass transition and not the longest relaxation time. The notion of the fragility is explored in further detail next.

7.5.1 Influence of Alkali Content on Heat Capacity and Activation Energy for Flow

The increasing fragility with increasing depolymerization was explained by Angell. Using Eqs. 7.1, 7.4, 7.9, and 7.11, it follows that (Problem 3)

$$B \propto \frac{1}{(T_g/T_\infty)\Delta c_p(T_g)} \qquad 7.12$$

This equation indicates that changes in $\Delta c_p(T_g)$ and T_g/T_∞ are connected to changes of the fragility of the glass melt.

The data in Fig. 7.6a indicate that with increasing depolymerization (increasing modifier content), at least for the tellurites and silicates, $\Delta c_p(T_g)$ increases monotonically. Moreover, the data in Fig. 7.6b indicates that the trends in the fragility index with x are similar to those exhibited by the dependence of $\Delta c_p(T_g)$ on x. This strongly suggests that the fragility increases

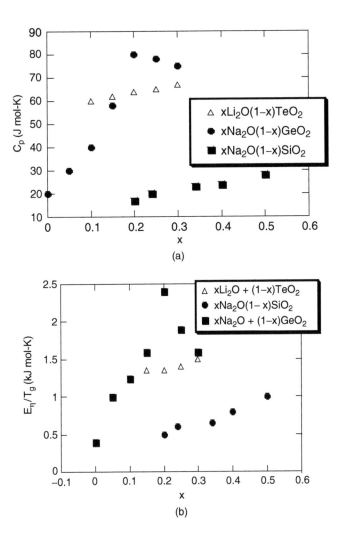

FIG. 7.6
a) Changes in the heat capacity at T_g, $\Delta c_p(T_g)$ are shown here for three different types of glasses.
b) E_η/T_g, which is proportional to the fragility index, is plotted here for the glasses whose heat capacity changes at T_g are plotted in Fig. 7.5a. (Data in these figures were extracted from S.K Lee, M. Tatsumisago and T. Minami, 1993, 1994, 1995).

with increasing modifier content, consistent with depolymerization of the network (Angell).

The trends in the data representing the germinates exhibit maxima in Fig. 7.6(a) and 7.6(b). These maxima are the result of what is commonly referred to as the germinate anomaly (wherein germanium oxide undergoes a coordination change from 4 to 6 in the absence of NBO formation). The increasing coordination results in an increase in the covalent network connectivity. Therefore the change in fragility in single alkali glasses is associated with changes in the network connectivity. In the vicinity of T_g, c^{liq}/c^{glass} should increase with m (see Problem 4).

The overall utility of the foregoing analysis to capture the behavior of other glass forming systems (polymers, small molecule liquids) remains uncertain. A study of a wide range of systems—small molecule, polymers, and other inorganic glass formers—by Huang and McKenna show no correlation between Δc_p and m. In summary while the situation is clear for some classes of materials, a complete picture that describes a wide class of materials is yet to emerge.

7.5.2 The Viscosity of Mixed Alkali Glass Melts

In mixed alkali phosphates and silicates containing sodium and lithium ions, changes in fragility are observed by changing the ratio of sodium to lithium ions while keeping the total alkali ion concentration constant. The fragility exhibits a minimum when the number of dissimilar cations is comparable. Figure 7.7 shows the temperature dependencies of various

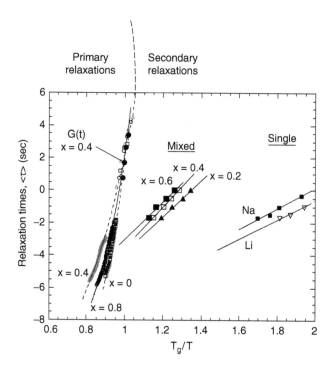

FIG. 7.7

This is a relaxation plot for a mixed alkali metaphosphate glass, containing a total of 50% Na_2O and Li_2O and 50% P_2O_5. The primary relaxations associated with the viscosity are shown to the left for different mole fractions of Na_2O in the sample. The data identified with $G(t)$ were obtained from stress relaxation measurements. The so-called secondary relaxations, associated with the hopping of alkali ions are shown to the right of the diagram (Figure appeared in Green et al 1999).

relaxation processes that occur in a mixed alkali metaphosphate glass, containing 50 mol% (Na_2O and Li_2O) and 50% P_2O_5. The primary relaxation times determined from viscosity data of samples with Na_2O weight fractions of $x = 0$, $x = 0.4$ and $x = 0.8$, as shown ($x_{Na_2O} + x_{Li_2O} = 1$ and $x = x_{Na_2O}$). It is noteworthy that the fragilities are larger at $x = 0$ and $x = 0.8$ than at $x = 0.4$. These changes in melt fragility are evidently associated with changes in the configurational heat capacity change at T_g and reconcilable (see Problems 4 and 15) with m (Putz and Green 2004; Green et al. 1998).

The other data points in Fig. 7.7, at lower T, with the lines drawn through them, represent secondary relaxation processes associated with ion hopping below the glass transition (Green et al 1998, 1994, Bucheneau, 2001, Roling et al, 1998, 1998). A broader discussion of the analysis of primary and secondary relaxation processes will be addressed in Section 7.7. In the meantime the discussion of the effect of alkali oxides on the properties of network glasses continues.

7.5.3   Effect of Alkali Composition on T_g

Like $\Delta C_p(T_g)$, commensurate changes of the glass transition temperature also occur with the increasing in NBO content in single alkali glasses, as illustrated in Fig. 7.8 for alkali germinates and alkali tellurites. The maximum in the GeO_2 data is associated with the germinate anomaly. The T_g for alkali tellurites and alkali phosphates tend to be much lower than alkali silicates and alkali germinates. The reasons are unclear.

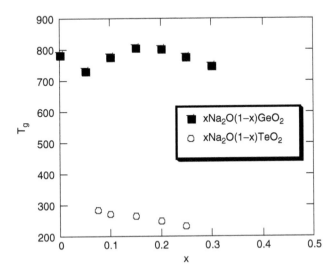

FIG. 7.8
The compositional dependence of T_g (Kelvin) is shown here for the alkali germinates and alkali tellurites. (T_g data from tellurites is taken from Zhu et al, 2003.) The decrease of T_g accompanies the increase of the NBO content.

7.5.4   The Energy "Landscape" Approach

It was suggested by Goldstein in 1969 that flow processes in glass melts should be connected to a so-called potential energy "landscape." The potential energy landscape picture was introduced to provide a rationale for the configurational entropy scenario for understanding the glass transition. The idea is that if the potential energy of an n-particle system is known as a function of the $3n$-particle coordinates, then the structure (configurational integral from Chapter 2) and the dynamics could in principle be calculated. The landscape of a glass, in contrast to a crystal, is characterized by distribution of potential energy barrier heights and minima. The potential energy barriers are large compared to the available thermal energy and the probability of surmounting a barrier is, of course, determined by the Boltzmann factor. Hence the probability that barriers are surmounted at low temperatures is low and at higher T the process would occur with higher probability.

The connection to the Adam-Gibbs model is that Ω_c (Section 7.3.2) is interpreted as the number of basins in the landscape available to the system. The landscape for fragile glasses is characterized by a large number of minima with a distribution of barrier heights, whereas for strong glasses, there exist fewer (yet deeper) extrema. The existence of fewer barriers in systems that exhibit strong behavior evidently reflects the smaller number configurations (lower configurational entropy) available to them as they vitrify.

7.6   Relaxation Functions

The upcoming discussions in this chapter will involve the time-dependent response of glass formers $(T > T_g)$ to external influences. So the goal of this section is to introduce procedures that will enable the analysis of relaxation processes. We begin with a description of linear response theory. Generally, one may consider the imposition of a perturbation $\Im_1$ at time t_1 for a short duration δt, where δt is short compared to the relaxation response of the material. The perturbation could be due to the imposition of a mechanical stress, an electric displacement, or a magnetic field, for example. The response, $\zeta_1(t)$, of the material would be proportional to $\phi_1 \delta t$, provided the perturbation in sufficiently small. In the case of an applied mechanical force, $\phi_1 \delta t$ would represent the displacement. The specifics of the response at time $t > t_1$ depends on the material and on the measurable property of interest and would be $\xi_1(t) = \Im_1(t)\phi(t - t_1)\delta t$. Note the similarity of this discussion with the discussion in Section 6.4.5 regarding the mechanical response. It is left as an exercise to show that if the superposition principle is applied to a large number of such perturbations, then the response is reasonably well approximated by

$$\xi(t) = \int_{-\infty}^{\infty} \Im(t')\phi(t-t')dt' \qquad 7.13$$

Often the perturbations are of an oscillatory nature and Fourier Transform methods are relied on to perform the analysis. Equation 7.13 may be rewritten (Problem 9)

$$\xi(\omega) = \phi^*(\omega)\Im(\omega) \qquad 7.14$$

$\phi^*(\omega)$ is typically identified as the subsusceptibility of the system; it is a material parameter. With regard to the discussion in Section 6.4, $\phi^*(\omega)$ would be the compliance and with regard to a dielectric measurement, $\phi^*(\omega)$ would be the susceptibility. $\phi^*(\omega)$ has real and imaginary parts, $\phi'(\omega)$ and $\phi''(\omega)$, respectively, largely because the response of the system is not instantaneous. The perturbation and the response are out of phase, with a phase difference δ,

$$\phi^*(\omega) = \phi_0 + \phi'(\omega) - i\phi''(\omega) \qquad 7.15$$

By definition,

$$\phi^*(\omega) = \int \phi(t)e^{i\omega t}dt = \phi_0 + \phi'(\omega) + \phi''(\omega) \qquad 7.16$$

from which it follows that

$$\phi'(\omega) + \phi_0 = \int_{-\infty}^{\infty} \phi(t)\cos\omega t dt \qquad 7.17$$

and

$$\phi''(\omega) = \int_{-\infty}^{\infty} \phi(t)\sin\omega t dt \qquad 7.18$$

In addition, $\text{Tan}(\delta) = \phi''(\omega)/\phi'(\omega)$. Recall that for a purely elastic or instantaneous process, $\delta = 0$ and $\phi^*(\omega) = \phi(0) = \phi_0$ is independent of frequency and $\phi_0 = \xi_0/\Im_0$.

With regard to the time-dependent response, it is often convenient to analyze the response in terms of the Kohlrausch-Williams-Watts (KWW) function

$$\phi(t) = \phi_0 \exp[-(t/\tau)^\beta] \qquad 7.19$$

where $0 < \beta < 1$ describes the deviation of the response from exponential dependence (Williams and Watts 1970). Specifically, the value of β provides information regarding the distribution of relaxation times, as we saw earlier. An alternate view point is that, a value of $\beta < 1$ also reflects evidence of cooperativity in the dynamics (see for example Ngai 1996, Ngai and Martin 1996).

If the external influence is a constant mechanical strain, then the time dependence of the response (stress relaxation) of the material throughout the duration of the strain is described in terms of $\phi(t) = \frac{G(t)}{G_0} = e^{-(t/\tau_p)^\beta}$, where G_0 is the high frequency modulus and τ_p can be identified as a characteristic relaxation time associated with the process. $G(t)$, as we saw in Chapter 6 on polymer dynamics, is related to the viscosity. In the case of viscoelastic materials, a temperature independent β implies a thermorheologically simple response. In this case the only effect of temperature is to increase the rate of the relaxation processes; the mechanism of transport remains the same.

A final example involves an experiment in which the temperature of the glass is rapidly changed from an initial temperature, T_1, at which a sample has equilibrated, to a new temperature, T_2. This is a *structural relaxation* experiment (Moynehan 1995). The relaxation time associated with equilibration of properties such as the enthalpy, h, and the specific volume, v, at the new temperature exhibits a highly nonlinear dependence on time (Moynehan 1995, Scheer 1990, Narananaaswamy 1971, 1988). The time-scale depends on the final temperature, the initial temperature, and fictive temperature T_f, such that

$$\phi(t) = \frac{v - v_{eqbm}(T_2)}{v_1 - v_{eqbm}(T_2)} = \frac{h - h_{eqbm}(T_2)}{h_1 - h_{eqbm}(T_2)} = \frac{T_f - T_2}{T_1 - T_2} = \exp[-(t/\tau_{str})^\beta] \qquad 7.20$$

The fictive temperature provides a measure of the contribution of structural relaxations to properties such as the enthalpy and the specific volume of the glass forming material under nonequilibrium conditions after the temperature is suddenly changed from T_1 to T_2.

7.7 Mechanical Relaxations

Thus far we have discussed various primary and secondary relaxation processes that occur in network glasses and melts. Ion hopping processes and associated local network relaxations occur below the glass transition temperature and are of technological importance. Ionic hopping processes are important for various electrochemical applications, including batteries and various smart cards and sensors. Taking advantage of the information in Section 7.6 above, the analysis of primary and secondary relaxation processes due to mechanical deformations is now discussed.

7.7.1 Primary Relaxations

A mixed alkali metaphosphate glass is now considered, wherein both the stress relaxation modulus and the real and imaginary moduli are analyzed.

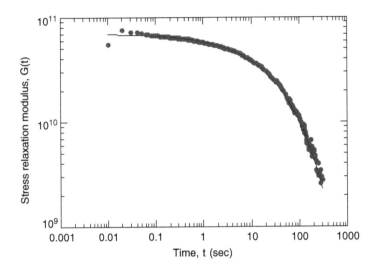

FIG. 7.9
The stress relaxation modulus, $G(t)$ is plotted here for a mixed alkali metaphosphate glass.
These data were taken at a temperature of 250°C.
(Figure originally appeared in Green et al 1999). Based on these data, $\beta = 0.5 \pm 0.1$, $t = 27 \pm 2$ secs
and $G_0 = 7 \times 10^{10}$ dynes/cm².

The composition of the material is: 20 mol% Li_2O, 30 mol% Na_2O, and 50%P_2O_5 ($x = 0.4$). Plotted in Fig. 7.9 is the stress relaxation modulus for this material at a temperature of 250°C. The line drawn through the data was computed using equation 7.19 with values of $\beta = 0.5 \pm 0.01$, $G_0 = 7.0 \times 10^{10}$ dyn/cm², and $t = 27 \pm 2$ seconds. This value of β indicates that the distribution is rather broad. A value of $\beta = 1$ corresponds to a breadth (full width at half maximum, FWHM) of 1.144 decades, see Table 7.2. Based on the information in this table, the distribution is evidently approximately 2.2 decades broad.

In order to get an appreciation for the breadth of this distribution, oscillatory shear (rheological) experiments were performed on the same glass and these data, $G'(\omega)$ and $G''(\omega)$, are plotted in Fig. 7.10. The same values of

TABLE 7.2

Relation between β and the FWHM.

β	FWHM (decades)
.8	1.4
.6	1.8
.5	2.2
.4	2.75
.3	3.6

(Data extracted from Sidebottom et al (1995).)

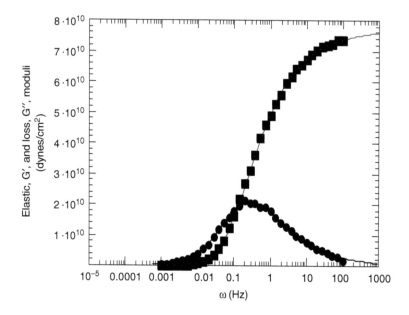

FIG. 7.10

The frequency dependent moduli are shown here for the same sample whose $G(t)$ is shown in Fig. 7.9. (Figure originally appeared in Green et al 1999).

β and G_0 were used to fit these data using Eqs. 7.15 to 7.19 (bearing in mind that $G^*(\omega) = G'(\omega) + iG''(\omega)$).

The data in Figs. 7.9 and 7.10 experimentally illustrate the connection between the responses in the time and frequency domains.

7.7.2   Secondary Mechanical Relaxations ($T < T_g$)

In this section, relaxations largely associated with cation dynamics are examined. Mechanical relaxation measurements of $\tan(\delta)$ are effective means to identify local mechanical relaxation processes in a variety of materials. These measurements are routinely used to examine short-range molecular motions, such as rotations of chemical side groups, in polymers. As discussed earlier, $\tan(\delta)$ is the ratio of the energy dissipated per cycle due to a periodic external perturbation to the energy stored. In the literature, $\tan(\delta) = 1/Q = D$ and D is called the dissipation function, which increases with increasing internal friction. It is noteworthy that internal friction measurements are also used to study diffusion of small interstitial ions in the lattice of BCC crystals (Flynn, 1972). In these experiments, the sample is perturbed at a given frequency and the temperature of the sample varied until a peak appears. The characteristic frequency is related to the hopping rate of the interstitial atom. With knowledge of the jump distances and coordination number the diffusion coefficient may be calculated.

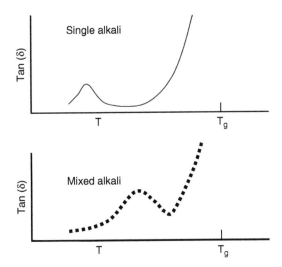

FIG. 7.11
A typical schematic of the temperature dependence of tan(δ) for a) single and b) mixed alkali glasses is shown. In a typical experiment, the sample is perturbed at a fixed frequency while the temperature is varied. The locations of the maxima identify characteristic frequencies that reflect the dynamical processes of interest.

Internal friction measurements have been performed on single and mixed alkali glasses (Day 1976, van Ass and Sievels 1974; Green et al 1998; 1994; Rollings and Ingram 1998; Buchenau 2001). Measurements of tan(δ) of phosphates and of silicates reveal distinct differences between the behavior of single and mixed alkali glasses. The first series of experiments were performed in thin wire samples which were oscillated at a given frequency while the temperature was changed. Today dynamic mechanical analyzers capable of performing measurements over a wide range of frequencies and temperatures are utilized. (Green et al 1998; 1994; Rollings and Ingram 1998). In single alkali glasses a low temperature peak appears in the spectrum, representing relaxations that accommodate the motions of the single type of cations. The relaxations are believed to be local network relaxations and ionic hopping relaxations. A schematic of typical data is shown in Fig. 7.11(a).

In the mixed alkali analogs, a high temperature maximum appears Fig. 7.11b, even at low concentrations of the second alkali cation. This suggests the occurrence of larger reconfigurations in the system to accommodate the dynamics of dissimilar cations. Theory predicts that for single alkali glasses there should be one internal friction peak and for mixed (2 alkali ions) alkali glasses there should be three, one for each single cation and a third representing interactions (see appendix A). Experimentally, in mixed alkali glasses the single alkali peaks are severely diminished and in most cases masked by the large mixed alkali peak.

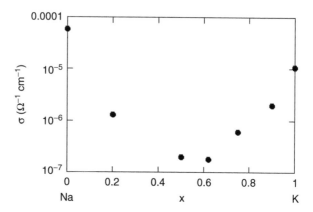

FIG. 7.12
The ionic conductivity is shown here for a mixed alkali sodium potassium silicate glass. The ionic conductivity in shown to exhibit a minimum when the fraction of dissimilar cations is comparable. These data reveal that the effect increases with decreasing temperature. (Data extracted from G.N. Greaves and K.L. Ngai, (1995)).

In internal friction experiments, the peak position is also known to be sensitive to the disparity in the size of the cations. The peak size increases and the peak position shifts to higher T with increasing disparity of the dissimilar cations (Van Ass and Stevels 1974; Day 1976). The data representing the secondary relaxations in Fig. 7.7, reveal that with regard to the mixed alkali dynamics, the relaxation times are longer near $x \sim 0.5$ than at $x = 0.2$. These data, moreover, reveal that the conductivity relaxation rates of mixed alkali ions are much lower than for the single alkali counterparts.

Additional ionic conductivity measurements of a mixed alkali glass are shown in Fig. 7.12. Typically in these experiments a series of glasses, each containing two dissimilar cations, A^+ and B^+, are prepared; the total number of cations in each glass is fixed but the relative number is varied. The ionic conductivity exhibits a deep minimum in the middle of the composition regime where the number of A-type and B-type cations is comparable. This so-called mixed alkali effect is a well-documented phenomenon and is still not completely resolved. The discussion in Section 7.11 will provide insight into the origins of this phenomenon.

Dynamic phenomena associated with the mixed alkali glasses are known as mixed alkali effects (MAE). The MAE has an interesting history and was initially associated with the thermometer effect, discovered over a century ago! At the time, thermometers were made with SiO_2 as the glass former. The glass composition also included two dissimilar alkali oxides, Na_2O and Li_2O. If, during calibration, the thermometer was placed in boiling water and subsequently placed in ice water, the temperature of the thermometer would read $-0.5°C$, not $0°C$. The only way that this problem could be circumvented would be to employ a single type of alkali oxide to modify the structure.

7.8 Phenomenology of Secondary Relaxations: Ionic Conductivity

Unlike metals and other crystalline materials, a mechanism describing the transport of ionic species within the disordered glass has been elusive. This has had far-reaching implications on the desire to provide a fully satisfactory explanation for MAE. This section discusses ionic conductivity and diffusion in network glasses. The simple random walk analysis introduced in earlier chapters to describe diffusion in crystals is not effective at describing the transport process in network glasses. The reasons are, in part, related to: 1) ion-ion correlations induced by Coulombic interactions, and 2) M—O bond distances are characterized by a distribution of jump distances due to the disorder.

7.9 Ionic Conductivity and Diffusion

The total current density is in general determined by the bound and the free charges. In typical experiments a disc-shaped sample (cross-sectional area A and thickness l_0) of the material is prepared and metal electrodes are deposited on its two surfaces, thereby creating a capacitor. An a.c. bridge is often used to measure the conductance and the capacitance as a function of frequency, $G(\omega)$ and $C(\omega)$, respectively. Ionic transport can be expressed in terms of a complex conductivity

$$\sigma^*(\omega) = \sigma'(\omega) + i\sigma''(\omega) \qquad 7.21$$

where the real and imaginary parts are $\sigma' = G(l_0/A)$ and σ'', respectively (Scher and Lax 1973, Dietrich et al. 2002). Alternatively, particularly in experiments that examine dipolar relaxations, the complex permittivity, $\varepsilon^*(\omega)$, is measured

$$\varepsilon^*(\omega) = \varepsilon'(\omega) - i\varepsilon''(\omega) \qquad 7.22$$

($\varepsilon' = CL/A\varepsilon_0$). The following two points are noted. The cation and the NBO, in principle, constitute a dipole (transient), so in this regard the use of the complex dielectric constant may be rationalized. Second, the relative permittivity, ε_∞ (dielectric constant) of the bound charges is independent of frequency, i.e., the bound charge response is instantaneous (Sidebottom et al, 1997, 1995).

$\varepsilon^*(\omega)$ are $\sigma^*(\omega)$ are related such that

$$\sigma^*(\omega) - \sigma(0) = i\omega\{\varepsilon^*(\omega) - \varepsilon_\infty\}\varepsilon_0 \qquad 7.23$$

The conductivity is determined by the auto-correlation function of the current $J(t)$, due to moving charges (Scher and Lax 1973),

$$\sigma^*(\omega) = \frac{V}{3kT} \lim_{\delta \to +0} \int_0^{\infty} \langle \vec{J}(t)\vec{J}(0) \rangle e^{i\omega t - \delta t} dt \qquad 7.24$$

where

$$\langle \vec{J}(t)\vec{J}(0) \rangle = \frac{Nq^2}{V^2} \left\{ \langle \vec{v}(t)\vec{v}(0) \rangle + \frac{\sum_{i \neq j}^{N} \langle \vec{v}_i(t)\vec{v}_j(0) \rangle}{N} \right\} \qquad 7.25$$

Note that the cross-correlation terms for the velocity are represented by the second term. The diffusion coefficient can be identified with this expression because, as discussed in Chapter 2, it can be written in terms of the auto-correlation function of the velocities. The mean-square displacement, of course, can be obtained from the velocity autocorrelation functions. To show the connection explicitly, we consider writing down a general expression for the frequency dependence of the diffusion coefficient (Dietrich 2002, Scher and Lax 1973). Specifically, we write down the Fourier Transform of the mean square displacement,

$$D^*(\omega) = -\frac{\omega^2}{6} \lim_{\delta \to +0} \int_0^{\infty} \langle [\vec{r}(t) - \vec{r}(0)]^2 \rangle e^{i\omega t - \delta t} dt \qquad 7.26$$

(we need the limits associated with δ in the above equation otherwise the integrand is not bounded). As a reality check, please note that if the particles undergo a simple random hopping process, then

$$\langle [r(t) - r(0)]^2 \rangle = \langle r^2(t) \rangle = 6Dt \qquad 7.27$$

If this result is substituted into the equation for $D^*(\omega)$, it is evident that indeed in the limit where δ approaches zero, and the frequency approaches zero, we recover the relation that (see Problem 10)

$$D^*(\omega) = D \qquad 7.28$$

In the meantime we return to the expression for the autocorrelation of the velocities and recognize that from eqn. 2.90

$$D = \frac{1}{3} \lim_{\delta \to 0} \int_0^{\infty} e^{-\delta t} \langle v(0)v(t) \rangle dt \qquad 7.29$$

With this, we can consider two cases regarding the frequency dependence of the conductivity

7.9.1 Case I

If we ignore the cross-correlation terms in equation 7.25 and together with eqns 7.27 and 7.29 the complex conductivity becomes,

$$\sigma^*(\omega) = \frac{-\omega^2 Nq^2}{6VkT} \lim_{\delta \to +0} \int_0^\infty \langle r^2(t) \rangle e^{i\omega t - \delta t} dt \qquad 7.30$$

This may, in turn, be written as,

$$\sigma^*(\omega) = \frac{\rho q^2}{kT} D^*(\omega) \qquad 7.31$$

where ρ is the density of charge carriers and q is the charge. As another reality check, please note that in the case of a simple random walk, this equation reverts to the Nernst-Einstein equation discussed in Chapters 2 and 4.

7.9.2 Case II

If the cross correlation terms are not excluded then the conductivity becomes

$$\sigma^*(\omega) = \frac{\rho q^2}{kT} \frac{D^*(\omega)}{H_R^*(\omega)} \qquad 7.32$$

where $H_R^*(\omega)$ is the Haven ratio and ρ is the density of charge carriers. The Haven ratio is a measure of the effect of ion-ion correlations on the diffusivity of the ions. In the case where all cations contribute to diffusion the Haven ratio is unity.

One might anticipate from the foregoing equation that independent measurements of diffusion and conductivity reveal that the values of diffusion determined from conductivity D_σ or from direct diffusion experiments D can be reconciled with the introduction of the Haven ratio $D = H_R D_\sigma$ (note that correlation factors are used to describe the atomic diffusion process in metals (Chapter 3) but their origins are fundamentally different, as should be clear context).

7.9.3 Comments Regarding Ionic Conductivity in Network Glasses

Two important facts about the conductivity in oxide glasses are clear (see for example Ngai 1996). First it is observed that below T_g, the $\sigma_{dc}T$ exhibits an Arrhenius dependence on temperature,

$$\sigma_{dc}T \propto e^{-E^{dc}/kT} \qquad 7.33$$

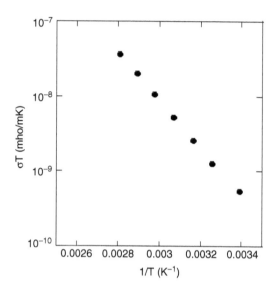

FIG. 7.13

The ratio of the dc conductivity to inverse temperature, σT is plotted as a function of $1/T$ for a lithium metaphosphate $LiPO_3$ glass ($E = 0.66$ eV). Data extracted from Ngai 1996.

A plot revealing typical temperature dependencies of the conductivity is shown for a lithium metaphosphate glass in Fig. 7.13. The diffusion coefficient is also Arrhenius.

Second, both the prefactor and the activation energy in equation 7.33 are functions of modifier content, x. The ionic conductivity increases with increasing alkali modifier content because the activation energy decreases with increasing modifier content. The cation mole fraction dependence of the activation energies associated with transport in ionic network glasses shown in Fig. 7.14 for a lithium metaphosphate glass ($LiPO_3$) (i.e., the glass with the specific composition 50% P_2O_5 and 50% Li_2O) illustrate this point (Ngai and Martin 1989).

One of the most well-known attributes of the ionic conductivity is that it exhibits a universal frequency dependence, wherein at low frequencies the conductivity is constant and at high frequencies it exhibits a power-law dependence (Ngai, Roling et al 1998; Sidebottom et al 2000, Sidebottom 1999). The real part of the conductivity is often written as (Joncher 1983, 1996)

$$\sigma'(\omega) = \sigma(0)[1 + (\omega/\omega_0)^n] \qquad 7.34$$

where $\sigma(0) \equiv \sigma_{dc}$. Joncher first recognized that the exponent, n, is a universal exponent, possessing values varying between $n = 0.6$ and 1 for all ionic materials. Since then, the value appears to be 0.6 for these materials.

Figure 7.15 shows the general shapes of the conductivity and the permittivity as a function of frequency which is typical for these systems. The dc

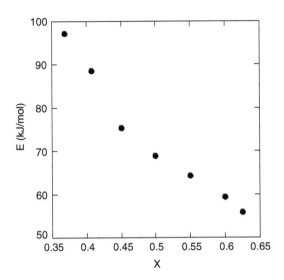

FIG. 7.14
The activation energy for electrical conductivity decreases as the amount of modifier increases in this lithium phosphate system $x\text{Li}_2\text{O} + (1 - x)\text{P}_2\text{O}_5$. The composition $x = 0.5$ corresponds to lithium metaphosphate (Ngai and Martin 1989).

conductivity is determined by the long-range ionic diffusion where the mean square displacement is proportional to t. At higher frequencies, the conductivity exhibits power-law behavior and this is due to the increasing influence of correlations associated with forward and backward hops. In this regime the mean square displacement scales approximately as $\langle r^2(t) \rangle \propto t^{1-n}$. The

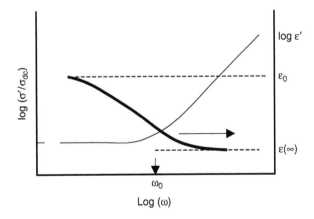

FIG. 7.15
Typical frequency dependencies of the conductivity and relative permittivities of network glasses are illustrated here.

backward correlated hops are possibly due to the Coulombic interactions associated with the NBO. However this remains an open issue. In a more formal sense one can define a length scale, ξ, associated with a cross over from a traditional random walk to the correlated backward and forward hoping process (B. Roling et al 1999; Sidebottom et al 1999; Funke 1993).

$$\langle r^2(t) \rangle \propto \begin{cases} t, & r > \xi \\ t^{1-n}, & r < \xi \end{cases} \qquad 7.35$$

In the above equation n is approximately 1/3. These dependencies represent limiting values of the time dependencies. The dependence of $\langle r^2(t) \rangle$ on t has been examined in the $x\mathrm{Na_2O}(1 - x)\mathrm{GeO_2}$ system (Roling et al 1998). At short times, $\langle r^2(t) \rangle \propto t^{2/3}$, and at longer times, $\langle r^2(t) \rangle \propto t$. This length scale is a function of the alkali ion fraction, becoming smaller with increasing x.

There is now growing evidence that a master equation might be written to describe the complete temperature and compositional dependence of the d.c. conductivity. The activation energy can be written as

$$E = E_0 \ln(x_0/x) \qquad 7.36$$

with

$$\sigma_{dc}T = A_0 e^{-(E_0 \ln(x_0/x))/kT} \qquad 7.37$$

The importance of length scales in this problem is underscored with regard to identifying a universal scaling picture for the conductivity. The master curve can be written as,

$$\frac{\sigma}{\sigma_0} = F\left(\frac{f}{f_0}\right) \qquad 7.38$$

where f is the frequency and

$$F(x) \approx 1 = x^n \qquad 7.39$$

If $f_0 = \frac{\sigma_0 T}{x}$ is used to scale the data, the scaling works over a narrow compositional range. However, the use of the relation, as suggested by Sidebottom,

$$\frac{\sigma}{\sigma_0} = F\left(\frac{f\varepsilon_0 \Delta\varepsilon}{\sigma_0}\right) \qquad 7.40$$

where $\Delta\varepsilon = e_0 - e_\infty$, is proportional to the d^2, where d is the average separation between NBOs may be more appropriate (Sidebottom 1999 and Roling et al. 1998).

7.9.3.1   The Electrical Modulus Representation

Some time ago (see, for example, Macedo et al. 1972, Ngai 1996) it was shown that the complex permittivity, $\varepsilon^*(\omega)$, could be expressed in terms of an electrical modulus,

$$M^*(\omega) = 1/\varepsilon^*(\omega), \qquad 7.41$$

and

$$M^*(\omega) = M'(\omega) + M''(\omega) = M_\infty \left[1 - \int_0^\infty dt e^{-i\omega t} \left(-\frac{d\phi}{dt} \right) \right]$$

where $M_\infty = \lim_{\omega \to \infty} M' = 1/\varepsilon_\infty$ is a measure of the magnitude of the electric field relaxation. The frequency at which the peak height appears in the $M''_\sigma(\omega)$ dispersion curve is proportional to the d.c. conductivity, $\sigma_0 \sim \omega_\sigma$. For network glasses ε_∞ typically possesses values between 4 and 20; $M_0 = 1/\varepsilon_0$ and ε_0 is the permittivity of free space (dielectric constant), $\varepsilon_0 = 8.854 \times 10^{-14}$ F/cm. The KWW relaxation function describing the electrical field relaxation is

$$\phi(t) = \exp[-(t/\tau_\sigma)^{\beta_\sigma}] \qquad 7.42$$

The average relaxation time can be extracted directly from the d.c. conductivity,

$$\langle \tau_\sigma \rangle = 1/(M_0 M_\infty)\sigma_0, \qquad 7.43$$

It may also be shown that

$$\langle \tau_\sigma \rangle = \tau_\sigma \Gamma(1/\beta_\sigma)/\beta_\sigma \qquad 7.44$$

Values of β_σ for ionic conductivity typically range 0.5–0.75 for most glass formers. For the electrical conductivity, one can identify the characteristic rate as $\nu_\sigma = 1/\tau_\sigma$.

7.10 Secondary Relaxations in ECR and MR Experiments

The secondary relaxation rates in single and mixed alkali glasses measured using electrical conductivity relaxation, mechanical relaxation, and NMR relaxation experiments are now discussed. The data are shown in Fig. 7.16 for single and mixed alkali metaphosphates. The total alkali ion mole fraction is 0.5 in each glass; in the mixed alkali samples the Li : Na ratio is unity.

The following observations may be made from these data. 1) The mixed alkali relaxation times are much longer than the single alkali relaxations (Angell 1990; 1991). This point was mentioned earlier in relation to the data in Fig. 7.7) 2) The activation energies of the single alkali materials are slightly smaller than those of the mixed alkali materials, $E_\sigma(Na) = 70 \pm 2$ J/mol, $E_\sigma(Li) = 67 \pm 2$ J/mol and $E_\sigma(Na, Li) = 11 \pm 12$ J/mol (Green et al 1998). The implication is that as the temperature increases, the relative relaxation times between

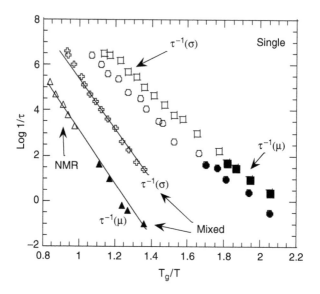

FIG. 7.16

Relaxation times obtained from ECR, MR, and NMR measurements in single and mixed alkali metaphosphate glasses and melts are shown here. Single alkali: The filled circles and squares represent MR data whereas the open circles and squares represent ECR data. Mixed alkali: The + symbols and the filled triangles ere obtained from ECR and MR data, respectively. The open triangles were obtained from $T_{1\rho}$ NMR data of phosphorous [31]P.

the single and mixed alkali materials become closer. They should eventually become comparable to the viscosity relaxation times at sufficiently high temperatures. 3) The relaxation times measured by the MR experiments of the mixed alkali glasses are longer than those measured by electrical conductivity relaxation ECR. The data from NMR are shown and the NMR experiments are sensitive to the phosphorous, not the alkali cations. The characteristic NMR rates are comparable to the characteristic MR rates. Moreover, the breadth of the MR spectrum is typically decades wide, encompassing the spectrum measured by the ECR measurements (Angell 1990; Green et al 1994). The MR experiments are evidently sensitive to local network relaxations as well as to the ion hopping dynamics.

    In summary, for the single alkali glasses the electrical conductivity relaxation rates (ECR) are comparable to the mechanical relaxation rates (MR). With regard to differences between single and mixed alkali glasses containing the same number of alkali ions, dynamic processes that occur in mixed alkali glasses are much slower than those in their single alkali analogs. In addition, the distribution of relaxation times observed in the MR experiments is much broader than those measured in ECR experiments. This continues to be an active area of research and much is yet to be said about relaxations in a wide range of systems.

7.11 Mechanism of Cation Transport in Ionic Glasses

In this section we describe a mechanism of ionic transport that has been given serious consideration though not universally accepted. It is loosely described here as a unified site relaxation model (Bunde, Funke, and Ingram 1996). The description in this section is qualitative and intended to provide some physical intuition into the issue.

One should recall that the local environment around each type of alkali ion in a network glass possesses a distinct configuration in order to accommodate that cation (Section 7.2). Cations are generally located in the vicinity of NBOs because of the charge neutrality condition. The ions interact via long-range Coulombic forces and this imposes constraints on their spatial locations. In this sense, their locations are correlated and their dynamics are, by extension, correlated as well, particularly at high concentrations. In what follows the discussion for transport in single and mixed alkali glasses is separated.

7.11.1 Single Alkali Glasses

Now consider a single alkali network glass. One can envision that alkali ions vibrate with a given frequency in their equilibrium positions where their energies are minimized. Statistically, an ion can gain sufficient energy such that a hop may occur. Sites that are vacated can relax over some time scale τ_{site}. For an ion, say A^+, to hop it is confronted with one of two options.

7.11.1.1 Option I

It can hop into a region previously occupied by another A^+ ion, vacated only momentarily, $t < \tau_{site}$. We will identify this as an A-site. Bearing in mind that this is a continuously dynamic process, the new environment must reconfigure to accommodate the arrival of the A^+ cation. The extent of the adjustment depends on the relative magnitudes of t and τ_{site}. This new site, as suggested by Funke, possesses a somewhat higher energy because it would have evolved with the departure of the previous A^+ cation. Figure 7.17 shows a potential energy diagram for this process. Initially, the ion, located at position 1, is at the minimum of the free energy.

The wings of the curves indicate that the ion has a very negligible probability of escape. In a typical crystal, you will recall, the potential is periodic and each site that the atom can visit possesses the same energy. In this case the new site, location 2, is of higher energy. This type of diagram is typical of some diagrams used to illustrate hopping of charged species and is an essential component of the jump relaxation model by Funke. This asymmetric double well potential is a consequence of the Coulombic interactions and not a function of the disorder.

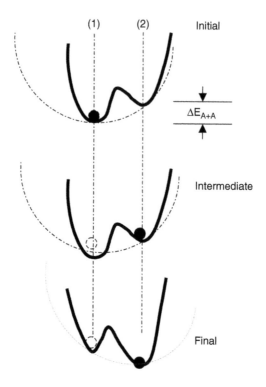

FIG. 7.17

Potential energy diagram for an ion making a hop from location 1 to location 2. The reorganization of the local environment to accept the ion in location 2 is depicted here.

We now return to the fate of the A^+ cation. The rearrangement of the potential in the final stage reflects to some degree the process required to accommodate the arrival of the A^+ cation. Note that during the reconfiguration process the ion could hop back to its original location. So the longer the ion spends in its new location, the higher the probability that it will remain there and the hop would be successful.

7.11.1.2 Option II

What happens if the site is vacant for a long time ($t > \tau_{site}$) and the configuration of the environment is near final? This final "equilibrium" configuration is identified in the dynamics structure model as a C-site. If the A^+ ion arrives at the C-site, the C-site has to be reconfigured to accommodate the A^+ ion. If this relaxation process is too long, the $A+$ ion could hop back into its original position, because it is experiencing a somewhat less accommodating environment. If it is sufficiently fast, then the hop is successful. Note, however, that the activation energy in this case is larger than a hop into the A-site, $\Delta E_{A+C} > \Delta E_{A+A}$. It follows that the relaxation time associated with a hop into the C-site is much longer, since the relaxation time increases exponentially with the activation barrier (these barriers are larger than $k_B T$). Indeed the A^+ cations can hop in many directions and, statistically, a small number will hop into the C-sites, since the probability is related to the activation energy barrier.

The theories have remained silent on the specific nature of the sites. One could surmise two scenarios that would create C-sites. Statistically, bonds break, creating an opportunity for the A+ ion to hop. One would anticipate that the probability that bond breaking occurs increases with temperature. Another scenario is that the bridge (BOs) does not have sufficient time to completely reform before the arrival of the next A^+ ion, thereby making a site available. In either event, the arrival of the A^+ ion triggers a relaxation of the new environment so it becomes more hospitable. If the concentration of alkali ions is large, then many A-sites are available. As we saw earlier, experiments on ionic conductivity and diffusion indicate that the conductivity, and hence diffusivity (Nearnst equation) increases rapidly, with increasing alkali fraction x. One could venture to suggest that with increasing cation mole fraction, the jump distances decrease since there exists a larger number of NBOs that are necessarily distributed throughout the system because of the Coulombic constraints. Since the diffusivity increases as the square of the jump distance, then both the diffusion coefficient and the conductivity should increase as $x^{2/3}$.

7.11.2   Mixed Alkali Glasses

We now consider the mixed alkali case where there are in principle three types of sites A, B, and C to accommodate the A^+ and B^+ cations. The activation energy associated with the hop of an A^+ cation and its subsequent accommodation by a B-site (or a B^+ ion into a A-site) is large compared to hops of either ion into their respective sites (A^+ into A-sites and B^+ into B-sites) due to the site configurations. It also follows, by extension that the associated relaxation times are much longer. As the disparity in the size of the ions increases, the extent of the reorganization increases. The aforementioned is the basis of the "unified site relaxation model" for ionic transport.

Computer simulations based on the notion that: 1) the alkali cations occupy and maintain distinct environments, A-sites and B-sites for A^+ and B^+ cations, respectively; 2) the existence of C-sites that result from an A-site or B-site remaining vacant for a sufficiently long period of time; 3) dissimilar sites must reconfigure to accommodate a different cation and as a result the activation energy is much higher, and 4) the existence of conducting pathways, had reasonable success. Such simulations provide a rationale for the changes in conductivity with the changing concentration of modifier cations and the mixed alkali effect in the conductivity (Maass, 1998). Moreover, the large mixed alkali peak was suggested to be the result of cations exchanging sites. This model was, or course an oversimplification, but nevertheless contained features worthy of further examination (Bunde et al 1991; Maass et al 1992, 1996).

More recent simulations indicate that, fundamentally, the reason for the mixed alkali effect is that the dissimilar cations, while occupying and maintaining distinct pathways, are responsible for blocking the paths of other types of cations (Swenson and Adams 2003). The simulations indicate that the cations, while randomly mixed, are not statistically distributed. Each type of cation follows distinct low dimensional pathways. Each cation totally or partially blocks the other from entering the pathway. These simulations suggest that the MAE can be explained without invoking a structural relaxation process, particularly at low temperatures. Nevertheless, at high temperature, the relaxation processes are sufficiently fast that the effectiveness of blocking is reduced, hence the effect diminishes, in accordance with experimentation.

This will probably not be the last word on the issue of ionic conduction, but it represents important progress toward a rigorous understanding of a problem many decades old. Ironically, this problem must have appeared somewhat straightforward at the outset, but it was not until the advent of sophisticated spectroscopic tools late during the previous century that reliable progress was finally made. This topic is still actively examined by a number of researchers as a more realistic connection to structure and interactions is sought.

7.12 Final Remarks

The study of relaxations in network glasses has been and is currently a very active area of research. This chapter could not necessarily address all aspects of the problem. However the intent was to address some of the general features that are central to understanding relaxations and dynamics in network forming glasses. One topic we did not discuss specifically is the dynamics of halide glasses, e.g, $AgI - AgPO_3$. Alkali halides are important for applications associated with electrochemical sensors. These materials posses lower glass transition temperatures than alkali oxide network glasses and their ionic transport rates far exceed those of alkali oxides. The rapid ion transport in halide glasses is believed to be due to the fact that the free volume available to the cations is larger in these systems than in pure alkali oxide network glasses (Wicks, 1995). In the next chapter aspects of dynamics in the supercooled state are discussed.

7.13 Problems for Chapter 7

1. The short-range tetrahedral structure of SiO_2 is shown in Fig. 7.1. Draw the short-range structure of TeO_2 and show how the addition of an alkali oxide would affect the structure.

2. Starting with expressions for the entropy of the liquid and crystalline phases, show that

$$\frac{\Delta h_m}{T_m} = \int_{T_K}^{T_m} \frac{c_p^{liq} - c_p^{cryst}}{T'} dT'$$

where Δh_m is the difference between the specific enthalpies. Now, using
eqn. 7.10 derive 7.11. state any assumptions.

3. Determine the constant of proportionality in equation 7.9,
$B \propto \frac{1}{(T_g/T_\infty)\Delta c_p(T_g)}$.

4. Show that the fragility index may be written as $m = m_{min} + \frac{m_{min}^2}{C}$
$T_g \Delta C_p(T_g)$, where $C = B(T_g/T_\infty)\Delta C_p(T_g)$ and $m_{min} = \log \tau(T_g)/\tau_0$; think
of τ_0 as a Debye frequency).

5. The data in Fig. 7.8 show the dependence of T_g on the modifier
content for single alkali germinates and tellurites. Sketch and explain
the relative magnitude and trends of the T_g dependence of the modifier content for silicates.

6. Starting with the Vogel-Fulcher equation, show that

$$\log \frac{\eta(T)}{\eta(T_g)} = -(1-r)m \left[\frac{(1-T_\infty/T_g)}{(1-\alpha T_\infty/T_g)} \right],$$

where $r = T_g/T$ and

$$m = \frac{BT_g}{[T_g(T_g/T_\infty - 1)]^2}.$$

In addition, derive m starting with

$$m = \frac{1}{2.3T_g} \frac{d_\eta}{d(T_g/T)} \bigg|_{T=T_g}$$

7. Starting with the WLF equation, show that it is mathematically
equivalent to the Vogel-Tammann-Fulcher equation. Identify the
relations between the constants c_1 and c_2 from the WLF equations
and the constants in the VTF equation, B and T_∞.

8. Show based on the super position principle that $\zeta(t) = \int_{-\infty}^{\infty} \Im(t')\phi(t-t')dt'$,
assuming that the response to as series of minor perturbations per-
turbation, $\Im_i(t)$, is $\zeta_i(t) = \Im_i(t)\phi(t-t_i)\delta t$.

9. Show that $\xi(\omega) = \phi(\omega) * \Im(\omega)$ may be derived from $\xi(t) = \int_{-\infty}^{\infty} \Im(t')$
$\phi(t-t')dt'$ using Fourier transforms. Determine explicit expressions
for ϕ' and ϕ'' in terms of ϕ_0, ω and t.

10. Show that based on a random hopping process, $\langle[r(t)-r(0)]^2\rangle = \langle r^2(t)\rangle = 6Dt$ is obtained from

$$D^*(\omega) = -\frac{\omega^2}{6}\lim_{\delta\to+0}\int_0^\infty \langle[\bar{r}(t)-\bar{r}(0)]^2\rangle e^{i\omega t-\delta t}dt$$

11. If $\phi^*(\omega) = \frac{\varepsilon_0\varepsilon_\infty}{\sigma^*(\omega)}$, calculate and plot expressions for $\sigma'(\omega)$ and $\sigma''(\omega)$.

12. Show that, based on the Modulus formalism, $\langle\tau_\sigma\rangle = 1/(M_0 M_\infty)\sigma_0$. In addition, based on the KWW function, show that $\langle\tau_\sigma\rangle = \tau_\sigma\Gamma(1/\beta_\sigma)1/\beta_\sigma$.

13. Discuss the essential differences between the Modulus formalism and the mechanistic formalism based on the ion hopping dynamics.

14. Explain why the values of the stretching exponent, β, would not be the same from a stress relaxation experiment and an electrical conductivity relaxation experiment.

15. The table below contains data for mixed alkali (Na, Li) metaphosphate glasses of varying composition of sodium, x. The parameters T_0, m, and T_g are shown here. Compare the temperature dependences of the viscosity at the compositions in the table. Comment on your results in terms of expected trends in the heat capacity change at T_g. Further, comment on any possible connections to the stretching exponent, β.

x	$T_g(°C)$	$T_0(°C)$	m
1	606	488	90
0.8	562	422	55
0.4	521	328	45

15. Based on an anomalous hopping model (Sidebottom et al 1995), it has been shown that the ionic conductivity of a glass may be written in terms of a length scale, ξ,

$$\sigma = K\xi^2\omega_c\left[1+\Gamma(2-n)\cos\left(\frac{n\pi}{2}\right)\left(\frac{\omega}{\omega_c}\right)^n\right]$$

where $\sigma_0 = K\xi^2\omega_c$ and $\Gamma(2-n)$ is the Gamma function. The table that follows shows values of the parameters that describe the ionic conductivity of a lithium metaphosphate glass.

T (°C)	σ_0 (mho/m)	n	ω_c (Hz)
22	1.6×10^{-7}	0.67	2.6×10^3
53	1.7×10^{-6}	0.67	2.7×10^4
83	8.3×10^{-5}	0.67	2.1×10^5

a) Compute the average jump distance.

b) Compute the frequency dependence of the conductivity for each temperature.

c) Draw M' and M'' (use $\varepsilon_\infty = 8.4$ for each temperature). Discuss any approximations.

16. Consider a sample subject to a periodic stress. Show that the total energy per unit volume stored per cycle is $E_0 = \pi \sigma_0^2 J'$, where σ_0 is the maximum stress. Second, show that the maximum energy per unit volume stored is $E_{max} = \frac{J' \sigma_0^2}{2}$. Now show that $\tan\delta = \frac{1}{2\pi} \frac{E_0}{E_{max}}$.

17. Derive a relationship between m/m_{min}, $\Delta c_p(T_g)$ and $S(T_g)$. Using the data in Table 7.1, comment on any trends in $S(T_g)$.

7.14 References

Adam, G. and Gibbs, J.H., "On the temperature dependence of cooperative relaxation properties in glass forming liquids," *Journal of Chemical Physics*, 43, 139 (1965).

Angell, C.A., *Journal of Non Crystalline Solids*, 131, 13 (1991).

Angell, C.A., "Correlation of mechanical and electrical relaxation phenomena in superionic conducting glasses," *Materials Chemistry and Physics*, 23, 143 (1989).

Angell, C.A., "Dynamic Processes in Ionic Glasses," *Chem. Rev.*, 90, 523 (1990).

Angell, C.A., Ngai, K.L, McKenna, G.B., Martin, S.W., "Relaxation in glass forming liquids and amorphous solids," *Journal of Applied Physics: Applied Physics Rev.*, 88, 3113 (2000).

Angell, C.A., *Science*, 267, 1924 (1995).

Böhmer, R., Senapati, H., Angell, C.A., *J. Non-Crystalline Solids* 131, 182 (1991).

Brow, R.K., Editor, "Structure, properties and applications of phosphate and phosphate containing glasses" Proceedings of the Fifteenth University Glass Conference on Glass Science, North-Holland, Elsiver (2000).

Buchenau, U., "Dynamics of Glasses," *J. Phys. Cond. Matter.*, 13, 7827 (2001).

Bunde, A., Funke, K., Ingram, M.D., "A unified site relaxation model for ion mobility in glassy materials," *Solid State Ionics*, 86, 1311, (1996).

Bunde, A., Funke, K., Ingram, M.D., "Ionic glasses: History and Challenges," *Solid State Ionics*, 105, 1, (1998).

Busch, R., "The thermophysical properties of bulk metallic glass forming liquids," *Journal of Materials*, 52, 39 (2000).

Day, D.E., "Mixed alkali glass: Their properties and uses," *J. Non. Cryst. Solids*, 21, 343 (1976).

Debenedetti, P., Metastable Liquids, Princeton University Press, NJ, 1996.

Dieterich, W. and Maass, P., "Non-Debye relaxations in disordered ionic solids," *Chemical Physics*, 284, 439 (2002).

Doremus, R.H., Glass Science, John Wiley and Sons, NY (1973).

Flynn, C.P., Point Defects and Diffusion, Clarendon Press, Oxford, 1972.

Greaves, G.N. and Ngai, K.L., "Reconciling ionic-transport properties with atomic structure in oxide glasses," *Phys. Rev. B* 52, 6358 (1995).

Green, P.F., Hudgens, J.J., Brow, R.K., "Specific Heat and Transport Anomalies in Mixed Alkali Glass," *J. Chem. Phys.*, 109, 7907 (1998).

Green, P.F., Sidebottom, D. and Brow, R.K., "Scaling Parallels in the Non-Debye Dielectric Relaxation of Ionic Glasses and Dipolar Supercooled Liquids," *Physical Review B* 56, p. 170–177 (1997).

Green, P.F., Sidebottom, D. and Brow, R.K., "Structural Correlations in the AC Conductivity of Ion Containing Glasses," *Journal of Non-Crystalline Solids* 222, 354–360 (1997).

Green, P.F., Sidebottom, D. and Brow, R.K., "Anomalous Diffusion Model of Ionic Transport in Oxide Glasses," *Physical Review B* 51, p. 2770–2776 (1995).

Green, P.F., Sidebottom, D. and Brow, R.K., "Scaling Behavior in the Conductivity of Alkali Oxide Glasses," *Journal of Non-Crystalline Solids* 203, p. 300–305 (1996).

Green, P.F., Sidebottom, D. and Brow, R.K., "Two Contributions to the AC Conductivity of Alkali Oxide Glasses," *Physical Review Letters* 74, p. 5068–5071 (1995).

Green, P.F., Sidebottom, D. and Brow, R.K., "Dynamics of Mixed Alkali Metaphosphate Glasses and Liquids," *Journal of Non-Crystalline Solids* 255, 87 (1999).

Green, P.F., Sidebottom, D. and Brow, R.K., Hudgens, J.J., "Mechanical Relaxation Anomalies in Mixed Alkali Glasses," *Journal of Non-Crystalline Solids* 231, 89–99, (1998).

Green, P.F., Sidebottom, D. and Brow, R.K., "Relaxations in Mixed Alkali Metal Phosphates," *J. Non-Crystalline Solids*, 172–174, 1352 (1994).

Goldstein, M., "Viscous liquids and the glass transition: a potential energy barrier picture," *Journal of Chemical Physics*, 51, 3728 (1969).

Hodge, I.M., "Strong Fragile liquids- A brief critique" *Journal of Non-Crystalline Solids*, 202, 164 (1996). see also Roland, C.M. and Ngai, K.L., "Commentary on Strong and Fragile liquids-A brief Critique," *Journal of Noncrystalline Solids*, 212, 74 (1997).

Houde-Walter S. and Green, P.F., The New Functionality of Glass, Materials Research Society Bulletin, ed. S. Nov. (1998).

Huang D. and McKenna, G.B., "New insights into the fragility dilemma in liquids," *J. Chem. Phys.*, 114, 5621 (2001).

Huang, W.C., Jain, H., Meitzner, G., "The structure of potassium germinate glasses by EXAFS," *Journal of Non-Crystalline Solids*, 196, 155 (1996).

Inagaki, Y., Maekawa, H., Yokokawa, T., "Nuclear magnetic resonance study of the dynamics of network glass forming systems," $xNa_2O(1-x)B_2O_3$," *Physical Review B*, 47, 674 (1993).

Jackle, J., "Theory of glass transitions, new thoughts and old facts," *Philosophical Magazine B* 56, #2, 113 (1087).

Joncher, A.K., *Dielectric Relaxation in Solids,* Chelsea Dielectrics Press, London 1983.

Joncher, A.K., *Universal Relaxation Law,* Chelsea Dielectrics Press, London 1996.

Kieffer, J., Masnik, J.E. and Nickolayev, O., "Structural developments in supercooled alkali tellurite melts," 58, 694 (1998).

Lee, S-K., Tatsumisago, M., and Minami, T., "Fragility of liquids in the system, $Li_2O\text{-}TeO_2$," *Phys. Chem. Glasses*, 35, 226 (1994).

Lee, S-K., Tatsumisago, M. and Minami, T., "Relationship between average coordination number and fragility of sodium borate glasses," *J. Ceramic Society Japan*, 103, 398 (1995).

Lee, S-K., Tatsumisago, M., and Minami, T., "Transformation range viscosity and thermal properties of sodium silicate glasses," *J. Ceramic Society Japan*, 101, 1018 (1993).

Macedo, P.B., Moynihan, C.T. and Bose, R., "The role of ionic diffusion in polarization in vitreous ionic conductors," *Phys. Chem. Glasses*, 13, 171 (1972).

Maass, P., "Towards a theory for the mixed alkali effect in glasses," *Journal of Non-Crystalline Solids*, Volume 255, 35–46 (1998).

Maass, P., Meyer, M., Bunde, A., Dieterich, W., *Physical Review Letters*, 77, 1528 (1996).

Moynihan, C.T., "Structure relaxation and the glass transition," in Struture, Dynamics and Properties of Silicate Malts, Eds. Stebbins, J.F., McMillan, P.F. and Dingwell, D.B., Minerological Society of America, Series Editor,Volume 32, Washington D.C., Ribbe, P.H., 1995.

Narananaaswamy, O.S., "A model for structural relaxation in glass," *J. Am. Ceram. Soc.* 54, 491 (1971).

Narananaaswamy, O.S., "Thermorheological simplicity in the glass transition," *J. Am. Ceram. Soc.*, 71, 900 (1988).

Ngai, K.L, Martin, S.W., "Correlation between the activation enthalpy and Kohlrausch exponent for ionic conductivity in oxide glasses," *Physical Review B.*, 40, 10550 (1989).

Ngai, K.L. "A review of critical experimental facts in electrical relaxation and ionic diffusion I ionically conducting glasses and melts," *Journal of Non-Crystalline Solids*, 203, 232 (1996).

Paul, P., Chemistry of Glass, Chapman and Hall, 2nd edition New York (1990).

Putz, K. and Green, P.F., "Fragility in mixed alkali glasses," *Journal of Non-Crystalline Solids*, 337, 254 (2004).

Richert, R. and Angell, C.A., "Dynamics of glass forming liquids: On the link between molecular dynamics and configurational entropy," *Journal of Chamical Physics*, 108, 9016 (1998).

Roling, B., Martiny, C. and Bruckner, S., "Analysis of mechanical losses due to ion-transport processes in silicate glasses," *Physical Review B.*, 57, 14192 (1998).

Roling, B., Meyer, M., Bunde, A., Funke, K., "Ionic conductivities of glasses with varying modifier content," *Journal of Non-Crystalline Solids*, 226, 138 (1998).

Roling, B., Martiny, C., Funke, K., "Information on the absolute length scales of ion transport processes in glasses from electrical conductivity and tracer diffusion data," *Journal of Non-Crystalline Solids*, 249, 201 (1998).

Scheer, G.W., "Theories of Relaxation," *Journal of Non-Crystalline Solids*, 123, 75 (1990).

Scher, H. and Lax, M., "Stochastic Transport in a Disordered Solid. I. Theory," *Physical Review B*, 7, 4491 (1973).

Scher, H. and Lax, M., "Stochastic Transport in a Disordered Solid. II. Impurity Conduction," *Physical Review B.* 7, 4502 (1973).

Sidebottom, D.L., Green, P.F., Brow, R.K., "Anomalous Diffusion Model of Ionic Transport in Oxide Glasses," *Physical Review B* 51, p. 2770–2776 (1995).

Sidebottom, D.L., Green, P.F., Brow, R.K., "Comparison of KWW and Power Law Analyses of an Ion-Conducting Glass," *Journal of Non-Crystalline Solids*, 183, p. 151–160 (1995).

Sidebottom, D.L., Roling, B. and Funke, K., "Ionic conduction in solids: Computing conductivity and modulus representations with regard to the scaling properties," *Physical Review B*, 63, 024301 (2000).

Sidebottom, D.L., "Universal approach for scaling the ac conductivity in ionic glasses," *Physical Review Letters*, 18, 3653 (1999).

Sidebottom, D.L., Green, P.F., Brow, R.K., "Scaling Parallels in the Non-Debye Dielectric Relaxation of Ionic Glasses and Dipolar Supercooled Liquids," *Physical Review B* 56, p. 170 (1997).

Stebbins, J.F., McMillan, P.F., Dingwell, D.B., Editors, Structure, Dynamics and Properties of Silicate melts, J. Reviews in Minerology, v. 32 Minerological Society of America, Washington DC (1995).

Stebbins, J.F., Sen. S., George, A.M., *J. Non-Crystalline Solids*, 192 , 298 (1995).

Stillinger, F.H., Debenedetti, P.G., and Truskett, T., "The Kauzmann Paradox Revisited," *Journal of Physical Chemistry B*, 105, 11809 (2001).

Swenson, J. and Adams, S., "Mixed alkali effect in glass," *Physical Review Letters*, 90, 155507-1 (2003).

Tanaka, H., "Relation between thermodynamics and kinetics of glass-forming liquids," *Physical Review Letters*, 90, 055701-1 (2003).

Uhlmann D.R. and Kreidl, N.J., Editors, Glass: Science and Technology, Vol. 3 Viscosity and Relaxation Academic Press, NY, 1986.

Van Ass, H.J.M. and Stevels, J.M., "Internal friction of mixed alkali metaphosphate glasses," *Journal of Non-Crystalline Solids*, 16, 27 (1974).

Webb, S.L., "Silicate melts: Relaxation, rheology, and the glass transition," *Rev. Geophys.* 35, 191–218 (1997).

Wicks, J.D., Borjesson, L., Buschnell-Wye, G., Howells, W.S., McGreevy, "Structure and ionic conduction in $(AgI)_x(AgPO_3)_{1-x}$ glasses," *Physical Review Letters*, 74, 726 (1995).

Williams, G. and Watts, D.C., "Non-symmetrical dielectric relaxation behavior arising from a simple empirical deay function," *Trans. Faraday Soc.*, 66, 80 (1970).

Zhu, D., Ray, C.S., Zhou, W. and Day, D.E., "Glass transition and fragility of $Na_2O–TeO_2$ glasses," *Journal of Non-Crystalline Solids*, 317, 247 (2003).

7.15 Appendix

The following is a basic sketch of the analysis due to Maass (1998) suggesting the existence of three peaks in the internal friction spectrum of a mixed alkali (two cations) glass. In the earlier chapter we discussed the internal friction problem. However it worthwhile to revisit this issue with regards to ion hopping. The ions hop into sites where their energy is minimized. As discussed earlier, each site has a distinct configuration and the environment. As suggested by Maass, each site i may be described by a set of structural variables, $\bar{s}_i$. These vectors define the positions of each nearest neighbor atom around each mobile ion. With this in mind, each A and B ion will possess energies $\varepsilon^A(\bar{s}_i)$ and $\varepsilon^B(\bar{s}_i)$, respectively. Each empty site (or site that is available to accept an ion) possesses energy $\varepsilon^C(\bar{s}_i)$. Now, consider an experiment in which a periodic shear field, ζ, is applied to the sample,

$$\xi = u_o \operatorname{Re}\left[e^{\omega t} \right] \qquad\qquad A1$$

The deformation of the local environments of the ions and sites is given by

$$\Delta \bar{s}_j = u_o \operatorname{Re}\left[\phi_j e^{\omega t} \right] \qquad\qquad A2$$

where ζ_j are coupling parameters. Changes in the local energy of the sites necessarily occur due to this deformation. Consequently, the ions are induced to undergo hops allowing some of the energy to be dissipated. The internal friction spectrum for the mixed alkali glass is predicted to be

$$Q^{-1}(\omega,T) \propto \frac{\omega}{kT} \mathrm{Re}\{\gamma_A S_A(\omega,T) + \gamma_B S_B(\omega,T) + \gamma_{AB} S_{AB}(\omega,T)\} = \frac{1}{Q_A} + \frac{1}{Q_B} + \frac{1}{Q_{AB}}$$

A3

The reader is referred to Maass (1999). Here the structure factors are specified in terms of the correlation functions of the site occupations, n_j^A, n_j^B and n_j^C. These n's possess values of 1 when a site is occupied and 0 otherwise, so $n_j^A = 1$ when an A atom occupies a site. The correlation functions are

$$S_A = \int_0^\infty \langle n_j^A(t) n_j^C(0) \rangle e^{i\omega i}$$

A4

$$S_B = \int_0^\infty \langle n_j^B(t) n_j^C(0) \rangle e^{i\omega i}$$

A5

and

$$S_{AB} = \int_0^\infty \langle n_j^A(t) n_j^B(0) \rangle e^{i\omega i}$$

A6

The γ_A, γ_B and γ_{AB} functions describe the degree to which the energy of a site changes in response to the applied field. The essential point is that the foregoing equation predicts that for a single alkali glass, one peak should exist and for a mixed alkali glass, there should be three peaks, one for each alkali cation and a third representing interactions.

8

Comments on Heterogeneous Dynamics in the Disordered State

8.1 Introduction

The last two chapters dealt largely with specific details regarding mechanisms of transport in two systems, polymers and network glasses, which possess structures that lack long-range structural order. The discussion in this chapter is more generally applicable to dynamics in a disordered environment, particularly in the supercooled regime.

At temperatures sufficiently far above the temperature range where the glass transition occurs, the atomic or molecular entities of the system move at sufficiently rapid rates that the system achieves equilibrium on reasonable time scales. With decreasing temperature, glass-forming liquids exhibit increasingly slow dynamics in the absence of the formation of long-range order. The relaxation times increase in a non-Arrhenius manner with decreasing temperature, and, eventually, the rate of cooling exceeds the ability of the system to reach equilibrium on a reasonable time-scale. Under such circumstances, the system is nonergodic. It is considered to be frozen on the time scale of observation, denoting glass formation.

Generally, details of the relaxation dynamics exhibited by a molecular entity are determined by the size, the architecture, and the environment (interactions with neighboring entities, to which it may or may not be bonded) of the entity and by the temperature. The dynamics of long-chain polymers are characterized by a translational (snake-like) center of mass motions facilitated by more rapid time-scale relaxations, such as motions of groups of monomer segments and rotations and vibrations of chemical side groups. In the liquid, or melt, the reptative motions determine the viscous relaxations (primary relaxations). Below T_g, the secondary relaxations are associated with local segmental dynamics. In network alkali oxide glasses, the main network relaxations largely determine the viscous melt dynamics. Below T_g, secondary relaxations, ionic hopping dynamics, are primarily responsible for internal friction and conductivity. In small molecule liquids, vibrational, rotational, and translational motions of the molecules characterize the dynamics.

A universal feature of the disordered state is that correlation functions that describe the material response in the linear response regime, are invariably characterized by a distribution of relaxation times, regardless of the architecture of the molecular entities (polymers, organic liquids, metallic glasses, ionic liquids, etc.). In this chapter, this issue is explored in further detail with regard to the existence of a spatially dynamic heterogeneous environment wherein local regions of the sample relax at rapid rates while others relax slowly. The size of the regions are on the order of nanometers (Ediger 2000, Richert, 2002). In the supercooled regime, a failure of the Stokes-Einstein relation has been documented in some system; this failure is connected to the notion that particles undergo transport in a dynamically heterogeneous environment. Finally, further insight into fragility of visions liquids, introduced in Ch. 7, is provided here.

8.2   Temperature Dependencies of Relaxations

In the previous chapter, the temperature dependence of the viscous relaxations were discussed in light of the Vogel-Tammann-Fulcher equation and of the Adam-Gibbs equations. The former was originally derived based strictly on free volume considerations, whereas the temperature dependence of the latter was rationalized in terms of a decrease of configurations (configurational entropy) available to the system with decreasing temperature. There are other functional forms based on different criteria used to describe the temperature dependence of relaxation processes. Cohen and Grest (1979) introduced another equation which, although based in part on free volume considerations, incorporates thermodynamic aspects of the system. The temperature dependence of the viscous relaxation time is predicted to be

$$\log_{10} \tau = A + \frac{2B}{\{T - T_\infty + [(T - T_\infty)^2 + CT]\}^{1/2}} \qquad 8.1$$

This equation is known to provide a better fit to the data for a number of systems over a wider temperature range. Ferry (1956) and, later, Richert and Bässler (1990) suggested the following based on random walk dynamics in a disordered medium,

$$\tau \propto e^{(T_0/T)^2} \qquad 8.2$$

A separate proposal by Stillinger (1988) involving flow associated with slip of densely packed regions and in contrast to Adam and Gibbs, predicts that

$$\tau \propto e^{const/TS_c^{2/3}} \qquad 8.3$$

The aforementioned predictions described are based on various notions of how the dynamic processes proceed in a disordered environment and, in fact, they underscore the complexity of the problem. The list of predictions is by no means exhaustive.

8.2.1 Dispersive Dynamics Associated with Disorder

Regardless of the mechanism of transport, the dynamic processes that occur in all these disordered systems are dispersive. In the previous chapter, it was shown that stress relaxation, enthalpy relaxation, and dielectric spectroscopy measurements were useful probes for studying the time dependences of the materials response. Other important techniques include neutron scattering measurements of the intermediate scattering function and nuclear magnetic resonance. Within the linear response regime, where the fluctuation dissipation theorem holds (Chapter 2), the correlation functions measured by these techniques exhibit a universal trend; they are all reasonably well described by the KWW relaxation function

$$\phi(t) = \phi_0 e^{-(t/\tau)^\beta} \tag{8.4}$$

This function, alternatively, may be expressed in terms of a relaxation time probability distribution function $P(t)$,

$$\phi(t) = \int e^{-t/\tau} P(\tau) d\tau \tag{8.5}$$

The universality of the time dependencies of the correlation functions suggests a commonality associated with the dynamics of the constituents. β, which represents the distribution of relaxation times, as mentioned in the last chapter, appears to be connected to the fragility of the system. In fact, an empirical relation between m and β was suggested by Böhmer et al. based on an assessment of data from ~60 substances,

$$m = a_1 - a_2 \beta \tag{8.6}$$

where a_1 and a_2 are constants. This correlation indicates that the fragility decreases as β increases; i.e., the dynamics of fragile systems are characterized by a larger distribution of relaxation times than those of strong systems. Table 8.1 contains values of m, β, and heat capacity ratios of the liquid to the glass at T_g for a wide range of systems. As discussed in the previous chapter, connections between m and $\Delta c_p(T_g)$ are well established in inorganic network glasses but are less certain in other systems. On the other hand, the connection between m and β appears to be more general.

Generally, the distribution of relaxation times broadens with decreasing temperature. At sufficiently high temperatures, β approaches unity. In homopolymeric melts, the distribution of relaxation times is typically broad, $\beta \sim 0.5$, and relatively insensitive to temperature. The effect of increasing temperature on the material is largely associated with decreasing the relaxation time of the structural entities. Nevertheless, the distribution of relaxation times increases with decreasing temperature, as T_g is approached, in a large number of glass-forming systems.

The dispersion that characterizes the dynamics may be understood based on the notion of heterogeneous dynamics. Under these conditions, the system

TABLE 8.1

Table of glass formers: fragility index (m), glass transition temperature (T_g), β, and ξ_{het} (m and T_g data adopted from Huang and McKenna, (T) data from Tanaka and (B) data from Böhmer et al., (E) data from Qiu and Ediger).

Material	m	T_g	$\xi_{het}(T_g + 10)$	$\beta(T_g)$	$C_p^l/C_p^{cryst,glass}$
GeO$_2$	24	818			1.073
SiO$_2$	20	1446			1.005
B$_2$O$_3$	32	521			1.449
Soda lime	40	536(B)	0.55(B)		
0.25Na$_2$O-0.75SiO$_2$	37	739	0.68(B)		
	30	764(B)			1.216
K$_3$Ca(NO$_3$)$_7$	93	332	0.45(B)		1.568
	93	343(B)			
2BiCl$_3$-KCl	85	306			1.647
ZnCl$_2$	30	370			1.279
BF$_2$	24	590			1.000
Se	87	307	0.42(B)		1.498
Salol	63	281	0.6, 0.53(B)		1.714
ethylene glycol	325	181			
glycerol	53	190	1.3 ± 0.5	0.65, 0.7(B)	1.847
sorbitol	93	274	2.5 ± 1.2	0.5, 0.41(E)	1.886
m-cresol		57	198.5		1.837
o-terphenyl	76	241	2.3 ± 1.0	0.52(E)	1.472
toluene	59	110			1.512
α-phenyl-o-cresol	83	220			1.457
dibutylphthalate	69	179			1.527
polystyrene	116	373		0.35(B)	1.188
polypropylene	137	260		0.35(B)	1.268
PDMS	79	149.5		0.35(B)	1.387
Polycarbonate	132	423			1.113
PVC	191	354			1.278
Polysulfone	141	459			1.142
PMMA	103	367			1.231
PEMA	81	344			1.189
PBMA	56	305			1.187
PVA	73(E)	305(E)	3.7 ± 1	0.43(E)	

is believed to be composed of domains in which the dynamics are very different (spatial heterogeneity). As pointed out by Ediger (2000), near T_g, the dynamics of one region of a sample could be orders of magnitude slower than the dynamics in an adjacent region a few nanometers away. To date, a structural basis for this phenomenon of fast and slow dynamics is unclear; at least no variations of structure or density that would be connected to the heterogeneity of the dynamics are directly observable experimentally.

The heterogeneity of the time scales is believed to be associated with the spatial heterogeneity of the system. If, during the diffusion process, particles experience domains whose dynamics are slow and, in other neighboring locations, domains in which the dynamics are fast, then the stretched exponential (the single exponential relaxation rate in each domain is different) behavior may be understood, whether or not there exists a distribution of domain sizes. At much longer time scales, in which a particle would have an opportunity to sample many domains, the dynamics appear to be homogeneous. In other words, correlation functions that reflect a sufficiently large center of mass displacements of a molecule are single exponential because the observable of interest averages over length scales that are many times the dimensions of the dynamic domain size.

Molecular dynamic simulations indicate that, on time scales much shorter than the primary relaxations, the dynamics are dispersive. The length scale of the domain dimensions is believed to be on the order of nanometers. Dynamic correlation functions capable of sampling dynamics on the millisecond time scale provide information about the length scales of the domains (Qiu and Ediger 2003). To this end, techniques such as nuclear magnetic resonance are used to measure the dimensions of these domains. Data from such experiments indicate that the length scale associated with the dimensions of the regions is $\xi_{het} \approx 1\ nm$, as shown in Table 8.1. A number of theories provide insight into the length scale of these domains. For example, the size of the cooperative relaxing regions in the Adams Gibbs model is one such proposal. For a recent assessment of dynamic heterogeneity, the reader should consult reviews by Ediger (2000) and by Richert (2002) and references therein as well as papers by Qiu and Ediger and by Colby (2000).

A recent model by Xia and Wolynes (2000, 2001) based on the dynamic heterogeneity picture provides a direct connection between β and the fragility. The system is composed of dynamically fluctuating states and the transition between one metastable state to another is associated with a free energy cost, ΔF, largely associated with the configurational entropy of a state into which it could subsequently reside. Since the free energy barrier is ΔF, then it follows that the relaxation time

$$\tau = \tau(\Delta F) \propto e^{\Delta F/kT} \qquad 8.7$$

In developing the model for β, a distribution function of free energy barriers is constructed. For simplicity this function was assumed to be Gaussian, so

$$P(\Delta F) = \frac{1}{\sqrt{2\pi(\delta\Delta F)^2}} e^{-(\Delta F - \Delta F_0)^2/2(\delta\Delta F)^2} \qquad 8.8$$

The correlation function (ef. Eq. 8.5) would be

$$\phi(t) = \int e^{-t/\tau(\Delta F)} P(\Delta F) d\Delta F \qquad 8.9$$

The remainder of the calculation by Xia and Wolynes is not repeated here, but the essential prediction is that

$$\beta = \left[1 + \left(\frac{\Delta F_0(T)}{2kTD^{1/2}}\right)^2\right]^{-1/2} \qquad 8.10$$

In Eq. 8.11, the parameter D is the fragility defined through the VTF equation

$$D = \frac{T - T_\infty}{T_\infty} \ln \frac{\tau}{\tau_0}$$

8.11

It is left as an exercise for the reader to derive a relationship between D and m. Based on the microscopic theory for D, it is shown to increase as Δc_p decreases,

$$D = \frac{32R}{\Delta c_p}$$

8.12

where R is the universal gas constant. This result is consistent with experiment. Equation 8.10 is somewhat approximate because of the Gaussian approximation, but it shows that a more accurate distribution function yields a functional form of β on D that is in agreement with experiment. The microscopic model provides some fundamental insight into the trends relating the distribution of relaxation rates to the fragility and heat capacity change at T_g.

8.3   Comments on Dynamics in the Supercooled State

Studies of the temperature dependencies of viscous flow and of the long-range diffusional transport of small molecule organic liquids in the super-cooled state reveal strong evidence that the Stokes-Einstein equation is not obeyed. This is largely because the appropriate hydrodynamic boundary conditions are not met in this regime. In Chapter 2, it was shown that the Stokes-Einstein law indicates that the diffusion coefficient of a particle of radius a in a medium of viscosity η is given by

$$D = \frac{k_B T}{6\pi\eta a}$$

8.13

This result indicates that $\eta D/T$ should be constant, independent of temperature. However, studies of the 1,3-bis-(1-napthyl)-5-(2-napthyl)benzene (TNB) system show that, with decreasing temperature, toward T_g, $\eta D/T$ can increase by over two orders of magnitude (Swallen et al). In this case, the translational diffusion coefficient exhibits a weaker temperature dependence than the viscosity near T_g. A similar observation was made by Fujara et al. in studies of OTP, although the difference was less dramatic. In these experiments, the temperature dependence of the rotational diffusional transport remains consistent with that of the viscosity; it is only dynamics connected to the translational diffusion that is implicated with the violation of the Stokes-Einstein prediction. A number of theoretical studies reveal that these discrepancies may be rationalized in terms of diffusional transport in a dynamically heterogeneous environment. (Ediger, 2002, Yamamoto et al. (1998),

Stillinger and Hodgdorn (1994), Fujara et al., 1992, Tarjus and Kivelson 1995, Rah and Eu 2003, Berthier 2004).

Finally, it is worthwhile to consider that heterogeneity might be responsible for the decoupling of the translation and rotational modes, which would be associated with an increase of the distribution of relaxation times as T_g is approached. However, the decoupling is also observed in systems in which the distribution is independent of temperature (time temperature superposition). In this regard, the fundamental origins of the decoupling do not appear to be clear cut at this time. (Richert et al., 2003, Swallen et al. 2003).

8.4 Comments on the Stokes-Einstein Relationship

It is worthwhile to conclude this chapter by revisiting the Stokes-Einstein prediction to examine its origins. The Navier-Stokes equation is the basic dynamical equation that governs the dynamics of a sphere in a liquid medium. Under steady state, incompressible, flow conditions, the equation may be rewritten as

$$\rho \bar{u} \cdot \nabla \bar{u} = -\nabla p + \eta \nabla^2 \bar{u} \qquad 8.14$$

where $\bar{u}$ is a fluid "particle" velocity, ρ is the fluid density, and p is the pressure. This equation is a force balance equation (Newton's Law for a fluid) where the first term represents inertial forces, the second term represents forces due to pressure gradients, and the third are forces associated with the action of the viscosity. Stokes solved this equation under conditions where the viscous forces are much larger than the intertial forces. This is the low Reynolds number (Re) regime, so Eq. 8.14 becomes

$$\nabla p = \eta \nabla^2 \bar{u} \qquad 8.15$$

The Reynolds number, Re, is physically the ratio of the inertial force to the viscous force, $Re \ll 1$. With no-slip boundary conditions (velocity components of the fluid at the surface of the sphere are zero, $u_r = u_\theta = 0$), the viscous drag force, ζ, on the sphere is

$$\zeta = 6 \pi \eta a V \qquad 8.16$$

where $u = V$ is the far field velocity (Leal 1992). Einstein used the relationship $D = k_B T / \zeta$, and the fact that the velocity is the product of the mobility and the force, to obtain what we have come to know as the Stokes-Einstein relation, equation, eqn. 2.62.

It has been known for some time that this equation fails under conditions of high Reynold's numbers. Interestingly, the equation has not only proven to be reliable for macroscopic spheres falling through a liquid, but it also works well empirically for particles of molecular size in many systems. In some systems with macroscopic Brownian particles, a failure has been noted, but the equation works well as long as the radius is interpreted as a hydrodynamic radius. (Schmidt and Skinner (2003), Tarjus and Kivelson, 1995, and Hodgdorn and Stillinger, 1992)

8.5 Final Comments

This is a vast area of research and therefore a number of topics have been omitted. The topic of mode coupling theory was not discussed. Mode coupling theory is based on the notion that each particle in a supercooled liquid is temporarily trapped by its surrounding neighbors. The theory attempts to describe the dynamics of the particle within the "cage" over time when the "cage" eventually breaks up, enabling the escape of the particle. Mathematically, a series of coupled, hydrodynamic, nonlinear integral differential equations of the intermediate scattering function are developed and solved. The reader is referred to one of a number of references to learn much more about this topic. (See, for example, Götze and Sjögren 1992.)

8.6 References

Bendler, J.T. and Slesinger, M.F., "Generalized Vogel law for glass forming liquids," *J. Stat. Phys.* 53, 531 (1988).

Berthier, L., "Time and length scales in super cooled liquids," *Phys. Rev. E.* 69, 020201(R) (2004).

Cohen, M.H. and Grest, G.S., "Liquid-glass transition, a free volume approach," *Phys. Rev. B* 20, 1077 (1979).

Colby, R.H., "Dynamic scaling approach to glass formation," *Phys. Rev. E,* 61, 1783 (2000).

Debenedetti, P.G. and Stillinger, F.H., "Supercooled liquids and the Glass transition," *Nature,* 410, 259 (2001).

Ediger, M.D., Angell, C.A. and S.R. Nagel, S.R.,"Supercooled liquids and glasses," *J. Phys. Chem.* 100, 13200 (1996).

Ediger, M.D., "Spatial heterogeneous dynamics in supercooled liquids," *Annu. Rev. Phys. Chem.* 51, 99 (2000).

Ferry, J.D., Grandine, L.D., Fitzgerald, E.R., "The relaxation distribution function of polyisobutylene in the transition from rubber-like to glass-like behavior," *J. Appl. Phys.* 24, 911 (1953).

Sillescu, H., "Heterogeneoty at the glass transition: a review," *Journal of Non-Crystalline Solids,* 243, 81 (1999).

Götze W. and Sjögren, L., "Relaxation processes in supercooled liquids," *Rep. Prog. Phys.* 55, 241 (1992).

Hodgdorn, J.A. and Stillinger, F.H., "Stokes-Einstein violation in glass forming liquids," *Phys. Rev. E.* 48, 207 (1993).

Johari, J.P. "Contributions of the entropy of a glass and liquid and the dielectric relaxation time," *Journal of Chemical Physics*, 112, 7518 (2000).

Leal, L.G., Laminar Flow and Convective Processes, Butterworth-Heinemann Series in Chemical Engineering, Newton, MA 1992.

Qiu, X. and Ediger, M.D., "Length Scale of Dynamic Heterogeneity in Supercooled D-Sorbitol: Comparison to Model Predictions," *J. Phys. Chem. B*, 107, 459 (2003).

Rah, K. and Eu, B.C., "Theory of the viscosity of supercooled liquids and the glass transition: Fragile liquids," *Phys. Rev. E.* 68, 051205 (2003).

Richert, R., "Heterogeneous Dynamics in Liquids: fluctuations in space and time," *J. Phys.: Condens. Matter* 14, R703 (2002).

Richert, R. and Bässler, H., "Dynamics of supercooled melts treated in terms of the random walk concept," *J. Phys.: Condens. Matter*, 2, 2273 (1990).

Richert, R., Duvvuri, K., Duong, L.-T., "Dynamics of glass-forming liquids. VII. Dielectric relaxation of supercooled *tris*-naphthylbenzene, squalane, and decahydroisoquinoline," *J. Chem. Phys.* 118, 1828 (2003).

Stillinger, F.H. and Hodgdon, J.A., "Translation-rotation paradox for diffusion in fragile glass-forming liquids," *Phys. Rev. E.* 3, 2064 (1994).

Stillinger, F.H., "Relaxation and flow mechanisms in "fragile" glass-forming liquids." *J. Chem. Phys.*, 89, 6461 (1988).

Swallen, S.F., Bonvallet, P.A., McMahon, R.J., Ediger, M.D., "Self-diffusion of *tris*-Napthylbenzene near the glass transition temperature," *Phys. Rev. Lett.*, 90, 015901, (2003).

Tarjus, G. and Kivelson, D., "Breakdown on the Stokes-Einstein relationin supercooled liquids," *J. Chem. Phys.*, 103, 3071 (1995).

Xia, X. and Wolynes, P.G., "Microscopic Theory of Heterogeneity and Nonexponential Relaxations in Supercooled Liquids," *Phys. Rev. Lett.*, 86, 5526 (2001).

Xia, X. and Wolynes, P.G., "Fragilities of liquids predictd from the random first order transition theory of glasses," *Proceedings of the Natl. Acad. Sciences* 97, 2990 (2000).

Part IV

Instabilities and Pattern Formation in Materials

Minor disturbances within a spatially and temporally homogeneous system can become amplified (unstable) by external influences leading to the formation of patterns characterized by a new level of complexity. Such events are fundamentally connected to a wide variety of events in nature such as the development of microsctuctural features in engineering materials, instabilities in the flow of liquid films, colonies of bacteria, and weather patterns. This section of the book explores the topics of spinodal decomposition in concentrated mixtures as well as instabilities that occur in moving interfaces (solid/melt interface of a solid in its supercooled melt or the solid/vapor interface of a solid in a supersaturated environment). The letter is known as the Mullins-Sekerka instability that engenders the formation of dendritic patterns found in a range of crystals (ice, metals, polymers, etc.).

9

Phase Separation in Binary Mixtures: Spinodal Decomposition and Nucleation

9.1 Introduction

Phase separation is ubiquitous, occurring in a diverse range of liquid-liquid mixtures, (metallic alloys, network glasses, polymer-polymer mixtures, rod-like molecule mixtures, surfactant solutions) regardless of the geometry and structure of the molecular constituents. Strategies that involve phase separation are often used to control the microstructural features and hence the properties of engineering materials. Phase diagrams are of particular utility since they identify the equilibrium structure (microstructural features, etc.) of a mixture at given temperature and composition ranges. The evolutionary, or transient, changes of the structure of a mixture due to changes in temperature that move it from one region of the phase diagram to another can be fascinating. One of the seminal papers on this topic describes the spinodal structure of a phase separated borosilicate network glass melt (Cahn 1969). The bicontinuous spinodal structures and, more importantly, signatures of spinodal decomposition (time dependencies of various characteristics of the structure) have been investigated in a wide range of systems in recent years, employing expe-rimental techniques such as microscopy (scanning and electron microscopy and scanning force microscopy) to examine cross-sections of samples (direct images) and scattering measurements (neutrons, X-rays, etc.) of the structure factor to learn about the spatial and temporal structure of the system. From a technological perspective, an understanding of this phenomenon is important. Spinodal decomposition processes influence the morphology of a wide range of alloys (metallic, polymeric, inorganic) and therefore determines a range of physical (mechanical, magnetic) and chemical (corrosion) properties. This phenomenon is often exploited to effectively tailor material microstructure.

The goals of this chapter are to describe the origins and dynamics of early stage phase separation (by nucleation and growth or by spinodal decomposition) of an initially homogeneous mixture thrust into the two-phase region of a phase diagram. While the topic is introduced here by describing the

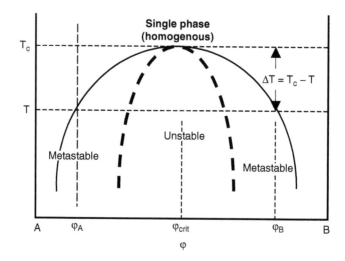

FIG. 9.1

A phase diagram for an A-B binary liquid that exhibits an upper critical solution temperature is shown here. The system demixes spinodally (unstable toward long wave length infinitesimally small amplitude fluctuations in composition) or by nucleation and growth in the metastable regime. ΔT is the depth of a quench below T_c.

phase diagram of a binary polymer-polymer mixture, the subsequent development is general.

A simple temperature versus composition phase diagram for a binary, A and B, mixture in which a single phase, homogeneous, region together with the two-phase, stable and unstable, regions are identified, is illustrated in Fig. 9.1. The mixture described here exhibits an upper critical solution temperature (UCST) below which it phase separates into A-rich and B-rich phases.

In the unstable region of the phase diagram, fluctuations of the composition of an initially homogeneous mixture would be amplified and the mixture would "spontaneously" phase separate into a bicontinuous structure, a *spinodal* pattern. The spinodal pattern is shown to the left of Fig. 9.2. A system that undergoes spinodal decomposition goes through different stages of structural evolution. The *early stage* is characterized by an exponential time-dependent growth of the amplitude of the fluctuations, while the wavelength of the fluctuations remains constant. As the compositional fluctuations grow, distinct phases begin to form. In a scattering experiment, during the *late stage*, the phases are fully formed and growth is characterized by asymptotic power-law time dependences of the scattering intensity and of the wave vector that characterizes the maximum mode of domain growth.

In the metastable region of the phase diagram, the system is unstable only toward large, localized fluctuations. Here a stable nucleus of critical size has to form. The size of a stable nucleus is determined by a balance between a volume free energy term, which favors formation, and an interfacial free energy term, which is unfavorable toward the formation of the cluster. If a

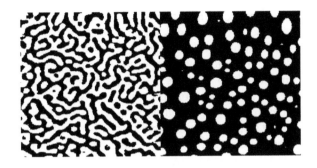

FIG. 9.2
A typical *spinodal* pattern is shown to the left of the figure whereas a pattern that develops during the early stages of the nucleation and growth mechanism of phase separation is shown to the right.

cluster is sufficiently large, the volume contribution dominates and the cluster would remain stable and increase in size. The late stage of structural evolution in this regime is also characterized by asymptotic power law growth and various coarsening mechanisms ensue depending on the system and growth conditions. The structure of the system is characterized by droplets of the minor phase dispersed throughout the major phase, as illustrated in Fig. 9.2.

9.2 Free Energy of Mixing of a Binary Polymer-Polymer Mixture

The free energy of mixing of an A/B polymer-polymer mixture is approximated by considering contributions from the entropy of mixing, associated with the mixing of n_A molecules of species A and n_B molecules of species B, and an energetic, or enthalpic, contribution. If it is assumed that mixing occurs on a lattice of n sites, then for an ideal mixture the free energy of mixing is due to entropy of mixing and

$$\Delta S_{mix} = -k(n_A \ln \phi_A + n_B \ln \phi_B) \qquad 9.1$$

where $\phi_A = V_A/(V_A + V_B)$ and $\phi_B = V_B/(V_A + V_B)$ are the volume fractions of species A and B, respectively, and ($\phi_A + \phi_B = 1$). V_A and V_B are the volumes occupied by species A and B, respectively. In this analysis, no volume change is assumed to occur upon mixing and the total number of sites is $n = \frac{V_A + V_B}{\Omega}$, where Ω is the volume of a unit cell of the lattice.

For the specific case of mixing a polymer of $N_A = n\phi_A/n_A$ monomers with another polymer of $N_B = n\phi_B/n_B$ monomers, the entropy of mixing per lattice site is

$$\Delta s_{mix} = \frac{\Delta S_{mix}}{n} = -k\left(\frac{\phi_A}{N_A}\ln \phi_A + \frac{\phi_B}{N_B}\ln \phi_B\right) \qquad 9.2$$

This equation has significant implications for the mixing of long-chain polymers because the entropy of mixing is small, it varies as $1/N$, where typically $N \sim 10^3$. This small entropy of mixing suggests that mixing of long-chain polymers would be dominated by the specific enthalpic interactions between the A and B segments of the mixture. The enthalpic contributions to the free energy of mixing are now considered. Only A-A, B-B, and A-B pair interactions are considered in this regular solution model. If the coordination number of the lattice is z, and each binary interaction is considered to be independent, then the interaction energy between the components before mixing is $E = \frac{zn}{2}\phi_A \varepsilon_{AA} + \frac{zn}{2}\phi_B \varepsilon_{BB}$, which is simply the sum of the contributions associated with the A-species, $(zn/2)\varepsilon_{AA}\phi_A$, and the B-species, $(zn/2)\varepsilon_{BB}\phi_B$. The factor of ½ avoids double counting. Upon mixing the interaction energy becomes

$$E = \frac{zn}{2}\phi_A[\phi_A \varepsilon_{AA} + \phi_B \varepsilon_{AB}] + \frac{zn}{2}\phi_B[\phi_B \varepsilon_{BB} + \phi_A \varepsilon_{AB}]$$ 9.3

The interaction energy per lattice site associated with mixing is

$$\Delta e = \frac{E - E'}{n} = \phi(1-\phi)\chi$$ 9.4

where

$$\chi = \frac{z}{2}\frac{(2\varepsilon_{AB} - \varepsilon_{AA} - \varepsilon_{BB})}{kT}$$ 9.5

is known as the Flory-Huggins interaction parameter. For a nonideal polymer-polymer mixture, the free energy of mixing per segment is approximated by the following expression

$$\Delta f_{mix} = kT\left(\frac{\phi}{N_A}\ln\phi + \frac{(1-\phi)}{N_B}\ln(1-\phi) + \phi(1-\phi)\chi\right)$$ 9.6

where $\varphi_A = \varphi$ and $\varphi_B = 1 - \varphi$. Note that for $N_A = N_B = 1$, eqn. 9.6 is generic regular solution free energy. Equation 9.6 has come to be known as the Flory-Huggins free energy function. It predicts that if $\chi > 0$, the relative contribution of the entropic component to the free energy decreases with decreasing T because $\chi \propto \frac{1}{T}$ and the mixture will phase separate at an upper critical solution temperature (UCST).

While in practice some A-B polymer-polymer mixtures exhibit UCST behavior, most A-B, polymer-polymer mixtures do not. Such mixtures exhibit lower critical solution temperatures (LCST) above which they would phase separate. The existence of the LCST is, in part, associated with changes in specific volume that accompany mixing the A and B components. This is entropic in origin. The Flory-Huggins formalism is not particularly well suited to describe LSCT behavior. Nevertheless, one ad hoc strategy used to address this issue is to write the χ-parameter such that

$$\chi \approx c_1 + \frac{c_2}{T}$$ 9.7

where the first term in this expression represents an entropic contribution and the second accounts for the enthalpic component. There is in fact empirical evidence for this temperature dependence. The existence of the LCST is reconciled by considering appropriate values of c_1 and c_2 ($c_2 < 0$). In general, the χ-parameter can be positive or negative wherein a negative value of χ favors mixing.

9.2.1 Phase Diagram of a Simple Binary Mixture

The Flory-Huggins free energy function is now used to calculate the spinodal and binodal regions of a phase diagram. For convenience, only the symmetric situation, $N_A = N_B = N$, is considered. Δf_{mix} is calculated for different values of χN, and plotted in Fig. 9.3. For the case where $\chi N = 0$, the free energy of the system is characterized by a single minimum and the system is homogeneous. For this situation, the entropy of mixing is the only contribution to the free energy, so the mixture resides in the single phase regime, $T > T_c$. When $\chi N = 3$, the free energy is characterized by two minima and the initially homogeneous mixture would reduce its free energy by separating into two different compositions, φ_1 and φ_2. This situation, clearly, corresponds to $T < T_c$.

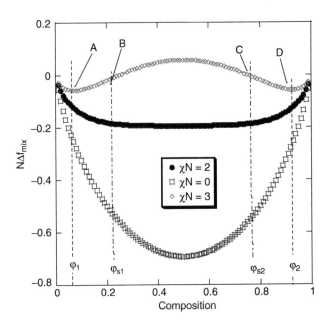

FIG. 9.3
The free energy is plotted as a function of composition for different values of χN, using the Flory-Huggins expression for the free energy of mixing per segment of an A-B polymer-polymer mixture. Note that the shapes of these curves are generic for any regular solution mixture.

There are two mechanisms by which such mixtures are known to phase separate, spinodally or by nucleation and growth. If the composition resides between boundaries B and C in Fig. 9.3, then the curvature of Δf_{mix} is negative ($\partial^2 \Delta f_{mix}/\partial \varphi^2 < 0$) and mixtures with compositions in this region will spontaneously demix in order to reach the lower free energy. This means that the mixture is unstable toward any weak compositional fluctuations from its mean values and will phase separate into compositions φ_1 and φ_2. This is *spinodal* decomposition. Compositions between A and B and between C and D at this temperature reside in a regime where the curvature is positive ($\partial^2 \Delta f_{mix}/\partial \varphi^2 > 0$). This is the *metastable* regime. In this situation, the composition fluctuations within this initially homogeneous sample would have to be large in order to demix into compositions φ_1 and φ_2. Demixing occurs via nucleation and subsequent growth. Specifically, in an A-rich environment, for example, a small nucleus of the B-component forms and, if it is beyond a critical size, it will grow.

The coexistence curve between the homogeneous and metastable regime (phase boundary) can be obtained by constructing a common tangent at compositions φ_1 and φ_2,

$$\left. \frac{\partial \Delta f_{mix}}{\partial \varphi} \right|_{\varphi=\varphi_1} = \left. \frac{\partial \Delta f_{mix}}{\partial \varphi} \right|_{\varphi=\varphi_2} \qquad 9.8$$

In other words, the chemical potentials are equal at the respective compositions. In the situation of interest ($N_A = N_B = N$), the slope is zero, $\partial \Delta f_{mix}/\partial \varphi = 0$, implying that the values of χN that define the phase boundary, or equivalently the binodal, are

$$(\chi N)_b = \frac{\ln[\varphi/(1-\varphi)]}{(2\varphi - 1)} \qquad 9.9$$

The curve which encloses the regime in which spinodal decomposition occurs is determined by the condition $\partial^2 \Delta f_{mix}/\partial \varphi^2 = 0$, and is

$$(\chi N)_s = \frac{1}{2}\left(\frac{1}{\varphi} + \frac{1}{1-\varphi} \right) \qquad 9.10$$

the spinodal. The critical point corresponds to the minimum in the spinodal ($\partial(\chi N)_s/\partial \varphi = 0$) which indicates for the symmetric case $\chi_c N = 2$. The condition defined by $\chi N = 2$ is or particular significance because a mixture characterized by $\chi N < 2$ is homogeneous whereas a mixture characterized by $\chi N > 2$ will exhibit a tendency to lower its free energy by phase separating. This result was obtained using equation 9.8 and the incompressibility requirement, $\phi_A + \phi_B = 1$ ($\varphi = \varphi_A$). In general, however, the condition for equilibrium is specified by equation 9.8 (equating the chemical potentials in each phase) and by equating the osmotic pressures, $\Pi = (\phi_i/v_i)\frac{\partial f(\phi_i)/\phi_i}{\partial \phi_i} (i = A, B)$, in each phase, A and B (see Safran 1994). Using the equations for $(\chi N)_b$ and $(\chi N)_s$, equations 9.9 and 9.10, respectively, the phase diagram in Fig. 9.4 was calculated for different values of χN. Since $\chi \sim 1/T$, then the critical temperature

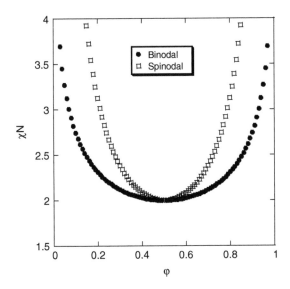

FIG. 9.4
The values of χN that identify the spinodal (Eq. 9.10) and binodal (Eq. 9.9) boundaries for a symmetric polymer-polymer mixture. The critical composition here (where $N_A = N_B = N$) corresponds to $\varphi_c = 1/2$, but as φ deviates toward the binodal the minor component cannot maintain a bicontinuous phase, due largely to surface tension effects, and becomes discontinuous.

T_c is a UCST when $\chi > 0$. At the critical point (φ_c, T_c) the spinodal and binodal coincide.

With the foregoing discussion, we have articulated the significance of the spinodal and binodal regimes of the phase diagram of a binary mixture. These phenomena are discussed in further detail below.

9.3 Spinodal Decomposition

9.3.1 Linearized Theory for the Early Stages of Spinodal Decomposition

In this section, the early stage dynamics and structural evolution of an A/B homogeneous mixture placed into the spinodal regime of the phase diagram are now discussed. The starting point for the dynamics is the diffusion equation

$$\frac{\partial \varphi(\bar{r},t)}{\partial t} = -\nabla \bullet \bar{J}(\bar{r},t) + \varsigma(\bar{r},t) \qquad 9.11$$

where

$$\bar{J}(\bar{r},t) = -L\nabla\bar{\mu} \qquad 9.12$$

with the local chemical potential,

$$\bar{\mu} = \frac{\partial F}{\partial \varphi(\bar{r})} \qquad\qquad 9.13$$

expressed in terms of a derivative of the free energy and a mobility term, *L*. The mobility term was introduced in Chapter 4, but will be discussed later, in Chapter 10 on Interdiffusion. The second term on the RHS in the diffusion equation is the noise term, discussed earlier in relation to the Langevin equation (Chapter 2). Briefly, it is due to phonon modes and, in principle, reflects the influence of thermal fluctuations on the dynamics of evolution.

When the single phase mixture is thrust into the two phase regime, compositional inhomogenieties develop and mean field or regular solution free energy functions, such as the Flory-Huggins free energy, which describe the system in the homogeneous regime are no longer appropriate. In the two-phase regime, the local composition gradients are dealt with in an approximate analytical manner by including a term proportional to the square of the gradient of the local composition, $|\nabla \varphi|^2$. An appropriate free energy functional $F\{\varphi\}$ is the Landau-Ginzburg functional, which has two contributions, a term describing the homogeneous mixture, $f(\varphi)$, and the square gradient term reflecting contributions due to the local compositional inhomogenities,

$$F\{\varphi\} = \int d\bar{r} \left[\frac{1}{2} K|\nabla \varphi|^2 + f(\varphi) \right] \qquad\qquad 9.14$$

where the $K = K(\varphi)$ is associated with the interfacial tension (see for example Safran, 1994, Brown and Chakrabarti 1993, Binder 1983, Gunton et al 1983). Generally, this free energy functional includes a series expansion of gradients of the local order parameter, but only the $|\nabla \varphi|^2$ term is retained here.

It has been emphasized that in the unstable, spinodal, region of the phase diagram, fluctuations of the composition are amplified. The current goal is to determine conditions under which compositional fluctuations are amplified or suppressed. Cahn solved a linearized version of the diffusion equation to describe the conditions under which this phenomenon would be possible in a mixture. The procedure involves expanding the free energy function $f(\varphi)$ in terms of a Taylor series, around the average, homogeneous, composition $\varphi = \varphi_0$

$$f(\varphi) = f(\varphi_0) + (\varphi - \varphi_0)\frac{\partial f}{\partial \varphi}\bigg|_{\varphi_0} + \frac{1}{2}(\varphi - \varphi_0)^2\frac{\partial^2 f}{\partial \varphi^2}\bigg|_{\varphi_0} + \cdots \qquad\qquad 9.15$$

from which it follows that Eq. 9.14 may be rewritten as

$$F\{\varphi\} \approx \int d\bar{r} \left[\frac{1}{2}\left(K|\nabla \varphi|^2 + \left(\frac{\partial^2 f}{\partial \varphi^2}\right)_{\varphi_0} (\varphi - \varphi_0)^2 \right) \right] \qquad\qquad 9.16$$

Note that since the volume integral $\int \varphi d\bar{r} = \int \varphi_0 d\bar{r} = M$ (*total* material), the term involving the first derivative of the free energy function and of $f(\varphi_0)$ vanishes.

In order to describe the local composition fluctuations in the mixture quantitatively, a new variable $u(\bar{r})$ was defined as the difference between the composition at location $\bar{r}$ and the average composition of the mixture, φ_0,

$$u(\bar{r}) = \varphi(\bar{r}) - \varphi_0 \qquad 9.17$$

$u(\bar{r})$ therefore represents the amplitude of a local fluctuation of the composition. Apart from linearizing (linear in amplitude) the equation, Cahn made an additional approximation by neglecting the noise term, which can be important under certain circumstances, as described later. The diffusion equation may be rewritten as,

$$\frac{\partial u(\bar{r})}{\partial t} = L\nabla^2 \left[-K\nabla^2 u(\bar{r}) + \left(\frac{\partial^2 f}{\partial \varphi^2} \right)_{\varphi_0} u(\bar{r}) \right] = L\nabla^2 \frac{\partial^2 f}{\partial \varphi^2} \qquad 9.18$$

A solution to this differential equation is now sought. Since the goal is to identify conditions under which the amplitude, $u(\bar{r})$, of any small fluctuations (infinitesimal disturbance) in composition would become unstable and grow, an effective strategy is to use the technique of linear stability analysis. One begins with a solution to the equation that describes the perturbations (in an idealized or, approximate, manner). The nature of the solution is such that it would reveal the conditions under which the parameter representing the amplitude of the disturbance would dampen or grow. The use of the word *linear* to describe the strategy reflects the fact that the differential equation includes terms linear in the amplitude of the fluctuation; terms of higher order in $u(\bar{r})$ are neglected. In this regard, the analysis describes only the initial stages of the instability.

In principle, the spatial and temporal dependence of the perturbation may be described analytically in terms of Fourier components,

$$u(\bar{r},t) = \hat{u}(t)e^{i\bar{q}\cdot\bar{r}} \qquad 9.19$$

where $\bar{q}$ is the wave vector. The time dependence is not directly specified at this point. Upon substituting Eq. 9.19 into the diffusion equation the following ordinary differential equation is obtained

$$\frac{d\hat{u}}{dt} = -\left\{ LKq^4 - Lq^2 \left(\frac{\partial^2 f}{\partial \varphi^2} \right)_{\varphi_0} \right\} \hat{u} \qquad 9.20$$

The solution to this equation is an exponential,

$$\hat{u}(q,t) = \hat{u}(q,0)e^{-\omega(q)t} \qquad 9.21$$

where

$$\omega(q) = LKq^2 \left(q^2 + \frac{1}{K} \left(\frac{\partial^2 f}{\partial \varphi^2} \right)_{\varphi_0} \right)$$
9.22

When $\omega(q) < 0$, Eq. 9.21 indicates that the amplitude of the fluctuation grows exponentially with time (amplification). On the other hand the fluctuations dampen when $\omega(q) > 0$. The sign of $\omega(q)$ is determined by the magnitude of q in the dispersion relation, Eq. 9.22. It is now possible to identify the modes which are unstable using Eq. 9.22. Recall from the earlier discussions that the curvature of the free energy in the spinodal regime is negative, $\left(\frac{\partial^2 f}{\partial \varphi^2} \right) < 0$. The critical value of the wave vector below which the disturbances will become unstable and grow is

$$q_c^2 = \frac{1}{K} \left| \left(\frac{\partial^2 f}{\partial \varphi^2} \right)_{\varphi_0} \right|$$
9.23

Otherwise, when $q > q_c$, and $\omega(q) > 0$ the amplitude decays. In other words, fluctuations in composition that are characterized by long wave lengths (small q) are unstable.

The significance of $\left(\frac{\partial^2 f}{\partial \varphi^2} \right)$ is now considered. If only long wavelength fluctuations are considered, the $KV^2\varphi$ term in Eq. 9.18 becomes negligible (concentration does not vary much in the long wavelength limit) and Eq. 9.18 becomes,

$$\frac{\partial u(\vec{r})}{\partial t} = LV^2 \left(\frac{\partial^2 f}{\partial \varphi^2} \right)_{\varphi_0} u(\vec{r})$$
9.24

implying that the diffusion coefficient, D, is

$$D = L \left(\frac{\partial^2 f}{\partial \varphi^2} \right)_{\varphi_0} < 0$$
9.25

The parameter $\left(\frac{\partial^2 f}{\partial \varphi^2} \right)_{\varphi_0}$ therefore represents a thermodynamic term that influences the magnitude and sign of the diffusion coefficient. The diffusion coefficient in Eq. 9.25 is identified as an interdiffusion (or mutual diffusion) coefficient and unlike tracer diffusion is determined by chemical potential gradients. Interdiffusion will be discussed in the next chapter. That $D < 0$ within the spinodal regime is expected since it indicates that the initially homogeneous system begins to *demix* when quenched into this regime. This, by the way, is the first time in this book that we have shown that diffusion can occur up a concentration gradient instead of down. Cahn referred to this phenomenon as "uphill diffusion."

By plotting the negative of $\omega(q)$ as a function of q in Fig. 9.5, it is evident that the (early stage) fluctuations grow for $q < q_c$, otherwise they dampen.

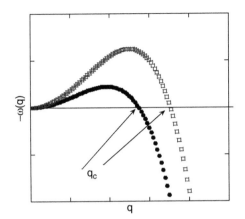

FIG. 9.5
The growth rate of the initial stage of the instability is shown above for a quench into the unstable state. Two different quench depths are shown here, with the squares representing the smaller quench depth.

Clearly, when $q^2 < -\frac{1}{K}|(\frac{\partial^2 f}{\partial \varphi^2})|_{\varphi_0}$, or equivalently $q < q_c$ and $\omega(q) < 0$, the fluctuations grow exponentially. This reinforces the notion that concentration fluctuations of longer wavelengths (smaller q) are associated with the destabilization of the homogeneous structure in this regime.

The structure of the phase separated mixture is dominated by the most rapidly growing wave, determined by the maximum in $\omega(q)$, and is characterized by an optimal wave vector

$$q_{max} = \frac{q_c}{\sqrt{2}} = \frac{1}{\sqrt{2}}\left(-\frac{1}{K}\frac{\partial^2 f}{\partial \varphi^2}\right)^{1/2}_0 \qquad 9.26$$

or equivalently a wavelength $\lambda_{max} = \sqrt{2}\lambda_c$, where $\lambda_c = \frac{2\pi}{q_c}$. The average dimension of the width of the features in the spinodal pattern (Fig. 9.3) is characterized by an optimal wavelength $\lambda_m \approx \sqrt{2}\lambda_c$.

A physical argument for an optimal wavelength will now be considered. In order to create narrow stripes one would naively surmise that the molecules would have to diffuse a short distance and for this reason narrow stripes (small wavelength pattern) would form rather quickly because the diffusion distance is short. By extension, a pattern with wide stripes would take a much longer time to develop because of the larger diffusion distance. This line of reasoning would argue that systems with narrow stripes would be optimal. It turns out that this argument would be true only for a homogeneous system (for a polymer-polymer mixture, the free energy would be specified by the Flory-Huggins equation). The situation in the inhomogeneous phase is qualitatively and quantitatively different. Notably, this is where the impact of the square gradient term comes in with regard to the inhomogeneous phase. The effect of the gradient term is to favor the formation of larger stripes to reduce the total A/B interfacial area. The result of this competition is the development of stripes characterized by an optimal wavelength.

9.3.2   Structure Factor

Scattering experiments have proven very effective at analyzing the structural evolution. Based on the discussions in Chapter 2, it should be evident that the correlation function, $S(|\vec{r} - \vec{r}_0|, t)$, of the concentration fluctuations, i.e.,

$$S(|\vec{r} - \vec{r}_0|, t) = \langle u(\vec{r}, t) u(r_0, 0) \rangle \qquad 9.27$$

would provide relevant information. The Fourier transform of $S(|\vec{r} - \vec{r}_0|, t)$ is the structure factor and the intensity of scattered light is proportional to the structure factor (Chapter 2)

$$I \propto S(\vec{q}, t) = \int e^{i\vec{q}\cdot\vec{r}} S(|\vec{r} - \vec{r}_0|, t) d\vec{r} \qquad 9.28$$

indicating that the intensity increases exponentially with time in a scattering experiment

$$I \propto |\hat{u}(q, 0)|^2 e^{-2\omega(\vec{q})t} \qquad 9.29$$

or alternatively,

$$S(q, t) = S(q, 0) e^{-\omega(\vec{q})t} \qquad 9.30$$

In the frequency domain the structure factor is readily rewritten as

$$S(\omega, q) \propto \frac{\omega^2 q^2}{\omega^2 + (\omega^2 q^2)^2} \qquad 9.31$$

In an actual experiment (in reciprocal space) the q-vector is related to the scattering angle θ, such that $q = (4\pi n/\Lambda)\sin(\theta/2)$, where n is the index of refraction of the medium and Λ is the wavelength of the beam of light. In a typical experiment, the peak occurs at q_{max} which is independent of time (in the linear regime). A plot of $S(q)$ versus q is shown in Fig. 9.6. The sketch reveals a maximum at q_{max}.

It is important to note that the amplitude grows while the wavelength remains constant (Fig. 9.6b) is a very important signature of the spinodal process during the early stage. An image of the morphology alone is insufficient because a scenario wherein a large density of droplets due to nucleation could evolve to create a pattern that looks spinodal is entirely possible.

9.4   An Example Involving a Polymer-Polymer Mixture

We are about to embark on a further analysis specifically involving polymers. For long-chain polymers the free energy per unit volume is approximated as

$$F\{\varphi\} = \frac{kT}{N}[\varphi \ln \varphi + (1 - \varphi)\ln(1 - \varphi)] + kT[\chi \varphi(1 - \varphi) + K(\nabla \varphi)^2] \qquad 9.32$$

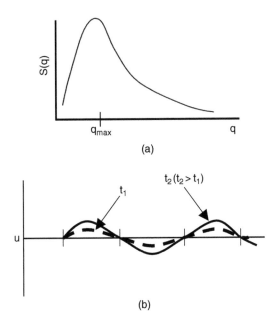

FIG. 9.6
a) A typical plot of $S(q)$ versus q is shown here for a an arbitrary mixture; b) early stage dynamics of mixture phase separating via spinodal decomposition. The amplitude increases while the wavelength remains constant.

with $K = \frac{b^2}{36\varphi(1-\varphi)}$. This term, as indicated by deGennes, is due to entropic constraints associated with the long-chain polymers. Nb^2 is the mean square end-to-end distance of the polymer. The diffusion equation now becomes

$$\frac{\partial \varphi}{\partial t} = kTL\nabla^2\left[\frac{1}{N}\ln\frac{\varphi}{1-\varphi} + \chi - 2\varphi\chi + \frac{(1-2\varphi)b^2}{36\varphi^2(1-\varphi)^2}(\nabla\varphi)^2 - \frac{b^2\nabla^2\varphi}{18\varphi(1-\varphi)}\right] + \zeta(\vec{r})$$

9.33

An expression for the relaxation time can be derived from this equation by following the procedure outlined above to get $\omega(q)$. The growth rate is described alternatively in terms of a relaxation time $\tau(q) = 1/\omega(q)$,

$$\tau(q) = \frac{1}{Dq^2\left[1 + \frac{K}{(\partial^2 f/\partial\varphi^2)_0}q^2\right]}$$

9.34

where

$$\frac{K}{(\partial^2 f/\partial\varphi^2)_0} = \frac{b^2}{36}\frac{\chi_s}{(\chi - \chi_s)}$$

9.35

This result is particularly significant since it represents a length scale, the correlation length, that characterizes the phase separation process. The correlation length is identified here as

$$\xi = \frac{b}{6}\sqrt{\frac{\chi_s}{(\chi_s - \chi)}}$$

9.36

Physically the correlation length tells us the length scale over which the fluctuations are correlated. As the system approaches the critical point, the correlation length diverges. Indeed the correlation length is much larger than the average dimension of a chain, which confirms the fact that the long wavelength modes are active and that the dimensions of the stripes in the spinodal pattern are larger than the size of the polymer chains.

Consequences of the divergence of the correlation length are now described. It is straight forward to show that the relaxation time now becomes

$$\tau(q) = \frac{1}{Dq^2[1 + \xi^2 q^2]}$$

9.37

The other point that should be emphasized is that the diffusion coefficient, D, undergoes "thermodynamic slowing down" as the temperature approaches the critical point, i.e., χ approaches χ_s. This should be obvious if we consider that D is the product of a mobility term and the driving force,

$$D = 2kTL(\chi_s - \chi) \propto \xi^{-2}$$

9.38

This prediction indicates that when the driving force decreases with T, the effect of the thermodynamic interactions is to reduce the magnitude of the effective D. When $\chi > \chi_s$, the system demixes, "uphill diffusion," we are now in the unstable regime, $D < 0$!

In the next chapter the topic of interdiffusion in materials is discussed. It is an important topic in its own right. The interdiffusion coefficient will be derived and used to describe dynamics of small molecule as well as long-chain mixtures.

9.5   Remarks Regarding Spinodal Decomposition

The theory described, heretofore, laid the groundwork for a large body of experiments and theory on the topic. This work of Cahn and Hillard (1958, 1965, 1968) was extended by Cook to include the noise term, revealing that an additional flux arises from the random thermal fluctuations. This became known as the linearized Cahn, Hillard, and Cook theory of spinodal decomposition. In some scientific communities, this is sometimes identified as the Ginzberg-Landau theory (Gunton et al. 1983). Cahn-Hillard theory has formed the basis for subsequent developments in the field. Simulations indicate that increasing the strength of the noise term has the effect of creating broader and more diffuse boundaries between the phases and rougher domain topography (Rogers et al 1988). The linear theory appears to provide a satisfactory description of the early stage phase separation for deep quenches in the unstable regime (see for example Jinnai et al 1993; and Kubota et al 1992).

The late stage dynamics have been the subject of appreciable research, theoretically and experimentally. Briefly, in cases where the volume fraction of one component is very small, the problem has been solved analytically by Lifshitz and Slyozov in 1961 and the predictions are that the structure can be characterized by one length scale! This length scale is the domain size, $R(t) \sim t^{1/3}$. An essential component of this theory is the absence of correlations in the locations and interactions between components of the dispersed minority phase (droplets). Experiments and computer simulations in the intervening years support this prediction both for polymer and for small molecule mixtures (Rogers et al 1988). Having mentioned this point, a significant distinction between polymers and the small molecule systems is the existence of the entropic term. This term influences the translational dynamics of chains. This has a further significance. The Lifshitz-Slyozov evaporation-condensation theory describes a mechanism by which molecules detach (evaporate), get transported from smaller droplets of the minority droplet A-phase and migrate through the major phase and subsequently become incorporated into (condensation) a larger A-droplet. Hence the large droplets grow at the expense of the smaller ones; we will address the fundamental reason for this in Chapter 11, Section 11.2. It is clear from foregoing chapters that the barrier to motion by long chains is prohibitively large. Therefore, with regard to polymers, there is evidence from simulations that this mechanism may not be highly favored (M.A. Kotnis and M. Muthukumar 1992). This is generally true for when the A-domains do not form a percolated network. We note briefly that when hydrodynamic effects become important the $t^{1/3}$ behavior no longer holds. Siggia examined the effects of hydrodynamic on the late stage coarsening indicating that for near-critical quenches, a coalescence mechanism, wherein droplets undergo Brownian motion and coalesce, is favored at finite concentrations (Siggia 1979 and Tanaka 1996).

9.6 Nucleation

Having discussed the phase separation process in the unstable regime, phase separation in the metastable region of the phase diagram is now discussed. The surface energy associated with the formation of a cluster of atoms, bound together, constituting a nucleus is an important energy barrier that needs to be surmounted before a nucleus can become stable and grow in size to form a droplet or a small crystal, depending on the system (liquid-vapor or solid-melt). The nucleation process may be *homogeneous* or *heterogeneous*. If nucleation is to proceed homogeneously then a sufficiently large driving force must exist. This implies that in a liquid-vapor or liquid-liquid system, the saturation vapor pressure, or the concentration of the relevant species, must be sufficiently high. Hence, the system must reside deep in a supercooled (or supersaturated) state for this to occur. For *heterogeneous nucleation*, the nucleation process, in principle, occurs on a foreign surface and the energy barrier for nucleation is lower than that required for homogeneous nucleation.

9.6.1   Nucleation in the A/B Mixture

The initially homogeneous mixture is unstable toward *large* localized fluctuations of composition when thrust into the metastable regime of the phase diagram. A nucleus, β-phase, forms in the majority α-phase provided the locally available energy is larger than a critical value, ε_c. This critical energy barrier is determined by a competition between a volume contribution to the free energy, F_v, of the system which favors cluster formation, and a surface free energy term, F_s, which opposes the formation of the cluster (Fig. 9.7). The implication is that this nucleus must possess a radius that is beyond a critical size, r_c, for $F_v > F_s$. When the phases are fully formed, a late stage growth process takes place.

The classical theory of nucleation and growth by Becker and Döring (1939); which is over 75 years old (!), provided reliable intuition regarding how the nucleation mechanism proceeds in simple mixtures. The intent of this section is to provide insight into the classical nucleation and growth process.

9.6.2   Elements of the Classical Theory of Nucleation

The critical energy ε_c for the formation of a nucleus, the critical radius, r_c, of a nucleus and the equilibrium number of nucler per unit volume area now calculated. We begin by considering the free energy change associated with the formation of a spherical nucleus of phase β (e.g., liquid) of radius r in phase α (e.g., vapor). The free energy has two primary contributions, a bulk and a surface/interfacial contribution,

$$\Delta G = \varepsilon(r) = 4\pi r^2 \gamma - \frac{4}{3}\pi r^3 \frac{\delta\mu}{v} \qquad\qquad 9.39$$

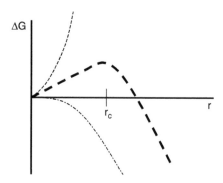

FIG. 9.7
A diagram of the Gibbs free energy as a function of cluster radius illustrating the relative contributions of the surface energy to the bulk free energy contribution toward formation of a stable nucleus with radius $r > r_c$.

The first term in this equation represents the interfacial contribution to the free energy ($\sim r^2$) that opposes formation of the nucleus of radius r; the interfacial energy between the α and β phases is γ. The second term is the volume contribution and is the driving force behind the formation of the nucleus ($\sim r^3$). $\delta\mu$ is the difference between the chemical potential at the coexistence curve (binodal) and the point within the metastable regime where the sample resides. In a liquid-vapor system, it would represent the difference between that of the liquid phase (β) and that of the vapor phase (α). The Gibbs free energy change $\delta\mu$ associated with the transfer of δk molecules to a cluster in the liquid β-phase from vapor phase (α) is $(\mu_\beta - \mu_\alpha)\delta k = \delta\mu\delta k$). Note that the number of molecules transferred to form the cluster is $k = \frac{4}{3}\pi r^3 \frac{1}{v}$, where the density is $\frac{1}{v} = \rho A_N/M$, is the number density (A_N is Avagardo's number and M is the molecular weight of a molecule). In a super saturated vapor environment, the second term in Eq. 9.39 is negative, which creates a driving force for nucleation.

Based on the condition $\partial\varepsilon/\partial r = 0$, the critical radius is

$$r_c = 2\gamma\frac{v}{\delta\mu} \qquad 9.40$$

or equivalently,

$$k_c = \frac{32}{3}\pi\frac{\gamma^3 v^2}{[\delta\mu]^3} \qquad 9.41$$

The associated critical energy is

$$\varepsilon_c = \gamma^3\frac{16\pi}{3}\frac{v^2}{[\delta\mu]^2} = \frac{4\pi r_c^2\gamma}{3} \qquad 9.42$$

The process is illustrated schematically in Fig. 9.8 where r_c and ε_c are identified. It should be anticipated that the number of nuclei, each composed of k particles, in the system would be determined by the Boltzmann factor,

$$\frac{n_k}{N} = e^{-\varepsilon_k/k_B T} \qquad 9.43$$

where $\varepsilon_k(\varepsilon_k > \varepsilon_c)$ is the free energy of formation of a nucleus of k particles and N is the total number of particles in the system. In the next section the steady state growth rate is considered.

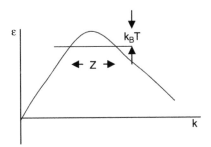

FIG. 9.8
The significance of the Zeldovich factor z is identified here.

9.6.3   Steady State Growth Rate

The number of nuclei of radius $r > r_c$ formed per unit volume under steady state conditions is of interest here. The basic assumption behind the classical theory of nucleation is that the average number of nuclei, $n_k(t)$, each composed of k particles, present at time t is determined by the rate at which a nucleus looses, or gains, a particle. In other words nuclei increase or decrease in size with the addition or removal of one particle at a time; coalescence of droplets is assumed to be prohibited. Therefore, for an embryo composed of n particles, the following growth law would apply

$$Q_n + Q_1 \rightarrow Q_{n+1} \qquad\qquad 9.44$$

For the case of shrinkage

$$Q_{n+1} - Q_1 \rightarrow Q_n \qquad\qquad 9.45$$

The rates of the processes, shrinkage/growth, are characterized by appropriate rate constants. Consequently, the rate per unit volume at which entities composed of k particles increase in size from $k-1$ to k is

$$I_k = R_{k-1}n_{k-1}(t) - R_k n_k(t) \qquad\qquad 9.46$$

where R_k and R_{k-1} are rate constants. In principle one can envision a flux of particles that impinge on the embryo Q_n. The flux Φ_0 might be given by Eq. 1.54 and the probability of an occurrence (scattering cross section) would be determined by the cross sectional area, A_k, of the embryo. It follows that $R_k \sim \Phi_0 A_k$. The other constant R_{k-1} would be determined by the detachment (of evaporation) of a particle from the embryo. The effective rate equation for $n_k(t)$ is therefore

$$\frac{\partial n_k(t)}{\partial t} = I_k - I_{k+1} \qquad\qquad 9.47$$

In an effort to find an explicit equation for the time-dependence of $n_k(t)$, we begin by eliminating R_{k-1} and recognizing that under equilibrium conditions $I_k = 0$. Hence Eq. 9.46 indicates that

$$\frac{R_k}{R_{k-1}} = e^{-[(\varepsilon_k - \varepsilon_{k+1})/k_BT]} \qquad\qquad 9.48$$

With eqn. 9.48 I_k to be rewritten as, (see problem 8)

$$I_k = R_k\left[-\frac{\partial n_k}{\partial k} - \frac{1}{kT}n_k\frac{\partial \varepsilon_k}{\partial k}\right] \qquad\qquad 9.49$$

In addition, since

$$I_k - I_{k+1} \approx \frac{\partial I_k}{\partial k} \qquad\qquad 9.50$$

and $\frac{\partial l_k}{\partial k} = -\frac{\partial n_k}{\partial t}$, a diffusion equation describing the time-dependence of the number of clusters or entities composed of k particles is, from eqn. 9.47,

$$\frac{\partial n_k}{\partial t} = \frac{\partial}{\partial k}\left\{ R_k \frac{\partial n_k}{\partial k} + \frac{1}{kT}\frac{\partial}{\partial k}\left(R_k n_k \frac{\partial \varepsilon_k}{\partial k} \right) \right\} \qquad 9.51$$

The equivalence to the diffusion equation is apparent if n_k is associated with the concentration, the term in parentheses is associated with the flux, and the diffusion coefficient with R_k. Various solutions to this equation can be obtained under different limiting conditions. This equation will enable calculation of the rate of production of entities (or nuclei or droplets) greater than a critical size under steady-state, and nonequilibrium, conditions (see problem 9)

$$I_S \approx \frac{N}{R_c} Z e^{-\varepsilon_c/k_B T} \qquad 9.52$$

where $Z \frac{1}{k_c}(\frac{2\varepsilon_c}{3\pi k_B T})^{1/2}$ is called the Zeldovich factor. The Zeldovich factor is a measure of the breadth of the ε versus k curve at a distance $k_B T$ below the peak position, as illustrated in Fig. 9.8. It would appear that the factor represents the relative influence of the size of a thermal fluctuation, $\sim k_B T$, for clusters composed of k_c molecules with free energy near ε_c, on the nucleation rate.

This result (cf. eqn. 9.52) is consistent with physical intuition. The steady state nucleation rate should be proportional to the number of molecules available in the system, the flux of molecules and a probability that a particle will stick ($R \sim 1/\text{flux} \bullet \text{scattering cross section}$). The result also indicates that if the critical number of molecules needed to create a nucleus is large, then the steady state growth rate should decrease. Moreover, the result indicates that the rate should decrease as the critical energy increases. The temperature dependence of the prefactor indicates that if the thermal energy is large, the nucleation rate should decease. This, too, is intuitive.

The aforementioned analysis is applicable to the liquid/vapor systems. Within the frame work of the model the situation on the other hand is in general different for the nucleation of a crystalline solid from a liquid or vapor. The important difference is that the critical energy for formation of the solid needs to include the shape factor (clusters are not necessarily spherical) and crystallographic orientations. In addition various thermally activated dynamic processes associated with the solid environment need to be considered. Specifically with respect to solid state nucleation events, strain energy, lattice mismatch, and coherency effects (coherency refers to the regularity of lattice planes) and various activated transport mechanisms involving defects need to be considered. In a crude way, one can make a first order approximation and lump all the strain energy effects in the form of a new free energy contribution. Equation 9.39 would then be modified by adding an additional term due to the effects of strain energy. In addition the new form of the free energy function (Eq. 9.39) would also be modified to include the shape factors since the clusters would not remain spherical (see for example K.C. Russell 1980).

9.7 Heterogeneous Nucleation

Brief comments are now made with regard to heterogeneous nucleation. The goal is to show that heterogeneous nucleation occurs with lower activation energy than homogeneous nucleation under appropriate conditions. Heterogeneous nucleation occurs on surfaces, impurities, or defects in crystals such as dislocations and grain boundaries.

Consider a case of nucleation of a liquid droplet on a substrate, the simplest case of heterogeneous nucleation, as illustrated in Fig. 9.9. In this figure, a droplet, β, is in contact with a substrate, s, in an environment, α. The free energy change associated with the formation of a nucleus has three contributions, a bulk term, the β-phase, and two interfacial energy terms, one associated with the α-phase/β-phase contact, $\alpha\beta$, and the other with the α-phase/substrate contact, αS,

$$\varepsilon = \frac{S^\beta r^3}{v^\beta}(\mu_\beta - \mu_\alpha) + r^2 S^{\alpha\beta}\gamma^{\alpha\beta} + S^{\alpha S}(\mu_{\beta s} - \mu_{\alpha s}) \qquad 9.53$$

where S^β, $S^{\alpha\beta}$ and $S^{\alpha S}$ are the shape factors. As an example a nucleus in the form of a hemispherical cap is considered. For a truly hemispherical cap, the shape factors are $S^\beta = \frac{\pi}{3}(2 - 3\cos\theta - \cos^3\theta)$, $S^{\alpha\beta} = 2\pi(1 - \cos\theta)$ and $S^{\alpha S} = \pi\sin^2\theta$.

The critical free energy necessary to form a nucleus is

$$\varepsilon_c^{het} = \frac{4\pi}{3}\frac{(\gamma^{\alpha\beta})^3(v^\beta)^2}{(\mu^\alpha - \mu^\beta)^3}(2 - 3\cos\theta - \cos^3\theta) \qquad 9.54$$

For an identical system the critical energy for formation of a nucleus via heterogeneous nucleation is smaller than that by homogeneous nucleation

$$\varepsilon_c^{het} = f\varepsilon_c \qquad 9.55$$

where $f = \frac{(2 - 3\cos\theta - \cos^3\theta)}{4} < 1$. Indeed it is easier for heterogeneous nucleation to occur in systems.

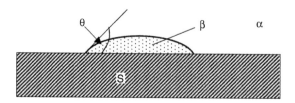

FIG. 9.9
Schematic of a heterogeneous nucleation process involving a nucleus (β-phase) in contact with a substrate in an $\alpha - q$.

9.8 Concluding Remarks on Nucleation and Growth

The late stage dynamics involves potentially a ripening process driven by chemical potential gradients associated with the curvature of the droplets. Large β-droplets grow at the expense of the smaller droplets. This is the theory of Ostwald ripening, first discussed by Lifshitz and Slyozov (1961). This theory is applicable to the case where one component is of small volume fraction so interactions between droplets of this species could be neglected (Alkemper et al. 1999). In later years this work was generalized by a number of authors (see, for example, Ratke and Voorhees 2002). Other coarsening mechanisms, such as coalescence of droplets, occur and volume fractions can form a percolated network. Various extensions and generalizations have been made in order to account for different correlations that this theory. Computer simulations have had an important impact on our overall understanding of this process. This topic has been examined in detail by a large number of authors in various texts and review articles (see, for example, Langer 1967; Coleman 1977; Binder 1983; Gunton et al. 1983).

We have discussed the topic of classical nucleation. In an A/B mixture, the late stage dynamics (the growth phase), which occurs after the phases have formed, can occur via a ripening process or by a coalescence process. The literature on this topic is vast and the topic is still an active area of research. Recent corrections to the classical theory involve corrections to the free energy (see, for example, Maksimov et al. 2000 and Ruth et al. 1988). Other recent theory examines the lag time associated with formation of the nucleus (Maksimov et al 2000; Lefebvre et al 2002). Others examine spatial correlations among droplets (Sagui et al 1999). This continues to be an active area of research. Very recent studies of nucleation in polymer-polymer mixtures have raised additional questions regarding aspects of the applicability of conventional theories of nucleation. Measurements of the critical nucleus and the its size dependence on quench depth suggest that aspects of the phase separation process may require closer scrutiny (Balsara et al 2004).

9.9 Problems for Chapter 9

1. The Flory-Huggins interaction parameter, c, for an A-B polymer-polymer has a value of 9×10^{-4}. a) If they possess the same degree of polymerization, $N_A = N_B$, calculate the molecular weights beyond which they would phase separate. b) If $N_A = 100$, determine the value of N_B for which they would not be compatible.

2. Using the expression for the Flory-Huggins free energy, a) show that $(\chi N)_b = \frac{\ln[\varphi/(1-\varphi)]}{(2\varphi-1)}$ and that $\chi_s = \frac{1}{2N}(\frac{1}{\varphi}+\frac{1}{1-\varphi})$ for a symmetric

blend, $N_A = N_B$. b) Calculate the relevant expressions for general values of N_A and N_B.

3. If $\chi = c_1 + c_2/T$, calculate an expression for the critical temperature T_c *in terms of* c_1 *and* c_2 for a symmetric blend. If $c_2 = 0.1$, calculate a range of values for c_1 that would predict an LCST.

4. The following is an expression that has been used for $f(\varphi)$ to approximate the free energy of binary small molecule mixtures

$$f(\varphi) = -\frac{1}{2}rx^2 + \frac{1}{4}ux^4$$

where r and u are phenomenological parameters. If $u = 1$, use appropriate values of r to illustrate the shape of the curve for $T > T_c$ and $T < T_c$ (x may have positive and negative values). Determine an expression for the spinodal and for the binodal.

5. Show that the dominant wave vector $q_{max} = \frac{q_c}{\sqrt{2}} = \frac{1}{2}(-\frac{1}{K}\frac{\partial^2 f}{\partial \varphi^2})_0^{1/2}$

6. Derive the expression for the relaxation time

$$\tau(q) = \frac{1}{Dq^2\left[1 + \frac{K}{(\partial^2 f/\partial \varphi^2)_0}q^2\right]}$$

and show that

$$\tau(q) = \frac{1}{Dq^2[1 + \xi^2 q^2]}$$

7. Show that the wavelength of the dimensions of the pattern is

$$\lambda_{max} \approx b\left(\frac{\chi_s}{\chi - \chi_s}\right)^{1/2}$$

Comment on the reason this is physically intuitive.

8. First show that eqn. 9.48 becomes $I_k = -R_{k-1}\{(R_k/R_{k-1})n_k(t) - n_{k-1}(t)\}$ second, with the use of eqn. 9.48 and the expansion $e^{-x} \approx 1 + x$ and $\varepsilon_k - \varepsilon_{k+1} \approx \frac{\partial \varepsilon}{\partial k}$ and $\frac{\partial n_k}{\partial k} = n_k(t) - n_{k-1}(t)$ derive eqn. 9.49.

9. First consider the steady state nucleation rate, $\frac{\partial n_k}{\partial t} = 0$, where $I_k = I_S =$ constant is considered. In the model by Becker and Doring, it is assumed that when the number of particles that compose a nucleus is greater than a large number k_c it is no longer considered part of the system. This is specified by the boundary condition $\lim\limits_{k\to\infty} n_k^s = 0$. In addition the source of droplets is associated with $k \to 0$, $\lim\limits_{k\to 0} n_k^s = n_k$. Herewith, Eq. 9.51 becomes

$$-I_S = R_k\frac{\partial n_k^s}{\partial k} + \frac{1}{kT}\frac{\partial}{\partial k}\left(R_k n_k^s\frac{\partial \varepsilon_k^s}{\partial k}\right) \qquad\qquad 9.55$$

The expression for the steady state current of clusters is show that

$$I_S = N\left[\int_0^\infty \frac{e^{-\varepsilon_{k'}/k_BT}}{R_{k'}}dk'\right]^{-1}$$
9.57

Second, a more explicit relation for the steady state nucleation rate may be obtained under a limiting condition. For small super saturations, ε_k possesses a sharp maximum in the vicinity of k_c. With that in mind, ε_k can be expanded in terms of a Taylor series around k_c to yield $\varepsilon_k \approx \varepsilon_{k_c} + \frac{1}{2}\frac{d^2\varepsilon}{dk^2}|_{k_c}(k-k_c)^2 + \cdots$. Recall that the second term associated with the first derivative of ε_k is zero at k_c. Moreover, because the second term is a maximum, $\frac{1}{2}\frac{d^2\varepsilon}{dk^2}|_{k_c} < 0$, necessarily an approximation may be made wherein R_k is replaced with a constant R_{k_c}. With these substitutions show that

$$I_S = \frac{N}{R_{k_c}}\left\{\left[\frac{2(|\partial^2\varepsilon/\partial k^2|_{k_c}|)}{\pi k_BT}\right]^{1/2}\left(1+erf(k_c\sqrt{\frac{(|\partial^2\varepsilon/\partial k^2|_{k_c}|)}{2k_BT}})\right)^{-1}\right\}e^{-\varepsilon_{k_c}/k_BT}$$
9.55

Finally show that:

$$I_S \approx \frac{N}{R_c}\frac{1}{k_c}\left(\frac{2\varepsilon_c}{3\pi k_BT}\right)^{1/2}e^{-\varepsilon_c/k_BT}$$

9.10   References for Spinodal Decomposition

Binder, K., "Collective diffusion, nucleation and spinodal decomposition in polymer mixtures," *J. Chem. Phys.* 79, 6387 (1983).

Brown, G. and Chakrabarti, "Phase separation dynamics in of critical polymer blends," *J. Chem. Phys.* 98, 2451 (1993).

Cahn, J.W, and Hilliard, J.E.,"Free energy of a nonuniform system. I. Interfacial free energy," *J. Chem. Phys.* 28, 258 (1958).

Cahn, J.W., "Phase separation by spnodal decomposition in isotropic systems," *J. Chem. Phys.* 42, 93 (1965).

Cahn, J.W., ""Spinodal Decomposition," The 1967 Institute of Metals Lecture," *TMS Trans, Metall. Soc. AIME* 242, 166 (1968).

de Gennes, P-G., "Dynamics of fluctuations and of spinodal decomposition in polymer blends," *J. Chem. Phys.* 72, 4756 (1980).

Gunton, J.D., San Miguel M. and Sahni, P.S., Phase Transition and Critical Phenomena, edited by C. Domb and J.L. Lebowitz (Academic Press N.Y. vol. 8, p. 269, (1983).

Hayashi, M., Jinnai, H., Hashioto, T., "Validity of linear analysis in early-stage spinodal decomposition of a polymer mixture," *J. Chem. Phys.*, 22, 3414 (2000).

Jinnai, H., Hasegawa, H., Hashimoto, T. and Han, C.C., "Time-resolved small-angle neutron scattering study of spinodal decomposition in deuterated and protonated polybutadiene blends. I. Effect of initial thermal fluctuations," *J. Chem. Phys.* 99, 4845 (1993).

Koga, T. and Kawasaki, K., "Late stage dynamics of spinodal decomposition in binary fluid mixtures," *Physica A*, 196, 389 (1993).

Kotnis, M. A. and Muthukumar, M., "Entropy-induced frozen morphology in unstable polymer blends," *Macromolecules*, 25, 1716 (1992).

Kubota, K., Kuwahara, N., Eda, H., Sakazume, M., and Takiwaki, K., "Dynamic scaling behavior of spinodal decomposition in a critical mixture of 2,5-hexanediol and benzene," *J. Chem. Phys.* 97, 9291 (1992).

Langer, J.S., Bar-on, M., and Miller, H.D., "New computational method in the theory of spinodal decomposition," *Physical Review A.* 11, 1417 (1975).

Lifshitz, I.M. and Slyozov, V.V., "The kinetics of precipitation from supersaturated solid solutions," *J. Phys. Chem. Solids* 19, 35 (1961).

Pincus. P., "Dynamics of fluctuations and spinodal decomposition in polymer blends. II," *J. Chem. Phys.* 75, 1996 (1981).

Rogers, T.M., Elder, K.R.and Desai, R.C., "Numerical study of the late stages of spinodal decomposition," *Physical Review B*, 37, 9638 (1988).

Safran, S.A., Statistical Thermodynamics of Surfaces, Interfaces and Membranes, Frontiers in Physics, Addison-Weslye Publishing Co. NY 1994.

Siggia, E.D., "Late stages of spinodal decomposition in binary mixtures," *Phys. Rev. A* 20, 595 (1979).

Tanaka, H., "Coarsening mechanisms of droplet spinodal decomposition in binary fluid mixtures," *J. Chem. Phys.*, 105, 10099 (1996).

9.11   References for Nucleation and Growth

Avrami, M., "Kinetics of Phase Change. I General Theory," *J. Chem. Phys.* 7, 1103 (1939).

Alkemper, J., Snyder, V.A., Akaiwa, N., and Vorhees, P.W., "Dynamics of late stage phase separation: A test of theory," *Physical Rev. Lett.*, 82, 2725 (1999).

Balsara, N.P., Rappl, T.J. and Lefebvre, A.A., "Does conventional nucleation occur during phase separation in polymer blends?" *Journal of Polymer Science: Polymer Physics ed.* 42, 1793 (2004).

Becker, R. and Döring, W., "Behandlung der Keimbildung in übersättigten Dämpfern," *Ann. Phys (Leipsig)* 24, 719 (1935).

Binder, K.J., "Collective diffusion, nucleation and spinodal decomposition in polymer mixtures," *J. Chem. Phys.* 79(12), 6387 (1983).

Langer, J.S. and Schwartz, A.J., "Kinetics of nucleation in near-critical fluids," *Physical Review A.* 21, 948 (1980).

Lefebvre, A.A., Lee J.H., Balsara, N.P. and Hammouda, B., "Critical length and time scales during the initial stages of nucleation in polymer blends," *J. Chem. Phys.* 116, 4777 (2002).

Maksimov, I.L., Sanada, M. and Nishioka, K., "Energy barrier effect on transient nucleation kinetics: Nucleation flux and lag-time calculation," *J. Chem. Phys.* 113, 3323 (2000).

Ratke, L. and Voorhees, P.W., Growth and Coarsening: Ripening in materials processing, Springer, New York (2002).

Russell, K.C., "Nucleation in solids: The induction and steady state effects," *Adv. In Colloid and Intf. Sce.* 13, 205 (1980).

Ruth, V. and Hirth, J.P. and Pound, G.M., "On the theory of homogeneous nucleation and spinodal decomposition in condensation from the vapor phase," *J. Chem. Phys.* 88, 7079 (1988).

Sagui, C. and Grant, M., "The theory of nucleation and growth during phase separation," *Phys. Rev. E* 59, 4175 (1999).

Langer, J.S. and Schwartz, A.J., "Kinetics of nucleation in near critical fields," *Physical Review A* 21, 948 (1980).

10

Interdiffusion: Diffusion in Chemical Potential Gradients

10.1 Introduction

Much of the emphasis throughout this book, thus far, has been on microscopic mechanisms of diffusional transport in different types of materials. In metals and ionic crystals the influence of the periodic lattice and the nature of the point defects that mediate diffusional transport were highlighted. In metals, different types of defects and lattices of varying geometric structures were responsible for a diverse range of transport mechanisms that occur in these systems. In network glasses the structural disorder, coupled with the transient nonbridging sites that accommodated cationic transport, imposed certain limitations on the nature of the dynamics. Spatial correlations were imposed on the mobile species due to long-range Coulombic effects. In long-chain polymers, translational diffusion of a chain is subject to topological constraints imposed by neighboring chains ("tubes") leading to one-dimensional motion along its own contour. Tracer diffusion and self-diffusion were discussed in detail in order to illustrate the effect of these processes on polymers. The driving force for tracer and self-diffusion is entropic, devoid of complications associated with enthalpic interactions that would influence the magnitude of the interdiffusion coefficient, which is of particular interest in this chapter.

Figure 10.1 shows a sketch of an A/B diffusion couple. Initially at $t = 0$, the concentration profile of each component changes abruptly at the interface. The profile of the A component is shown in the Fig. 10.1. After a sufficiently long period of time the A and B components interdiffuse across the interface and the concentration of each species in the central region is significant. The rate at which the A and B species diffuse is determined by the interdiffusion coefficient which is highly concentration dependent. Generally gradients in chemical potential drive interdiffusion and the profile across the interface is typically not symmetrical. In Chapter 9 an expression for the interdiffusion coefficient was derived and it was found that the interdiffusion coefficient

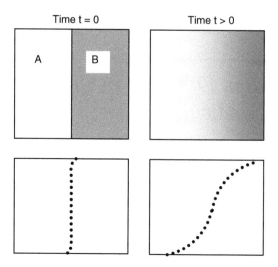

FIG. 10.1
An A/B diffusion couple is shown at time $t = 0$. Initially, the concentration profile is sharp. After a sufficiently long time, interdiffusion occurs and the profile broadens.

is dependent on a transport coefficient, L, and on the curvature of the free energy (Eq. 9.25),

$$D = L \left(\frac{\partial^2 f}{\partial \varphi^2} \right)_{\varphi_0} < 0$$

The goal of this chapter is to derive explicit expressions for the interdiffusion coefficient for both small molecule species and for long-chain polymers.

Technologically the topic of interdiffusion is important in its own right. In many situations, interdiffusion is desirable; it is responsible for the development and evolution of microstructure in materials. This is important because microstructure is intimately connected to physical properties (mechanical properties, magnetic, electronic, and optical) and by extension applications and reliability. Interdiffusion controls adhesion in multilayer systems, from polymers, to intermetallics, to compound semiconductors. Interestingly there are cases in which interdiffusion is not particularly desirable or at least needs to occur only under very limited conditions.

For convenience, most of the discussion in this chapter will be devoted to two-component systems. Enthalpic interactions between unlike components can have the effect of enhancing or decreasing the interdiffusion coefficient which is composition dependent. In Fig. 10.2 mutual diffusion (interdiffusion) data for two very different A/B systems are shown to illustrate the effects of enthalpic interactions on interdiffusion. Part 10.2a describes the compositional dependence of the interdiffusion of d-PS into PS at 174°C, above the $T_g = 100$°C for PS. The data indicate that the effect of enthalpy is to decrease

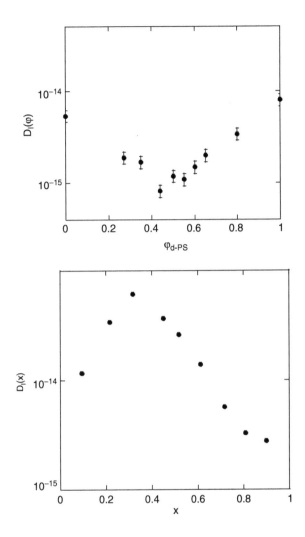

FIG. 10.2
The compositional dependence of the interdiffusion is shown here for a a) polymer-polymer, polystyrene-deuterated polystyrene (D_I versus volume fraction, replotted from Green and Doyle (1987)) and b) a metallic alloy, iron-palladium (D_I versus atomic fraction, replotted from Van Dal et al 2000).

the diffusivity relative to that of the pure species. In other words the magnitude of D_I resides appreciably below that which a rule of mixtures would predict (Green and Doyle 1986; 1987). Under normal circumstances the use of an isotopically labeled species should only have a minor effect on the diffusion rate. The situation with long-chain polymers is particularly interesting because mixing is determined primarily by the enthalpic interactions between the A and B components (the entropy of mixing in polymers varies as $1/N$ ($N\sim 10^3$-number of monomers per chain). The small enthalpic effects in isotopic mixtures are due to differences between the polarizabilities of the normal and duterated species. For metals, or other small molecule mixtures, this is not the case. The entropy of mixing contributes comparatively more to the intermixing process than in the case of polymer-polymer mixtures.

The other example of interdiffusion illustrated in Fig. 10.2b involves the Ni-Pd system. In this case diffusion is enhanced over that which a rule of mixtures would predict. Generally, if the interactions favor enhanced mixing, D_I will be larger than that of the relevant tracer diffusion coefficient, otherwise it will be lower.

10.2 Transport in Diffusion Couples

10.2.1 Onsager Analysis

Generally, the relevant flux equations may be written to include all diffusing species, including defects (Flynn, 1972). For example, if the diffusion of different chemical species is considered together with vacancies, v, present in the system, then the flux of species 1 would be written in terms of a generalized force X_i acting on species i

$$\vec{J}_1 = L_{11}\vec{X}_1 + L_{12}\vec{X}_2 + \cdots + L_{1s}\vec{X}_s + L_{1v}\vec{X}_v \qquad 10.1$$

where L_{ij} are phenomenological Onsager coefficients. The generalized force

$$\vec{X}_i = \vec{F}_i - TV\left(\frac{\mu_i}{T}\right) \qquad 10.2$$

where $\vec{F}_i$ is an external force, μ_i is the chemical potential and T is the temperature. The chemical potential is defined,

$$\mu_i = \mu_i^0(T, P) + kT \ln \gamma_i x_i \qquad 10.3$$

where $x_i = n_i/n$ is the mole fraction of species i (n is the total number of atoms per unit volume), γ_i is the activity coefficient of species i and $\mu^0(T, P)$ is the chemical potential at standard temperature and pressure. The other relevant fluxes in the system are

$$\vec{J}_2 = L_{21}\vec{X}_1 + L_{22}\vec{X}_2 + \cdots + L_{2s}\vec{X}_s + L_{2v}\vec{X}_v$$
$$\vec{J}_3 = L_{31}\vec{X}_1 + L_{32}\vec{X}_2 + \cdots + L_{3s}\vec{X}_s + L_{3v}\vec{X}_v$$
$$\cdot$$
$$\cdot \qquad\qquad 10.4$$
$$\cdot$$
$$\vec{J}_s = L_{s1}\vec{X}_1 + L_{s2}\vec{X}_2 + \cdots + L_{ss}\vec{X}_s + L_{sv}\vec{X}_v$$
$$\vec{J}_v = L_{v1}\vec{X}_1 + L_{v1}\vec{X}_2 + \cdots + L_{v1}\vec{X}_s + L_{ss}\vec{X}_v$$

If all the lattice sites in the system are to be conserved, elements of the foregoing equations can not all be linearly independent. In fact

$$\bar{J}_v = -\sum_{i=1}^{s} \bar{J}_i \qquad 10.5$$

and the matrix (for the Onsager coefficients) is symmetric, $L_{\alpha\beta} = L_{\beta\alpha}$.

We now simplify the situation considerably by considering a binary mixture, A and B (with vacancies present). Two conditions are imposed on this system. The chemical potential for vacancies $\mu_v = 0$. Physically, the vacancy concentration would be expected to be at equilibrium with available sources and sinks (dislocations and grain boundaries) such that pores and voids would not form in the system, thereby violating the $\mu_v = 0$ condition (Meyer et al 1969; Höglund 2001). In practice this can be achieved but the right conditions have to be present. The second condition is that there are no external driving forces so the individual fluxes are

$$J_A = L_{AA}\nabla\mu_A + L_{AB}\nabla\mu_B$$
$$J_B = L_{BA}\nabla\mu_A + L_{BB}\nabla\mu_B \qquad 10.6$$

The Gibbs-Duhem relation of thermodynamics indicates that,

$$x_A\nabla\mu_A + x_B\nabla\mu_B = 0, \qquad 10.7$$

It follows from 10.6 and 10.7 that

$$J_A = \left(L_{AA} - L_{AB}\frac{x_A}{x_B} \right)\nabla\mu_A \qquad 10.8$$

From Eq. 10.3,

$$\nabla\mu_i = \frac{kT}{x_i}\nabla x_i + \frac{kT}{\gamma_i}\frac{\partial\gamma_i}{\partial x_i}\nabla x_i \qquad 10.9$$

Equation 10.8 may now be rewritten as

$$J_A = kT\left(L_{AA} - L_{AB}\frac{x_A}{x_B} \right)\left(1 + \frac{\partial\ln\gamma_A}{\partial\ln x_A} \right)\frac{\nabla x_A}{x_A} \qquad 10.10$$

This expression looks like the more familiar form of the diffusion equation describing the flux of species A,

$$J_A = -D_A\nabla c_A \qquad 10.11$$

where $c_A = nx_A$. The diffusion coefficient is now

$$D_A = \frac{kT}{n}\left(\frac{L_{AA}}{x_A} - \frac{L_{AB}}{x_B} \right)\left(1 + \frac{\partial\ln\gamma_A}{\partial\ln x_A} \right) \qquad 10.12$$

similarly,

$$D_B = \frac{kT}{n}\left(\frac{L_{BB}}{x_B} + \frac{L_{AB}}{x_A}\right)\left(1 + \frac{\partial \ln \gamma_B}{\partial \ln x_B}\right)$$

10.13

Equations 10.12 and 10.13 are identified as intrinsic diffusion coefficients which describe the transport of species i in the environment of chemical potential gradients.

10.2.2   The Darken Equation

The tracer diffusion coefficients are written as

$$D_B^* = kT\left(\frac{L_{BB}}{c_B} + \frac{L_{AB}}{c_A}\right)$$

10.14

for the B species. The tracer diffusion coefficient for the A-species is

$$D_A^* = kT\left(\frac{L_{AA}}{c_A} + \frac{L_{BA}}{c_B}\right)$$

10.15

With the use of the Gibbs-Duhem equation,

$$\left(\frac{\partial \ln \gamma_B}{\partial \ln x_B}\right) = \left(\frac{\partial \ln \gamma_A}{\partial \ln x_A}\right) \equiv \left(\frac{\partial \ln \gamma}{\partial \ln x}\right)$$

10.16

and the expressions for tracer diffusion, the intrinsic diffusion coefficient for the A-species may be rewritten as

$$D_A = D_A^*\left(1 + \frac{\partial \ln \gamma}{\partial \ln x}\right)$$

10.17

similarly, for the B-species

$$D_B = D_B^*\left(1 + \frac{\partial \ln \gamma}{\partial \ln x}\right)$$

10.18

A difference between the diffusivities of the A and B species would lead to a net flow of material. The implications are that a flow of vacancies would counter balance the fluxes due to the A and B species. In fact, Eq. 10.5 becomes

$$\bar{J}_v = -(\bar{J}_A + \bar{J}_B)$$

10.19

Experimentally this would be manifested as follows. The flow of vacancies would result in the movement of lattice planes relative to a fixed point at the edge of the crystal. Consider the diagram below (Fig. 10.3), part a), where the flux of species A, to the left, is larger than the flux of species B, moving

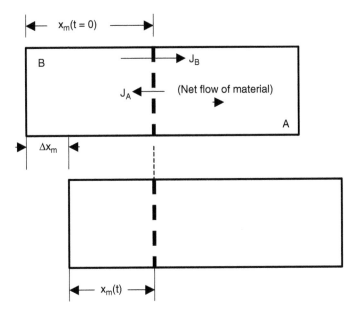

FIG. 10.3
a) A net flux of atoms move to the right and the region of the sample left of the marker decreases in size; b) The result is a change in the position of the marker relative to the ends of the sample.

to the right. This means that the region of the crystal to the left of the plane expands and the region to the left of the marker decreases in width due to the transfer of vacancies in that direction (vacancy mechanism). Therefore, if one measured the location of the marker with respect to the edge of the sample it would appear to have moved.

A reference plane can be identified such that the vacancy flux, $J_v = 0$. Such a plane would move with velocity $v = J_v/c$ with respect to the lattice. The plane would be sufficiently far from the interface where virtually no diffusion occurs. Under these conditions the fluxes J'_A and J'_B in this new coordinate system would satisfy the condition,

$$J'_A + J'_B = 0 \tag{10.20}$$

In other words,

$$J'_A = J_A - x_A(J_A + J_B)$$
$$J'_B = J_B - x_B(J_A + J_B) \tag{10.21}$$

Using these equations for J'_i,

$$J'_A = -(x_B D_A + x_A D_B)\nabla c_A \tag{10.22}$$

which is tantamount to writing

$$J'_A = -\tilde{D}\nabla c_A$$

$$J'_B = -\tilde{D}\nabla c_B$$

(10.23)

where the interdiffusion coefficient is

$$\tilde{D} = (x_B D_A + x_A D_B)$$

(10.24)

With this in mind we now have an expression that connects the interdiffusion coefficient with the tracer diffusion coefficient,

$$\tilde{D}(x) = (x_A D_B^* + x_B D_A^*)\left(1 + \frac{\partial \ln \gamma}{\partial \ln x}\right)$$

(10.25)

This is often referred to as the Darken equation. Other researchers identify it as the Hartley-Crank equation (Crank 1968).

10.2.3 Marker Velocity

We now return to the question of the marker movements. The velocity of the markers can be written as

$$v = \frac{dx_m}{dt} = -(D_A - D_B)\nabla c_A = -(D_A^* + D_B^*)\left(1 + \frac{\partial \ln \gamma}{\partial \ln c}\right)\nabla c_A$$

(10.26)

We now examine the time dependence of the marker displacement, as promised in the earlier section. The treatment is general in the sense that it will not matter whether the material is polymeric, metallic (see, for example, Kramer 1984 or Van Dal 2000). Begin by considering that the initial position of the interface is located at position x_m. Second, consider the second coordinate system with the marker located at the origin, x_0, of this moving coordinate system. Note that x_m is fixed in space and the displacement

$$x = x_0 + x_m$$

(10.27)

The marker displacement is shown to exhibit a parabolic time dependence

$$x_m = 2[D_A(\phi|_{x=x_0}) - D_B(\phi|_{x=x_0})]\frac{\partial \Phi}{\partial u}\bigg|_{u=0} t^{1/2}$$

(10.28)

Evidence of the marker movements was first shown by E.O. Kirkendall in 1942 and A.D. Smigelskas and Kirkendall in 1947. Today the effect is widely known as the Kirkendall effect. The mechanism by which the crystal grows, as suggested by Manning, is that the net flow of species in one direction creates a deficit of vacancies. New vacancies must be supplied by sources such as dislocations. An edge dislocation (an extra half plane of atoms wedged between two planes) that would be responsible for the source would have to grow, producing an extra half plane, the result of which is to expand the crystal. Equation 10.25 is now widely regarded as a very good approximation of the experimental situation, though more detailed calculations suggest that there is an additional correction term that changes the situation by less than a factor of 1.3. In some cases the correction is only a few percent (Flynn 1972). There is evidence of the migration of inclusions or the development of porosity and of internal stresses in the "diffusion zone" in some systems and these introduce additional complications in the analysis (see, for example, R.O. Meyer 1969; L. Höglund and J. Ågren 2001). Finally, not all experiments are performed using bilayer samples. In some experiments, multi layered samples are examined to determine the interdiffusion coefficient (see, for example, van Dal et al 2000 and Fedorov et al 2003). The data in Fig. 10.2b were obtained from Kirkendall marker experiments.

10.3 The Hartley-Crank Equation

Before concluding this section, it is worthwhile to revisit Eq. 10.25. This equation was derived based on the assumption that the two fluxes, J_A and J_B, in the diffusion couple were counterbalanced by a third flux of vacancies. It turns out that this equation is somewhat more general in that it may be obtained without the presence of a lattice. Bearman examined the question of diffusion of molecular liquid. In these systems a lattice obviously does not exist, however, as we saw in the section of polymers, one can understand diffusion on the basis of the frictional forces between molecules. The inter-diffusion coefficient may be expressed in terms of frictional coefficients ζ_{ij}, (Bearman, 1961)

$$\tilde{D} = \frac{\upsilon kT}{\zeta_{AB}}\left[1+\left(\frac{\partial \ln \gamma}{\partial \ln x}\right)\right]$$

10.29

where υ is a weighted average of the molecular volumes, $\upsilon = x_A \upsilon_A + x_B \upsilon_B$. The intrinsic diffusion coefficients are

$$D_A = \frac{\upsilon kT}{x_A \zeta_{AA} + x_B \zeta_{AB}}$$

10.30

and

$$D_B = \frac{\upsilon kT}{x_A \zeta_{AB} + x_B \zeta_{BB}}$$

10.31

Equation 10.13 is obtained if one assumes that

$$\zeta_{AB}^2 = \zeta_{AA}\zeta_{BB}$$

10.32

which is a geometric mean assumption. If on the other hand, if the friction factors are assumed to be represented by an arithmetic mean

$$\zeta_{AB} = \frac{\zeta_{AA} + \zeta_{BB}}{2}$$ (10.33)

then,

$$\tilde{D}(x) = kT\left(\frac{x_A}{D_B^*} + \frac{x_B}{D_A^*}\right)^{-1}\left(1 + \frac{\partial \ln \gamma}{\partial \ln x}\right)$$ (10.34)

This result for the interdiffusion maybe obtained from the Onsager analysis if one assumes that $J_v = 0$ and by extension $J_A + J_B = 0$. In other words there is no vacancy flow which would predict no marker movement. The two results are equivalent in the special case where the molecular sizes v_A and v_B are identical (see also G. Foley and C. Cohen). The assumption of the geometric mean is a good assumption for regular solutions and in this regard represents a special case, as pointed out by Bearman.

The foregoing discussion naturally leads to the discussion of the interdiffusion of polymers. Does a Kirkendall effect exist in such systems? How well do these equations describe the situation in long-chain polymers?

10.4 Interdiffusion in Polymers

In this section we determine an explicit expression for the interdiffusion coefficient between two species, A and B. With the use of Eqs. 10.8, 10.11, and the Flory-Huggins free energy expression from which the chemical potential is derived, the following equations for the intrinsic diffusion coefficients are, beginning with the A-species,

$$D_A = kTB_{AA}\left[\frac{1-\phi}{N_A} + \frac{\phi}{N_B} - 2\phi(1-\phi)\chi\right]$$ (10.35)

where

$$B_{AA} = \Omega\left(\frac{L_{AA}}{\phi} + \frac{L_{AB}}{1-\phi}\right)$$ (10.36)

For the B component,

$$D_B = kTB_B\left[\frac{1-\phi}{N_A} + \frac{\phi}{N_B} - 2\phi(1-\phi)\chi\right]$$ (10.37)

where

$$B_{BB} = \Omega\left(\frac{L_{BB}}{(1-\phi)} + \frac{L_{AB}}{\phi}\right)$$ 10.38

Note that in the foregoing $c_A = \phi/\Omega$ and $c_B = (1-\phi)/\Omega$.

The tracer diffusion coefficients would have to be defined appropriately for unentangled chains (Rouse dynamics) and for entangled chains (Reptation dynamics). The distinction arises in the mobility terms. For Rouse Dynamics,

$$D_i^* = D_i^{Ro} = kTB_{ii} = kT\frac{B_{i0}}{N_i}$$ 10.39

where the mobilities are equal to the inverse of the monomer friction factor, $B_{ii} = 1/\zeta_{ii} = B_{i0}/N = 1/N_i\zeta_{0i}$. For Reptation dynamics, the relevant mobility is identified with motion of the submolecules along the primitive path

$$D_i^* = D_i^{Rep} = kTB_{ii}\left(\frac{N_{e(i)}}{N_i}\right)$$ 10.40

where $N_{e(i)}$ is the degree of polymerization between entanglements for species i. The expression for the interdiffusion coefficient in terms of the intrinsic diffusion coefficients, you should recall, is $D_I = ((1-\phi)D_A + \phi D_B)$. For entangled chains, the interdiffusion coefficient becomes

$$\tilde{D}(\phi) = ((1-\phi)N_A D_A^* + \phi N_B D_B^*)\left[\frac{1-\phi}{N_A} + \frac{\phi}{N_B} - 2\phi(1-\phi)\chi\right]$$ 10.41

Recall that the condition $\partial^2 \Delta f_{mix}/\partial\phi^2 = 0$ dictates that the condition for the spinodal is

$$\chi_s(\phi) = \frac{1}{2}\left(\frac{1}{\phi N_A} + \frac{1}{(1-\phi)N_B}\right)$$ 10.42

from which it follows that the interdiffusion coefficient may now be written in a form similar to equation 10.25

$$\tilde{D}(\phi) = 2\phi(1-\phi)D_T(\chi_s(\phi) - \chi)$$ 10.43

where

$$D_T = [(1-\phi)N_A D_A^* + \phi N_B D_B^*]$$ 10.44

is the transport coefficient. Equation 10.43 is now a product of two quantities, a transport coefficient and a thermodynamic term which indicates how close the system is the spinodal boundary beyond which the mixture becomes unstable and phase separates ($\chi > \chi_s$).

The other transport coefficient corresponding to the situation in which

$$J_A + J_B = 0 \tag{10.45}$$

is

$$\frac{1}{D_{TS}} = \left[\frac{(1-\phi)}{N_A D_A^*} + \frac{\phi}{N_B D_B^*} \right] \tag{10.46}$$

A comparison of equations 10.44 and 10.46 indicates that in a mixture of short chains and long chains, the short chains determine the rate of inter-diffusion. For short chains the transport coefficient is given by 10.44. If the long chains control the rate of interdiffusion then the transport coefficient is given by Eq. 10.46.

The situation regarding which of these equations accurately describes interdiffusion in metallic alloys is obvious, particularly if a vacancy mechanism is operational. With regard to polymers the situation is not as clear-cut with regard to some unusual cases (Ackasu 1991; Brochard 1983; 1987). If one considers the polymer melt to consist of two components, A and B, and that the melt is incompressible $J_v = 0$, then there is no way to rationalize Eq. 10.44; the transport coefficient described by Eq. 10.46 would be valid. On the other hand, if one considers the melt to be compressible, with the presence of vacancies, then one recovers equation 10.44. An alternate view is to consider a three-component, incompressible system with vacancies as the third component. This would also lead to equation 10.44.

In an entangled mixture of long (L) and short (S) chains, the chains Reptate through their tubes. In essence the chains move through a network. Brochard (1987) points out that if a chain moves along its "tube" with a curvelinear velocity U, then the velocity of its center of mass through the network is

$$v_i = \frac{1}{L_i} U_i \langle r_i^2 \rangle^{1/2} \tag{10.47}$$

where L is a tube length of the long ($i = L$) or short ($i = S$) chain and $\langle r_i^2 \rangle$ is the mean square end-to-end vector of the chain. If the network moves with velocity v_T, then the center of mass velocity of component i becomes

$$v_i = \frac{1}{L_i} U_i \langle r_i^2 \rangle^{1/2} + v_T \tag{10.48}$$

Brochard points out that the relative friction between the L and S-chains and their local environment, ζ_{0S} and ζ_{0L}, respectively, contributes to the dissipation of energy. If the frictional forces on the network due to the L and S-chains is balanced then (Problem 7)

$$\frac{1}{\zeta_{LS}} = \frac{1}{\zeta_L} + \frac{1}{\zeta_S} \tag{10.49}$$

where $\zeta_i = \phi_i \frac{N_i}{N_e} \zeta_{0i}$ ($i = S, L$). In other words, the mobility (inverse of the friction factor) is represented by the sum of individual mobilities. This implies that

the faster moving short chains control the dynamics, as is the situation predicted by Eq. 10.44. Note that if $v_T = 0$, then the other prediction, which
indicates that the slower moving chains determine the rate of interdiffusion
(eqn. 10.46), would have been recovered as Brochard points out, recovering a
result in an earlier paper (see Brochard 1987 and Brochard et al 1983).

10.5 Measurements of Interdiffusion

Before discussing marker movement experiments a few additional comments could be made with regard to the use of other techniques to measure
interdiffusion. Apart from an obvious brute force method involving measurement and fitting of concentration profiles, scattering experiments such
as X-ray and neutron scattering could be employed.

To this end, we now briefly recapitulate points regarding the structure of
liquids. A radial distribution function, $g(r)$, plays an important role in
describing the structure of liquids. Imagine a situation in which a particle
is placed at the origin, $\bar{r}$, of a coordinate system. The quantity $\rho g(\bar{r})$, where
ρ is the density, provides information regarding the probability that a second
molecule is located within a distance $d\bar{r}$ of this first molecule. In a more
general sense, this function measures correlations between the concentrations at two points. Recall that while the function $g(r)$ is an oscillating
function of r, it rapidly diminishes in amplitude and approaches a value of
one for sufficiently large r. It is convenient to define a new function $h(\bar{r})$,
where $h(\bar{r}) = g(\bar{r}) - 1$, whose Fourier transform is the structure factor of the
liquid.

The structure factor due to concentration fluctuations in a binary polymer-
polymer mixture is given by (de Gennes, Binder)

$$S^{-1}(q) = \frac{1}{\phi g_D(N_A, q)} + \frac{1}{(1-\phi)g_D(N_B, q)} - 2\chi \qquad 10.50$$

g_D is the Debye scattering function for an ideal polymer chain of N monomers. Recall (section 2.3) that $g(r)4\pi r^2 dr$ is the number of molecules between
r and $r + dr$ about a central molecule, so

$$\int_0^\infty \rho g(r)4\pi r^2 dr = N - 1 \cong N = g(q = 0) \qquad 10.51$$

It can be shown (Problem 9) that for $qR_0 \gg 1$, where $R_0 = N^{1/2}a$ is the unperturbed dimension of a chain,

$$g_D(q) = \frac{12}{q^2 a^2} \qquad 10.52$$

For the case where $q = 0$,

$$S^{-1}(0) = \frac{1}{\phi N_A} + \frac{1}{(1-\phi)N_B} - 2\chi = 2(\chi_s - \chi) \qquad 10.53$$

It now becomes clear that the interdiffusion coefficient can be determined by the structure factor at $q = 0$,

$$D_I(\phi) = \phi(1-\phi)D_T S^{-1}(0) \qquad 10.54$$

These results reveal how scattering experiments can be used to measure the interdiffusion coefficient.

Photon correlation spectroscopy data for a poly(dimethyl siloxane) (PDMS)/poly(ethyl methyl siloxane) (PEMS) mixture are shown in Fig. 10.4 at 293K. Both $S(0)$ and $D_I(\phi)$ were determined from these measuremtnts and D_T extracted. These data support the notion that the faster diffusing species determine the rate of interdiffusion. Measurements by a number of authors strongly suggest that equation 10.43 adequately describes transport in many entangled polymer systems (Composto et al. Liu et al. Meier et al. Jordan et al.)

10.5.1 Marker Experiments

Recall that the marker velocity $v = (D_A - D_B)\nabla\phi$, is

$$v_{Ro} = kT(B_{AA} - B_{BB})\left[\frac{1-\phi}{N_A} + \frac{\phi}{N_B} - 2\phi(1-\phi)\chi\right]\nabla\phi \qquad 10.55$$

for *unentangled* chains and for *entangled* chains

$$v_{Rrp} = kT\left(\frac{N_{e(A)}B_{AA}}{N_A} - \frac{N_{e(B)}B_{BB}}{N_B}\right)\left[\frac{1-\phi}{N_A} + \frac{\phi}{N_B} - 2\phi(1-\phi)\chi\right]\nabla\phi \qquad 10.56$$

The first marker experiments in polymer systems were performed during the early 1980s with bilayers of polystyrene, supported by silicon substrates (Green et al 1985). Each layer had a different molecular weight, such that $N_A \gg N_B$. Gold particles were used as markers, as illustrated in Fig. 10.5. Rutherford backscattering was used to measure the location of the markers at different times in a bilayer of $N_A = 2 \times 10^5$ and $N_B = 320$, as shown in Fig. 10.5. The parabolic dependence of the marker displacement, ΔX_m is typical of the data determined with other couples of varying molecular weights (Fig. 10.6). In fact the values of D determined from the data are in excellent agreement with data using other techniques (Green 1995).

Measurements of marker displacements in PMMA were performed using X-ray reflectometry and these data also support the notion that the faster diffusing species determines interdiffusion, in support of a Kirkendall effect (Liu et al). The reader is asked to consult the original references for details.

While in metals the existence of a lattice and vacancies account for the marker movement, it is not immediately obvious for a polymer. Consider the

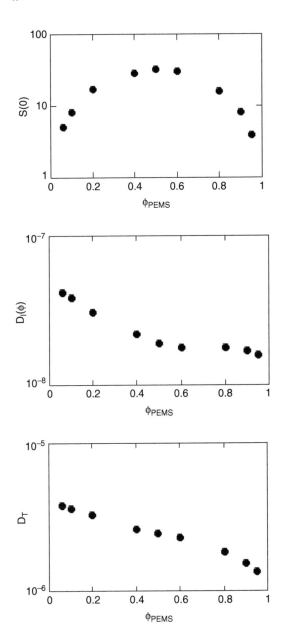

FIG. 10.4

Values of a) $S(0)$, b) $D_I(\phi)$ and c) D_T are shown here for a poly (dimethyl siloxane) (PDMA) of degree of polymerization $N_A = 80$ and poly (ethyl methyl siloxane) (PEMS) of degree of polymerization ($N_B = 90$) at 293 K. The glass transition temperatures for these polymers are $T_g(PDMS) = 148$ K and $T_g(PEMS) = 141$ K. This mixture has a lower critical solution temperature of approximately 393 K. Photon correlation spectroscopy was used to measure $S(0)$ and $D_I(\phi)$. These data were extracted from Table 1 of G. Meier et al. (1996).

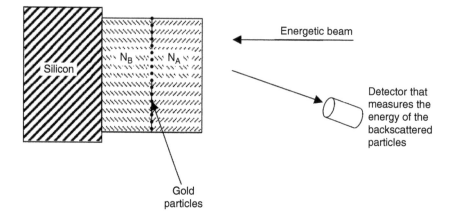

Energetic beam

Detector that
measures the
energy of the
backscattered
particles

Gold
particles

FIG. 10.5
The location of the marker with respect to the front surface is determined using Rutherford
backscattering spectrometry. Briefly, an energetic beam of particles is directed to the sample. A
fraction of the particles are backscattered and detected. The energy that the particles lose is
related to the depth below the surface from which it is backscattered (see Green et al for details)
and Doyle). The location of the gold particles beneath the surface is determined in a straightforward manner.

interdiffusion of two substances, A and B, initially separated by a permeable
boundary. If the rate of transfer of one species across the boundary is larger
than the other, a hydrostatic pressure will be generated in the region from
which the slower component emanates. The pressure would have to be

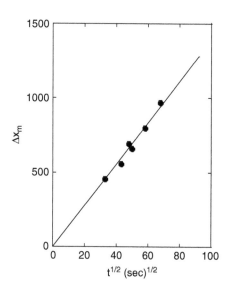

FIG. 10.6
Kirkendall marker shift is shown here
for a polymer-polymer system.

relieved by a compensating bulk flow of the entire solution. In this regard the phenomenology of interdiffusion of molten polymers is not different.

10.6 Concluding Remarks

What became known as the Darken equation (Eq. 10.25), developed based on the phenomenological Onsager analysis to describe interdiffusion in metals, was derived independently to describe diffusion in small molecule liquids. Such an equation is known as the Hartley Crank equation (Eq. 10.34) and has its foundations in Statistical Mechanics (Bearman 1961, Sillescu 1984, Foley 1987). Theoretical work (Ackasu, 1991, 1995, Jilge et al. 1990) suggest that the two transport equations 10.44 and 10.46 represent limiting situations of a more general relationship between the tracer diffusion coefficients used to describe interdiffusion (see Problem 1).

10.7 Problems for Chapter 10

1. Acasu et al. (1991) suggest the following form for the transport coefficient for interdiffusion. Here a three component incompressible system is considered: $D_T = \frac{D_A}{N_A} + \frac{D_B}{N_B} - \frac{(D_A^* - D_B)^2}{N_A D_A + N_B D_B + N_C D_C}$. Show that the transport coefficients in Eq. 10.44 and 10.46 represent limiting cases of this equation. Sketch the compositional dependence of D_T for both situations.

2. Enthalpic interactions also change the temperature dependence of interdiffusion compared to that of tracer diffusion. Explain what happens as the system approaches a phase boundary of a LCST and a UCST.

3. In the new coordinate system, with the markers located at x_0, Fick's second law for species A may be written as

$$\frac{\partial \phi}{\partial t} = \frac{\partial}{\partial x_0}\left(D_I(\phi)\frac{\partial \phi}{\partial x_0}\right)$$

a) Show that this equation can be transformed using the following equation

$$u = \frac{x_0}{\sqrt{t}}$$

to become an ordinary differential equation

$$-(u/2)\frac{d\phi}{du} = \frac{d}{du}\left(D_I(\phi)\frac{\partial\phi}{\partial u}\right)$$

b) Show that if the diffusion coefficient D_I is constant between compositions ϕ_L and ϕ_R, then compositions to the left and to the right of the interface, respectively, then the solution is

$$\phi(u) = \left(\frac{\phi_R - \phi_L}{2}\right)erfc\left(\frac{u}{2D_I^{1/2}}\right) + \phi_L$$

c) It is important to note that since the marker is always located at the origin ($x_0 = u = 0$) it remains at a constant composition for $t > 0$. Using the equation for the marker velocity,

$$v = [D_A(\phi|_{x=x_0}) - D_B(\phi|_{x=x_0})]\frac{\partial\phi}{\partial x}\Big|_{x=x_m}$$

$(\frac{\partial\phi}{\partial x}|_{x=x_m} = \frac{\partial\phi}{\partial x_0}|_{x=x_m} = \frac{\partial\Phi(u)}{\partial u}|_{u=0}t^{-1/2})$ show that the marker displacement exhibits a parabolic dependence on time.

4. Derive the relationships, Eq. 10.35 and 10.37, describing the intrinsic diffusion coefficients for polymers. Using realistic numbers, sketch their compositional dependencies. Discuss the differences between tracer diffusion and intrinsic diffusion.

5. Derive the following relationship $\chi_s = \frac{1}{2}(\frac{1}{\phi N_A} + \frac{1}{(1-\phi)N_B})$.

6. Show that the following expression $\frac{1}{D_{TS}} = [\frac{(1-\phi)}{N_A D_A^*} + \frac{\phi}{N_B D_B^*}]$ for the transport equation is valid for the condition $J_A + J_B = 0$.

7. Brochard shows that the frictional energy dissipated is $W = \int dr(\zeta_L(v_L - v_T)^2 - \zeta_S(v_S - v_T)^2)$. If a force balance equation, $\zeta_S(v_S - v_T) + \zeta_L(v_L - v_T) = 0$, is satisfied, then $W = \int\zeta(v_L - v_S)^2 dr$. Show that $\frac{1}{\zeta_{LS}} = \frac{1}{\zeta_L} + \frac{1}{\zeta_S}$.

8. Show that $S^{-1}(0) = \frac{1}{\phi N_A} + \frac{1}{(1-\phi)N_B} - 2\chi = 2(\chi_s - \chi)$.

9. In the limit $r \ll R_0$, $g_D \cong \frac{1}{a^2 r}$, show that $g_D(q) = \frac{12}{q^2 a^2}$. Sketch this function as a function of r.

10. Determine the value of the Flory interaction parameter, χ, from the data in Fig. 10.2a.

10.8 References

Ackasu, A.Z., Nagele G., and Klein, R., "Remarks on the fast ans slow mode theories of interdiffusion," *Macromolecules*, 28, 6680 (1995).

Ackasu, A.Z., Nagele G., and Klein, R.,"Identification of modes in dynamic scattering from ternary solutions," *Macromolecules*, 24, 4408, (1991).

Bearman, R.J., "On the molecular basis of some current theories of diffusion," *Journal of Physical Chemistry,* 65, 1961 (1961).

Binder, K., "Collective diffusion, nucleation and spinodal decomposition in polymer mixtures," *J. Chem. Phys.* 79, 6387 (1983).

Brochard, F., Jouffroy J. and Levinson, P., "Polymer-polymer diffusion in melts," *Macromolecules,* 16, 1638 (1983).

Brochard-Wyart, F., "Interdiffusion de polymèrs flexibles très ehchevêtrès," *C.R. Acad. Sci. Paris,* 305, 657 (1987).

Composto, R.J., Mayer, J.W., Kramer E.J. and White, D.M., "Fast mutual diffusion in polymer blends," *Phys. Rev. Lett.,* 57, 1312 (1987).

Crank, J., The mathematics of Diffusion, Clarendon Press, Oxford, 1979.

deGennes, P.G., Scaling Concepts in Polymer Physics, Cornell University Press, 1979.

Doi, M. and Onuki, A., "Dynamic coupling between stress and composition in polymer solutions andmelts," *J. Phys. II France,* 2, 1631 (1992).

Fedorov, A., Sipatov, A., Volobuev, V., "Diffusion and Kirkendall effect in PBSe-EuS multilayer," *Thin Solid Films,* 425, 287 (2003).

Feng, Y., Han, C.C., Takenada, M., Hashomoto, T., "Molecular weight dependence of mobility in polymer blends," *Polymer,* 33, 2729 (1992).

Flynn, C.P., Point Defects and Diffusion, Clarendon Press, Oxford, UK 1972.

Foley, G. and Cohen, C., "Diffusion in Polymer-Polymer Mixtures," *Journal of Polymer Science, Polymer Physics,* 25, 2027 (1987).

Green, P.F. and Doyle, B.L., "Ion Beam Analysis of Thin Polymer Films," Chapter in Characterization Techniques for Thin Polymer Films, pp. 1310–180, John Wiley & Sons, Inc., 1990 (eds. H. M. Tong and L. T. Ngugen).

Green, P.F. and Doyle, B.L., "Isotope Effects of Interdiffusion in Blends of Normal and Deuterated Polymers," *Physical Review Letters,* 57 (19), 2047, 1986.

Green, P.F. and Doyle, B.L., "Thermodynamic Slowing Down of Mutual Diffusion in Isotopic Polymer Mixtures," *Macromolecules,* 20, p. 2471–2474, 1987.

Green, P.F. and Doyle, B.L., "Application of Ion Beam Analysis Techniques to Polymer Science," Chapter in Scattering Methods in Polymer Science, Ellis Harwood (ed. Randal Richards). p. 193–215 (1995).

Green, P. F., "Translational Dynamics of Macromolecules in Melts" Diffusion in Polymers, Ed. P. Neogi, Mercell Dekker , p. 251–302, NY 1996.

Green, P.F., Palmstrom, C.J., Mayer J.W. and Kramer, E.J., "Marker displacement measurements of polymer-polymer interdiffusion," *Macromolecules,* 18, 501 (1985).

Höglund, L. and Ågren, J., "Analysis of the Kirkenall effect, marker migration and pore formation," *Acta Materialia.* 49, 1311 (2001).

Jilge, W., Carmesin, I., Kremer, K. and Binder, K., "A Monte Carlo simulation of polymer-polymer interdiffusion," *Macromolecules,* 23, 5001 (1990).

Jordan, E.A., Ball, R.C., Donald, A.M., Fetters, L.J., Jones, R.A.L., Klein, J., "Mutual diffusion in blends of long and short entangled chains," *Macromolecules,* 21, 235 (1988).

Kirkendall, E.O., *Trans Am. Inst. Min. Metall. Engrs,* 147, 104 (1942).

Kramer, E.J., Green, P.F. and Palmstrom, C.J., "Interdiffusion and marker movements in concentrated polymer-polymer diffusion couples," *Polymer,* 25, 473 (1984).

Liu, Y., Reiter, G., Kunz, K. and Stamm, M., "Investigation of the interdiffusion between poly(methyl methacrylate) films by marker movement," *Macromolecules,* 26, 2134 (1993).

Manning, J.R., "Diffusion in a concentration gradient," *Physical Review,* 124, 470 (1961).

Meier, G., Fytas, G., Mompler B. and Fleischer, G., "Interdiffusion in a homogeneous polymerblend far above its glass transition," *Macromolecules,* 26, 5310 (1996).

Meyer, R.O., "Pressure and vacancy flow effects on the Kirkendall shift in silver-gold alloys," *Physical Review* 181, 1086 (1969).

Sillescu, H., "Relation of interdiffusion and tracer diffusion in polymer blends," *Makromol. Chem., Rapid Communication*, 5, 393 (1987).

Sillescu, H., "Relation of interdiffusion and self-diffusion in polymer mixtures," *Makromol. Chem. Rapid Communication*, 5, 519 (1984).

Smigelskas, A.D. and Kirkendall, E.O., *Trans. Am. Inst. Min. Metall. Engrs.* 171, 130 (1947).

van Dal, H.J.M., Pleumeekers, M.C.L.P., Kodentsov, A.A., and Loo, F.J.J., "Intrinsic diffusion and Kirkendall effect in Ni-Pd and Fe-Pd solid solutions Ni-Pd system," *Acta Materialia*, 48, 385 (2000).

11

Growth: Moving Interfaces and Instabilities in Bulk Materials

11.1 Introduction

Thus far we have shown that the mechanism of transport influences the rate at which the atomic, or molecular, constituents migrate throughout materials at a given temperature. In multicomponent systems, thermodynamic driving forces may exert a strong influence on the migration process and determine the spatial organization of the material constituents (Chapters 9 and 10; cf. Eq. 10.19). Specifically, thermodynamic forces induce the constituents of a homogeneous, concentrated, A/B mixture to demix and to self-organize and form spinodal patterns. Mechanistically, fluctuations of the local composition of an otherwise homogeneous mixture, residing in the unstable region of the phase diagram, may become amplified, forcing the mixture to phase separate into A-rich and B-rich phases. This phenomenon, spinodal decomposition, is ubiquitous and is exhibited by a diverse range of mixtures, from metallic systems and polymeric melts to network glass melts. The demixing is an illustration of microstructural development in an alloy, engendered by a thermodynamic driving force. This instability is one of a number of instabilities that occur in condensed matter and is of practical and scientific interest.

The primary goal of this chapter is to introduce another important example of microstructural development (pattern formation) in material science. The problem of solidification is considered. Specifically, during the growth of a small crystal in its supercooled melt, the moving solid/melt interfaces may become unstable toward fluctuations (local deformations) in shape (Mullins and Sekerka, 1964, Langer, 1980, Gilcksman and Koss 1994) and form patterns on a macroscopic scale (Fig. 11.1). The patterns, as illustrated in the Fig., may include dendrites and columnar structures, depending on the growth velocity. The velocity of the interface is determined by the local temperature gradient, ∇T, responsible for removal of latent heat from the interface, which occurs via diffusion.

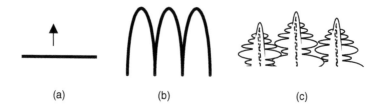

(a)   (b)   (c)

FIG. 11.1

Structures develop at the interface when the front becomes unstable during motion. a) stable interface, b) columnar structure, c) dendritic structure.

Dendritic structures are also developed during the growth of crystals in a supersaturated environment (Mullins and Sekerka 1963, Langer, 1980). The crystal increases in size due to the diffusion of solute from the environment to its interfaces. During growth, the solid/vapor interfaces which move due to the influence of a concentration gradient, ∇c, become unstable toward shape fluctuations and give rise to dendritic formation.

This instability associated with the moving interface is known as the Mullins-Sekerka instability; it is ubiquitous and occurs in all classes of materials, everything from long-chain polymers to simple liquids that form crystals upon solidification Bottinger, 2000, Mullis and Cochrane, 1997, Hunter and Bechhoefer 1997, Glicksman and Koss 1994. The most common-place example of dendritic formation is the formation of snow crystals. Dendritic structures are also commonplace in metals and metallic alloys subject to the *direct solidification* process. The dendritic pattern shown in Fig. 11.2 is that of long chain diblock copolymer chains capable of crystal-lizing from supercooled melts. The image taken using a scanning force microscope is shown to illustrate that dendrites can form in a diverse range of systems under appropriate conditions. The length scales are on the order of microns.

The surface tension plays an essential role in the development of such instabilities. Effects associated with capillarity create a force that opposes growth of the fluctuations and, as discussed further in this chapter, the competition between the driving forces that favor amplification of small shape fluctuations and the opposing forces associated with capillarity deter-mine a critical wave vector for fluctuations, below which the fluctuations are unstable and grow (Langer 1980, van Saarloos, 1998).

Capillarity has a profound influence on the properties of small particles in a medium. Due to the large curvature the chemical potential of a particle may become modified and physical properties, such as melting points, become size-dependent, differing appreciably from the bulk. Such capillarity effects are associated with the local variation in shape, due to the fluctuations, along the interface.

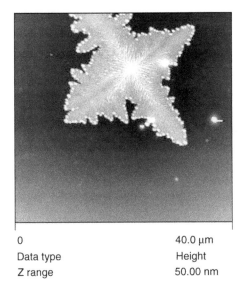

0   40.0 µm

Data type   Height

Z range   50.00 nm

FIG. 11.2
Dendritic pattern formed by a thin film diblock copolymer. (courtesy of Yuan Li)

This chapter begins with a discussion of the effects of capillarity on small systems in order to provide a context for the subsequent analysis of the instability. The model described here is simple in comparison to that which would be required to describe dendritic formation; but it is nevertheless instructive, as it provides conditions under which the instability would occur. The interested reader is referred to the end of the chapter, where references to more advanced and diverse treatments of the phenomenon are cited. This chapter concludes with a brief discussion of microstructure.

11.2 Effect of Curvature on the Properties of Small Particles

11.2.1 Elementary Concepts of Classical Capillarity

The wetting of a macroscopic liquid film on a substrate (Fig. 11.3) is determined entirely by capillarity. Whether a liquid forms droplets on a surface or spreads is determined by the spreading coefficient,

$$S = \gamma_{sv} - (\gamma_{lv} + \gamma_{ls}) \qquad\qquad 11.1$$

where the γ_{sv} is the substrate vapor interfacial energy, γ_{lv} is the liquid-vapor interfacial energy and γ_{ls} is the liquid-substrate interfacial energy.

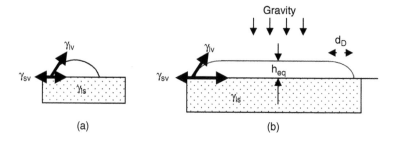

FIG. 11.3
Liquid droplet on a substrate. The angle of contact is θ_e. a) small droplet; b) large droplet flattened by gravity.

When $S > 0$ the liquid will wet (spread) the surface. The liquid will otherwise form droplets because the interfacial energy required to create two new interfaces, *liquid/vapor* and *liquid/solid*, is larger than the bare surface energy; consequently the droplet minimizes its area of contact.

If the droplet is sufficiently small, the effects due to gravity are negligible and the droplet assumes the shape of a spherical cap because the hydrostatic pressure within the droplet should equilibrate to conform to the Young-Laplace equation (Adamson and Gast 1997). Figure 11.4 shows a small droplet on a surface for which the two principal radii of curvature, $1/R_1$ and $1/R_2$ are identified. Generally, a pressure difference exists across a curved interface due largely to the existence of the surface tension, γ.

The pressure difference, Δp, across the liquid vapor interface is specified by the Young-Laplace equation

$$\Delta p = p_{in} - p_{out} = \gamma \left(\frac{1}{R_1} + \frac{1}{R_2} \right) \qquad 11.2$$

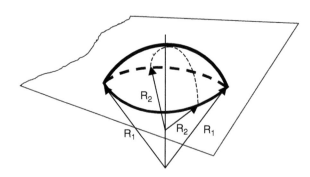

FIG. 11.4
The principal radii of curvature $1/R_1$ and $1/R_2$ for a liquid droplet on a substrate are shown here.

where p_{in} and p_{out} refer the pressures inside the droplet and outside the droplet, respectively. If the surface tension, γ, were zero, then there would be no pressure difference in such a hypothetical system. Physically, the effect of the surface tension is to induce a compressive stress on the droplet and this has to be balanced by the internal pressure within the droplet. The Young-Laplace equation is general. The curvatures $1/R_2$ and $1/R_1$ represent the principal radii of curvature of an arbitrary interface. The curvatures may be negative or positive; if the center of curvature resides within the region where p_{in} is located, then the curvature is taken as positive. There is, of course, no requirement for both curvatures to be of the same sign. It is typical to identify a mean radius of curvature $\kappa = (\frac{1}{R_1} + \frac{1}{R_2})$ which can always be defined independent of any coordinate system.

For a spherical object of radius r, the pressure difference is

$$\Delta p = \frac{2\gamma}{r} = \gamma\kappa \qquad 11.3$$

Equation 11.3 is readily understood by considering free energy of a bubble growing against the pressure from the external environment. The change of the free energy of the bubble is $dG = \gamma dA$, where the change in area $dA = 4\pi(r+dr)^2 - 4\pi r^2 \cong 8\pi r dr$. This is offset by the opposing work done by the environment, $dw = \Delta p dV$, where $dV = 4\pi r^2 dr$. The Young-Laplace equation follows from the condition $dG = dw$. The general case, Eq. 11.2, may also be proven from thermodynamic considerations (Hunter).

If the droplet is large, it is flattened by gravity in the center (Fig. 11.3b). However, near the line of contact the angle, θ_e, is determined by a mechanical balance of the horizontal components of the interfacial energies (forces per unit length), at the three-phase line of contact, $\cos\theta_e = 1 + \frac{S}{\gamma_{LV}}$ regardless of the size of the droplet. Here the capillary forces are dominant. Under the influence of gravity, the film assumes an equilibrium thickness of

$$h_{eq} = 2d_D \sin\frac{\theta_e}{2} \qquad 11.4$$

where

$$d_D = \left(\frac{\gamma_{LV}}{\rho g}\right)^{1/2} \qquad 11.5$$

is a capillary length. It characterizes the size of the droplet below which the gravitational effects can be ignored. g is the gravitational constant, ρ is the density of the film.

The capillary length may be understood by considering the situation in which a liquid wets the walls of a cylindrical container. At the walls of the container the hydrostatic pressure, ΔP, at depth d_D, $\Delta P = \rho g d_D$. The capillary pressure, from the Young-Laplace equation is γ/d_D. The capillary length (eqn. 11.5) is obtained by equating the two pressures. Physically, if a liquid wets a wall, d_D represents the distance over which capillarity effects perturb the shape of the liquid (gravitational effects are negligible) (de Gennes, Brochard-Wyart and Quéré 2004).

11.2.1.1 Effect of Curvature on the Properties of Small Systems

The pressure difference across the interface is particularly significant in systems where the radius is small (e.g., nanometer length-scales); it has the effect of altering the chemical potential and hence the properties of the material. Below, a basic set of equations is established to illustrate this point. Consider a spherical liquid droplet (β) immersed in an environment (α). Further, consider a small fluctuation in the system such that

$$\delta p_\beta - \delta p_\alpha = \delta(2\gamma/r) \tag{11.6}$$

At equilibrium, the chemical potentials in each phase are equal so

$$\delta\mu_\alpha = \delta\mu_\beta = \delta\mu \tag{11.7}$$

For an i-component system, the Gibbs-Duhem equation of thermodynamics stipulates that

$$S\delta T - V\delta p + \sum_i n_i \delta\mu_i = 0 \tag{11.8}$$

where n_i are the number of particles of species i. The extensive quantities, the entropy $S = \sum_i n_i s_i$ and the volume $V = \sum_i n_i v_i$ are written in terms of the partial molar entropy and partial molar volume $v_i = \frac{\partial V}{\partial n_i}\big|_{T,p,n_j}$ and $s_i = \frac{\partial S}{\partial n_i}\big|_{T,p,n_j}$ respectively. Within phase α and phase β (droplet phase), the appropriate forms of the Gibbs-Duhem relation are

$$s_\alpha dT - v_\alpha dp_\alpha + d\mu_\alpha = 0 \tag{11.9}$$

and

$$s_\beta dT - v_\beta dp_\beta + d\mu_\beta = 0 \tag{11.10}$$

respectively. Two situations are now described in order to illustrate the effect of curvature on the properties of materials.

Example 1: Effect of curvature in a constant temperature environment
At equilibrium, $v_\alpha dp_\alpha = v_\beta dp_\beta$ and from equations 11.6, 11.9 and 11.10,

$$\delta\left(\frac{2\gamma}{r}\right) = \left(\frac{v_\alpha - v_\beta}{v_\beta}\right)dp_\alpha \tag{11.11}$$

This equation is known as the Kelvin equation. If the situation is such that a droplet resides within a vapor phase, then $v_\alpha \gg v_\beta$ and the ideal gas law

$v_\alpha = RT/p_\alpha$ (R is the universal gas constant) may be applied. Under these conditions,

$$\delta\left(\frac{2\gamma}{r}\right) \cong \left(\frac{RT}{v_\beta p_\alpha}\right) dp_\alpha \qquad 11.12$$

If this equation is integrated from $r \to \infty$, where the vapor pressure is p_0 (vapor pressure of a flat surface) to a finite value of r where the vapor pressure is p_α, then

$$p_\alpha \cong p_0 e^{d_p \kappa} \qquad 11.13$$

where $d_p = \gamma v_\beta / RT$ would be the capillary length. This result indicates that the vapor pressure in the vicinity of a droplet of radius r is higher than that of the equivalent flat surface. Moreover, the vapor pressure increases as the radius of the droplet decreases. The implication is that if condensation of vapor occurs on one droplet, then the droplet will grow, driven by a decrease in the equilibrium vapor pressure. Likewise evaporation from one droplet results in a decrease of its radius and the vapor pressure increases, leading to a further reduction in size. Essentially, in a system composed of a distribution of droplets sizes in an infinite reservoir, the large droplets grow at the expense of the smaller ones. This situation was discussed in Chapter 10 in the section on nucleation and growth.

Similar calculations may be performed regarding the adsorption of particles to a curved interface. If the concentration of species in the environment is c_0 and the concentration of species in equilibrium at the surface of a particle of radius r is c_s, then

$$c_s = c_0 e^{d_c \kappa} \qquad 11.14$$

where $d_c = \gamma v_\beta / kT$ is a capillary length. The effect of curvature is an associated increase of the local solute concentration beyond that which would be encountered at a flat interface. The implication of this result is that in a large system, particles with larger radii are favored over particles of smaller radii. In other words, larger particles grow at the expense of small particles. This is the basis of the coarsening phenomena discussed in Chapter 9. It should also be evident from this discussion that shape fluctuations at an interface may be accompanied by local variations in the concentration of solute along the interface. This will be discussed further in Section 11.4.

Example 2: Effect of curvature in a constant external pressure,
  $dp_\alpha = 0$, *environment*
  If equation 11.9 is subtracted from 11.10, then

$$\Delta H_{vap} \frac{dT}{T} + v_\beta dp_\beta = 0 \qquad 11.15$$

where $s_\beta - s_\alpha$ is replaced with $\Delta H_{vap}/T$ (H_{vap} is the molar heat of vaporization). If a similar integration is performed, as was done above (from infinity to finite r), then the temperature at the interface is

$$T = T_0 e^{-d_0 \kappa} \qquad\qquad 11.16$$

where $d_0 = v_\beta \gamma / \Delta H_{vap}$ is a capillary length. Equation 11.16 is the Gibbs-Thompson equation. This result tells us that as the radius of the particle decreases, the local temperature decreases. The implication is that condensation of droplets from a vapor occurs at a lower temperature than the bulk. With regard to solid particles it is known that a small solid particle possess a lower melting point than the bulk (Hunter, 1995). This phenomenon has important consequences regarding solidification at a moving front, (discussed in section 11.3) because the local fluctuations at the interface lead to local changes of the interfacial melt temperature. This has important implications on the instability discussed in the next section.

11.3   Moving Front in a Supercooled Melt

The growth of a crystal phase within its own melt in the supercooled regime, $T < T_m$, is driven by local temperature gradients. The liquid to solid transition is 1st order, accompanied by the liberation of latent heat. If the latent heat remains in the vicinity of the interfacial region it has the effect of increasing the local temperature and this would retard growth. Hence there must exist a mechanism by which heat is transported away from the interface. Transport of heat may occur via diffusion or by a convection mechanism. In practical situations the transport of heat away from the interfaces by diffusion is accomplished by setting up an appropriate temperature gradient. As growth proceeds in the supercooled (*metastable*) regime, an initially planar interface becomes unstable toward long wave length fluctuations in shape and eventually breaks up into nonequilibrium patterns, which may be columnar, seaweed-like, or dendritic patterns, depending on the velocity of the moving front, as briefly mentioned above. (Langer, 1980, van Saarloos, 1998).

    The basic instability may be understood as follows. Consider a crystal immersed in its own supercooled melt, Fig. 11.5(a). The temperature of the supercooled melt is T_∞, which, of course, is lower than the melting temperature, T_m, of the solid and of the temperature at the solid/melt interface, Fig. 11.5(b). The latent heat produced due to solidification is removed at a sufficiently rapid rate by thermal diffusion.

    During motion, regions of the interface locally protrude outward, as illustrated in Fig. 11.6(a) for one such protrusion. When this occurs the effective temperature gradient at the tip of the protrusion is larger than that at the depression behind it (Fig. 11.6b). The difference between the local temperature gradients at the tip of the protrusion and behind the protrusion is a

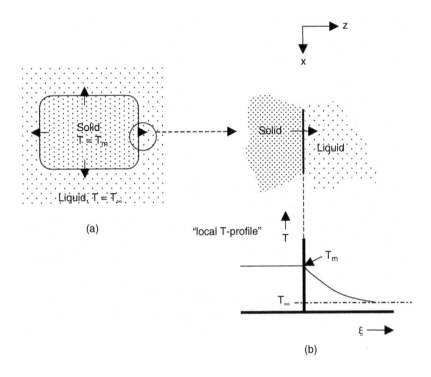

FIG. 11.5
Solid crystal immersed in its supercooled melt. The local temperature profile is shown here for this flat interface. The temperature sufficiently from the interface is $T_\infty < T_m$, thus forming a temperature gradient which will be responsible for growth.

result of the fact that the temperature isotherms are compressed in the vicinity of the protrusion. This larger temperature gradient at the tip of the protrusion creates a larger driving force for the protrusion to grow at a faster rate than the depression behind it because heat is removed at a faster rate (Fig. 11.6c).

The force that opposes growth is associated with local variations of the curvature of the interface; the Gibbs-Thompson effect. As mentioned earlier, the Gibbs-Thompson effect is the reason that sufficiently small crystals melt at lower temperatures than the bulk. The temperature at the interface due to this effect is

$$T_{intf} \cong T_m(1 - d_0\kappa) \qquad 11.17$$

where a Taylor series expansion of Eq. 11.16 has been performed. Equation 11.17 indicates that the curvature of the interface determines whether the melting temperature at the interface is higher or lower than the bulk T_m. The curvature of the protrusion in Fig. 11.6 is positive ($\kappa > 0$) so the Gibbs-Thompson effect suggests that the local melting temperature should be decreased by a factor $\gamma\kappa/L$ (for the remainder of this chapter $L = \Delta H_{vap}/v_b$ will be used instead

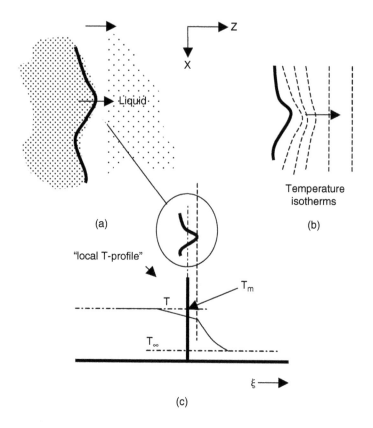

FIG. 11.6
Moving front during growth can become unstable. a) The interface moves with velocity v_n in a temperature gradient and becomes unstable. b) The temperature isotherms in the vicinity of a local protrusion are deformed. The local temperature at the tip of the protrusion is lower than the temperature at the interface due to the Gibbs-Thompson effect.

for convenience). It follows that because the curvature of the depression behind the protrusion is negative, the effective temperature at the depressions is higher than T_m. With regard to the instability, the implication is that growth of the protrusion is opposed by flow of heat from behind the protrusion to the front of the protrusion. If the temperature gradient in the vicinity of the protrusion is sufficiently large then the protrusion continues to grow provided that the surface tension effects associated with local curvature are sufficiently small. Indeed, when these small fluctuations become unstable, the system is driven into an entirely different morphological state.

This is naturally a complex dynamics problem and a self-consistent solution that accounts for the coupling between the interface velocity and the local fluctuations of shape and fluctuations of the temperature gradients would need to be sought. This is a daunting prospect. However, the fact that the rate of heat removal from the interface by diffusion is rapid compared to the shape dynamics at the liquid/solid interface, an approximate solution that provides a reasonable description of the situation suffices. As examined by a number of authors, the planar interface problem, wherein a steady state solution is sought in the absence of any fluctuations of the shape of the liquid/solid interface, is first sought. Linear stability analysis is subsequently exploited; a perturbation is applied to the interface thereby enabling the conditions under which the solution (modes) is unstable to be identified. Specifically we seek a dispersion relation, $\omega(q)$, that contains the parameters which characterize the instability. In this regard, the analysis is similar to that performed in Chapter 9.

The basic equation that governs the transfer of heat from the interface is the heat equation

$$\frac{\partial T}{\partial t} = D_T \nabla^2 T \qquad\qquad 11.18$$

where D_T is the thermal diffusion coefficient. This equation is analogous to Fick's second law of diffusion for mass transfer, except that the concentration $c(\vec{r}, t)$ is now replaced with the temperature, $T(\vec{r},t)$, which also exhibits a spatial and temporal dependence. The thermal diffusivity is related to the thermal conductivity, Λ (SI units of Joules/sec·m·Kelvin) and to the specific heat, c (SI units of J/m³·Kelvin·mole) such that

$$D_T = \frac{\Lambda}{\rho c}. \qquad\qquad 11.19$$

Therefore D_T, like the diffusivity for mass diffusion, has units of m²/s.

Because the interface moves with velocity v in the z-direction, it would be convenient to solve the equations in the moving coordinate frame of reference; i.e., the coordinates (x, y, z, t) need to be transformed to (x, y, ξ, t), see Fig. 11.6. This means that the z-coordinate in Eq. 11.18 needs to be replaced with

$$z = \xi + vt \qquad\qquad 11.20$$

The appropriate equation in the moving frame of reference is now

$$\frac{\partial T}{\partial t} = D_T \nabla^2 T + v\frac{\partial T}{\partial \xi} \qquad\qquad 11.21a$$

Because the temperature gradients and the thermal diffusivities are different in each phase, a separate equation needs to be considered for each phase. The superscripts (s) and (l) will identify variables associated with the solid and liquid phase, respectively. The equation that governs heat transport in the liquid phase is

$$\frac{\partial T^l}{\partial t} = D_T^l \nabla^2 T^l + v\frac{\partial T^l}{\partial \xi} \qquad\qquad 11.21b$$

and for the solid phase,

$$\frac{\partial T^s}{\partial t} = D_T^s \nabla^2 T^s + v \frac{\partial T^s}{\partial \xi}$$ 11.21c

The problem regarding the motion of a planar, stable, interface is now ready to be solved.

11.3.1 Stationary Solutions (planar interface, $\kappa = 0$)

The steady state solutions $(dT/dt = 0)$ of these equations for the planar interface in the solid phase, $T_{ss}^s(\xi)$, and in the liquid phase, $T_{ss}^l(\xi)$, may be obtained by considering the following boundary conditions. The boundary conditions, illustrated in Fig. 11.5, indicate that at the interface, $T_{ss}^l(\xi = 0) = T_{ss}^s(\xi = 0) = T_m$. The temperature in the solid, behind the interface, remains constant at $T = T_m$ and moreover,

$$T_{ss}^s = T_m$$ 11.22

On the side of the liquid the solution is

$$T_{ss}^l(\xi) = (T_m - T_\infty)e^{-\xi/l_D} + T_\infty$$ 11.23

where $l_D = D/v_n$ is the thermal diffusion length; the interface moves in the direction of the liquid phase with a component velocity, v_n, normal to the interface in the z-direction. Equation 11.23 represents the decay of the temperature ahead of the front (Fig. 11.5b). In this equation, the degree of undercooling is the difference between the temperatures T_m and T_∞;

$$\frac{L}{c} = T_m - T_\infty$$ 11.24

which specifies limitations on the value of T_∞.

Equation 11.24 is a natural consequence of the heat conservation boundary conditions associated with the moving interface. This may be seen as follows. Imagine a thin parallelpiped of cross sectional area A and thickness Δh, located at the interface, Fig. 11.7. Further, imagine that a flux of heat J_Q^s flows

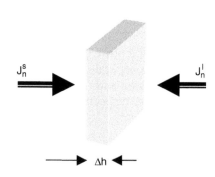

FIG. 11.7
Heat flux is shown to flow in and out of this box from the liquid and solid phases. The net heat per unit volume generated is determined by $J_Q^l - J_Q^s$.

from the solid phase in the direction of the liquid phase and a flux of heat J_Q^l flows from the liquid phase in the direction of the solid phase.

The flux of heat J_Q^i ($i = l, s$) can be expressed in terms of the heat capacity of the appropriate phase, c_p^i, the thermal diffusivity, D_T^i, and the temperature gradient, ∇T^i, across the slab,

$$J_Q^i = -cD_T^i\nabla T^i \qquad 11.25$$

The net heat per unit volume generated, or equivalently the latent heat per unit volume, L, is specified by $L/\Delta t = (J_n^s - J_n^l)/\Delta l$, where Δt is the time it takes to travel Δh, implying that

$$Lv_n = (D_T^s c_p^s \nabla T_n^s - D_T^l c_p^l \nabla T_n^l) \qquad 11.26$$

with $v_n = \Delta l/\Delta t$. It is stressed here that the temperature gradients in Eq. 11.26 are defined at the interface. If it is assumed, for simplicity, that the heat capacity at constant pressure is the same in the liquid as it is in the solid phase and that the diffusivities are equal in both phases, then we arrive at a simplified expression that defines an important boundary condition

$$Lv_n = Dc(\nabla T_n^s - \nabla T_n^l) \qquad 11.27$$

This equation becomes Eq. 11.24 upon substitution of Eqs. 11.22 and 11.23 into Eq. 11.27. This result is a restatement of the conservation of heat requirement to maintain equilibrium.

In fact, Eq. 11.24 may be rewritten as $v_n = (Dc/l_D L)(T_m - T_\infty)$, which identifies the interface velocity necessary to maintain the conservation condition. If the condition is not met, $L/c < (T_m - T)$, then the interface velocity decreases via a diffusive process and the displacement is $\propto t^{1/2}$. In practice, the experimental parameters can be chosen such that the condition is not violated.

11.3.2 Linear Stability Analysis

Having discussed the stationary solution ($dT/dt = 0$) for a planar boundary ($\kappa = 0$) moving with velocity v_n, perturbations to the boundary are now considered. This subsequent analysis enables the dispersion relation to be determined. We are now specifically interested in solutions to the time-dependent equations, 11.21b and 11.21c. The fluctuations of the interface along the x-direction are now considered such that $\xi = \zeta(x, t)$ where this function is approximated with the Fourier component

$$\zeta(x,t) = \zeta_q e^{\omega t + iqx} \qquad 11.28$$

The behavior of the system is assumed to be transformationally invariant along the y-direction, so only variations in the x-direction are considered for convenience. The temperature is now written as a sum of the steady state solution and a perturbation which depends on ζ, necessarily; hence in the liquid phase

$$T^l = T^l_{ss}(\xi) + \delta T^l(\xi)e^{\omega t + iqx} \qquad 11.29a$$

Similar to the discussion of linear stability analysis in Chapter 9, $\omega = \omega(q)$ is an amplification factor, or growth rate, whose sign determines whether the initial perturbation will become amplified or damped. The modes become amplified for $\omega = \omega(q) > 0$. In the solid phase,

$$T^s = T^s_{ss}(\xi) + \delta T^s(\xi)e^{\omega t + iqx} \qquad 11.29b$$

Upon substituting equations, Eq. 11.29(a) and 11.23 into Eq. 11.21(b), the following equations for the liquid is obtained

$$\frac{d^2 \delta T^l(\xi)}{d\xi^2} + \frac{1}{l_D}\frac{d\delta T^l(\xi)}{d\xi} = \left(\frac{\omega}{D} + q^2\right)\delta T^l(\xi) \qquad 11.30a$$

For the solid phase

$$\frac{d^2 \delta T^s(\xi)}{d\xi^2} + \frac{1}{l_D}\frac{d\delta T^s(\xi)}{d\xi} = \left(\frac{\omega}{D} - q^2\right)\delta T^s(\xi) \qquad 11.30b$$

is obtained after substituting eqn. 11.29(b) into 11.21(c). These are ordinary differential equations of the type discussed in Chapter 1 and the solution is in the form of exponentials. Bearing in mind that the temperature is not permitted to become large far away from the interfaces, it follows that for the solid phase, the solution is a single exponential

$$\delta T^s(\xi) = U^s e^{Q_s \xi} \qquad 11.31$$

and for the liquid phase

$$\delta T^l(\xi) = U^l e^{-Q_l \xi} \qquad 11.32$$

The Q's are solutions to the quadratic equations that arise from the solution to the differential equations

$$Q_l = \frac{1/l_D + \sqrt{1/l_D^2 - 4(q^2 - \omega/D)}}{2} \qquad 11.33$$

and

$$Q_s = \frac{-1/l_D + \sqrt{1/l_D^2 - 4(q^2 - \omega/D)}}{2} \qquad 11.34$$

The solutions describing the non-steady state case are now, beginning with the liquid phase

$$T^l = (T_m - T_\infty)e^{-\xi/l_D} + T_\infty + U^l e^{-Q_l \xi}e^{\omega t + iqx} \qquad 11.35$$

and for the solid phase

$$T^s = T_m + U^s e^{-Q^s_{\overline{s}}}e^{\omega t + iqx} \qquad 11.36$$

The coefficients U^l and U^s, in Eq. 11.31 and 11.32, respectively, are to be determined from the boundary conditions, which stipulate that the temperatures at both sides of the fluctuating interface should be equal to T_{intf},

$$T_{ss}^l(\xi=\zeta)+\delta T^l(\zeta)=T_{ss}^s+\delta T^s(\zeta)=T_{intf} \qquad 11.37$$

where T_{int} is specified by Eq. 11.17 and $T_{ss}=T_m$.

At this point, a number of approximations leading to the linearization of the equations are made. This may be justified since the perturbations are assumed to be small. The process begins by performing a Taylor series expansion of $T_{ss}^l(\xi=\zeta)$ about the origin and by keeping terms linear in the amplitude; Eq. 11.37 therefore becomes (Problem 8)

$$-\frac{L}{l_D c}\zeta_q+U^l=U^s \qquad 11.38$$

Furthermore, because the curvature in Eq. 11.17 may be approximated as $\kappa \approx -\partial^2\zeta/\partial x^2 = q^2\zeta_q$, one finds that $U^s = -(\gamma/L)T_m q^2\zeta_q$. It also follows that $U^l = \frac{L}{c l_D}\zeta_q - (\gamma/L)T_m q^2\zeta_q$ (see Problem 9).

Of interest here is the situation in which the velocity of the moving front is such that $v \ll D/l_D$. This corresponds to the limit where $q^2 \gg \omega/D$ and $q \gg 1/l_D$. Furthermore, with the aid of the conservation condition at the boundary (Eq. 11.27) and the fact that $v_n \approx v+\frac{d\zeta}{dt}$, the dispersion relation is obtained (see Problem 10)

$$\omega \approx vq(1-2d_0 l_D q^2) \qquad 11.39$$

Again, only terms linear in the amplitude were considered in the analysis that enabled Eq. 11.39 to be derived

Features of the instability are now examined. The liquid front moves with a velocity, v_n, due to a driving force proportional to the temperature gradient set up in advance of the tip. The growth of this front is characterized by a wavelength, λ, is longer than a critical wave length, λ_c, where $\lambda_c = \pi(8d_0 l_D)^{1/2}$, and is linear with q ($\sim qv$). Note that the long wave length modes are active here. Effects associated with the surface tension have a stabilizing, or restoring, effect as they attempt to dampen the fluctuations. Recall that effects associated with the surface tension tend to reduce the local melting temperature at the tip and this causes heat to be transferred from behind the protrusion to the tip of the protrusion. This has the effect of suppressing growth of the protrusions. However, if the protrusion is sufficiently large, associated with which are sufficiently large temperature gradients which enable heat to be effectively removed from the tip, then the protrusions grow. It is the competition between the growing protrusions and the stabilizing effect that gives rise to this instability.

The plot in Fig. 11.8 shows the dependence of ω on q, illustrating the two competing effects which determine a dominant wave vector, λ_{max}. A small wave vectors (long wave lengths), the instability grows as qv but becomes stabilized at large q by effects associated with capillarity. The growth rate is zero when the wave vector is $(2d_0 l_D)^{-1/2}$. Note that as the magnitude of the

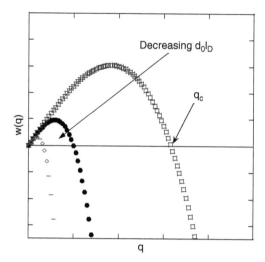

FIG. 11.8

The growth of the instability is shown here as a function of q. The instability is susceptible to long wavelength fluctuation at the surface. The critical wave vector decreases as $d_o l_D$ decreases, as indicated by the top arrow.

stabilizing interactions $d_o l_D$ decreases the dominant wavelength, λ_{max}, decreases. For values of q, beyond a critical wave vector, q_c, the growth is suppressed; effects associated with capillarity win. The most rapid growth rate possesses a wavelength $\lambda_{max} = \sqrt{3}\lambda_c$ (see problem 11). In essence the wavelength that characterizes the initial instability should be of order λ_{max} (λ_{max} evidently represents a length scale associated with the microstructure that develops as a result of the instability).

As a final note, we described the Mullins-Sekerka instability which is the underlying trigger for the formation of dendrites in various crystal forming systems. The analysis strongly indicates that the growth of any interface that is due to the diffusive transport of heat from an interface would be subject to such instabilities.

11.4   Instabilities at an Interface in a Supersaturated Environment

In the foregoing situation, the front moved in a supercooled melt as a result of solidification. In this upcoming the growth of a spherical particle is considered in a supersaturated environment (Fig. 11.9). Growth of the particle is due to the transport of a flux of solute from the environment where the concentration is c_α; the concentration at the surface of the particle is c_s.

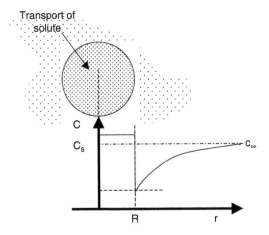

FIG. 11.9
Growth of a precipitate α-phase due to the transport of solute from the environment.

As the particle increases is size, the interface experiences shape fluctuations, protrusions and depressions, as illustrated in Fig. 11.10.

For a spherical object of radius R the driving force is associated with a concentration gradient, which is roughly proportional to $(c_s - c_\alpha)/R$. The concentration at the surface of the sphere is

$$c_s = c_0 e^{d_c \kappa} \tag{11.45}$$

If the perturbations are small, then the Eq. 11.45 may be expanded and

$$c_s \approx c_0 (1 + d_c \kappa) + \cdots \tag{11.46}$$

When the interface fluctuates in shape, the concentration of solute is higher at the protrusions than at the depressions (see problem 13). However, the

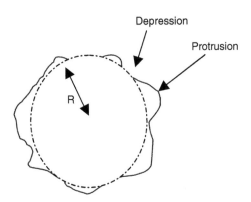

FIG. 11.10
Schematic of an unstable sphere.

fluctuating surface is associated with an increase in the interfacial free energy and the effect of the surface tension is to attempt to force the object back toward a spherical shape. The mechanism by which this occurs is with transport of matter from the protrusions to the depressions. The opposing force is proportional to d_c/R, as shown by Mullins and Sekerka (1963). Shape fluctuations in particles beyond a critical size R_c ($R > R_c$) become unstable (Problems 14 and 15).

11.5 Brief Comments on Microstructure

The Mullins-Sekerka instability is believed to be fundamentally responsible for dendritic formation in a wide class of materials. The velocity of the solid/ liquid interface of a crystal in its supercooled melt is determined by a temperature gradient, responsible for the removal of latent heat from the interface via diffusion. In the case of a solid in a supersaturated vapor environment, the velocity of the solid/supersaturated vapor interface is determined by a concentration gradient due to the diffusion of species. As such interfaces move, they experience local shape fluctuations. The shape fluctuations lead to differences in the local melting temperature via a Gibbs-Thompson effect. In the case of an alloy, there is an additional issue; redistribution of solute that accommodates these shape fluctuations leads to variations in local composition and associated changes in melting temperature because the liquidus temperatures (temperatures above which the sample is liquid) are composition dependent. This is known as constitutional undercooling. As shown earlier, capillarity effects are responsible for opposing the shape fluctuations. However, the long wavelength modes of the shape fluctuations are unstable and grow locally (protrusions). Since any interface is subject to the same instabilities, secondary branches develop at the interfaces of the growing protrusions. The result is the formation of dendrites.

Research in this area is directed at understanding the connection between the microstructural features (whether dendrites for of columnar structures of seaweed-like structures, etc.) velocity, degree of undercooling and isotropy. Various selection rules (connection between fundamental parameters that characterize details of a single structure in the pattern and velocity, etc.) are been examined for different types of patterns. (Kurz and Fisher, 1981, Li and Beckerman, 1999, Sekerka, 1995, Warren and Langer, 1993, Pochrau and Georgelin, 2003, Karma and Sarkissian, 1993).

Consider, for a moment the situation in which there is a high degree of anisotropy between the direction of the temperature gradient and the orientation of the crystallographic feature in the material. The velocity of the moving front has to be beyond a critical value before the front becomes unstable. With increasing velocity, the front becomes unstable and forms

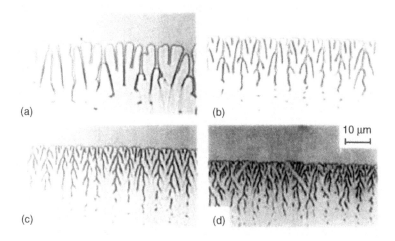

FIG. 11.11

Seaweed patterns formed in a sample where the anisotropy is low. In fact, simulations suggest that in the absence of anisotropy these patterns would develop. The velocity increases from a) through d) (Akamatsu et al, 1995, reproduced with permission).

cellular structures. At higher velocities dendritic structures develop. It is known that under these conditions the tip radius decreases with increasing velocity.

The relationship between crystalline anisotropy and pattern formation has been examined through modeling, theory, and experiments (Akamatsu et al 1995). If the material possesses a very low degree of anisotropy, then sea-weed-like structures develop instead of dendrites, assuming that the velocity is beyond a critical value. This situation is illustrated in Fig. 11.11 for a transparent (nonmetallic) sample. With increasing velocity, the features approach a smaller scale.

The reader is referred to a number of recent papers on this topic, combination of theory, simulations, and experiments. For a discussion of the recent status of solidification see a review by Boettinger et al.

11.6 Problems for Chapter 11

1. Prove that for a large liquid droplet on a substrate, subjected to a gravitational force, that its equilibrium thickness is $h_{eq} = 2d_D \sin \frac{\theta_e}{2}$. Calculate the capillary length for polystyrene on silicon oxide and for water on silicon oxide.

2. Consider a situation in which a liquid is in contact with a wall (schematic below).

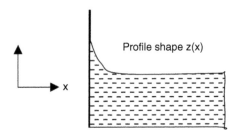

Profile shape z(x)

If the pressure at the wall is $p = p_0 - \gamma \frac{\partial^2 z}{\partial x^2}$ and the hydrostatic pressure is $p = p_0 - \rho g z$, show that the profile $z(x)$ decays exponentially from the wall.

3. Calculate the pressure difference across the interfaces, due to the curvature, of spherical particles of water, polystyrene, silica. Assume that the particles each have a radius $r = 50$ nm. Perform the calculation for particles of radius 20 nm and comment on the results.

4. Consider the surface energy of a hypothetical solid to be 2 J/m² and its density 1000 kg/m³. If the temperature is 370 K, how small would the particle have to be to increase its equilibrium vapor pressure by 10%? How small would the particle have to be to change its melting point by 15%?

5. Show that the capillary length may be written as $d_0 = \gamma T_m c_p / L^2$, where c_p is the heat capacity at constant pressure. Estimate the capillary length for a long chain polymer (e.g., polyethylene) and for a small molecule liquid.

6. Derive Eq. 11.23, the steady state solution to the heat equation.

7. Show that the heat equation becomes $\frac{\partial T}{\partial t} = D_T \nabla^2 T + v \frac{\partial T}{\partial \xi}$ with the transformation $z = \xi + vt$.

8.

a) Solve the following equation $\frac{d^2 \delta T^l(\xi)}{d\xi^2} + \frac{1}{l_D} \frac{d \delta T^l(\xi)}{d\xi} = (\frac{\omega}{D} - q^2) \delta T^l(\xi)$ and show that

$$Q_l = \frac{1/l_D + \sqrt{1/l_D^2 - 4(q^2 - \omega/D)}}{2}$$

b) In addition, show that in the short wavelength limit where $q^2 \gg \omega/D$ and $q \gg 1/l_D$ (This is accomplished when $D/l_D \gg v$, where the velocity of the front is assumed to be low.) $Q_s = Q_l q$.

9. Starting with the boundary condition $T_{ss}^0(\xi) + U^l = T_m + U^s$, show that $U^s = -(\gamma/L)T_m q^2 \zeta_k$ and $U^l = \frac{L}{d_D} \zeta_q - (\gamma/L)T_m q^2 \zeta_q$ (Hint: Rely on the

Taylor series expansion $T_{ss}^l(\xi) = T_{ss}^l(0) + \frac{\partial T_{ss}^l(\sigma)}{\partial \xi}|_{\xi=0}\zeta_q + \cdots$ the fact that $\kappa \approx -\partial^2\xi/\partial x^2 = q^2\xi_q + \cdots$).

10. With the aid of $Lv_n = D_sc_s\nabla T_{n,s} - D_lc_l\nabla T_{n,l}$ and $T_{int f} = T_m(1-\frac{\gamma\kappa}{L})$ show that $\omega = -\frac{v}{l_D} + Q_lv - \frac{Dv}{l_D}(Q_l + Q_s)\frac{\gamma}{L}q^2$. In the limit $q^2 \gg \omega/D$ and $q \gg 1/l_D$, show that $\omega \approx vq(1 - 2d_0l_Dq^2))$. Note that $dz/dt = v_n\cos\theta$, where θ is the angle between the z-direction and the vector normal to the interface. Discuss any assumptions.

11. Show that $\lambda_c = \pi(8d_0l_D)^{1/2}$ and that $\lambda_{max} = \sqrt{3}\lambda_c$.

12. With the use of Fick's 1st law, show that the velocity of the moving interface is $v = \frac{dr}{dt} = \frac{D}{(c-c_s)}\frac{\partial c}{\partial r}|_{r=R}$ for a material in a supersaturated environment.

13. Show that the concentration at the surface of a growing sphere is $c_s = c_0 e^{\Omega\gamma\kappa/kT} \approx c_0(1 + d_{cD}\kappa) + \cdots$ where δ_c is a capillary length and $d_c = \frac{\gamma\Omega}{RT}$ and Ω is the volume per particle and R is the universal gas constant and k is the curvature. (Hint: Begin by noting that the chemical potential at the surface of this spherical object is $\mu(curved) = \mu^0 + kT\ln k'c_s$, where $\mu^{(0)}$ is that of the standard state and k' is a constant. In the case of a planar surface, under otherwise identical conditions, $\mu(planar) = \mu^{(0)} + kT\ln k'c_0$. The chemical potential difference is therefore $\Delta\mu = kT\ln\frac{c_s}{c_0}$). Second, develop an expression for $\Delta\mu$ in terms of the curvature, κ, of the particle.

14. The general solution to the equation $\nabla^2 c = 0$ in spherical coordinates for the growing sphere shoes surface fluctuated is shown by Mullins and Sekerka to be

$$c(r,\theta,\phi) = c_\infty + \frac{(c_0 - c_\infty)R + 2c_c\Gamma_D}{r} - \frac{[(c_0 - c_\infty)R^l + c_c\Gamma_D l(l+1)R^{l-1}]\delta(t)Y_{lm}}{r^{l+1}}$$

where $Y_{lm}(\theta, \phi)$ are spherical harmonics of order l and m and a solution to Laplace's equation; $Y_{lm} = P_l^m(\cos\theta)e^{\pm m\phi}$ where P_l^m are Legendre polynomials). $\delta(\theta,\phi,t)Y_{lm}(\theta,\phi)$ represents the distortion at the surface of the sphere whose radius is given by

$$r(\theta,\phi,t) = R(t) + \delta(\theta,\phi,t)Y_{lm}(\theta,\phi)$$

subject to the boundary conditions specified in Section 11.5 for a growing sphere. The growth rate is

$$\frac{d\delta}{dt} = \frac{c_0D(l-1)}{(c-c_R)}\{f(R)\}\delta$$

where

$$f(R) = \frac{c_\infty - c_R}{R^2c_0} - \left[\frac{\Gamma_D^2}{R^3}[(l+1)(l+2)+2]\right]$$

Plot $f(R)$ as a function of R and comment on the physical significance of the results.

15. There exists a critical radius, R_c, beyond which the fluctuations of the surface of the sphere become unstable and this radius is

$$R_c = \left[\frac{1}{2}(l+1)(l+2)+1\right]R^*$$

where $R^* = 2\Gamma_D[(c_\infty - c_0)/c_0]$.

    a) Derive this result and determine R_c for $l = 1$ and 2.

    b) Further show that the maximum growth rate, is $\lambda = 2\pi R/l_{max} = \pi[6RR^*]^{1/2}$.

11.7 References and Further Reading

Adamson, A.W. and Gast, A.P., Physical Chemistry of Surfaces, 6th ed. Wiley, NY 1997.

Akamatsu, S., Faivre, G., Ihle, T., "Symmetry-broken double fingers and seaweed patterns in thin-film directional solidification of a nonfaceted cubic crystal," *Phys. Rev. E.* 51, 4751 (1995).

Bottinger, W.J., Coriell, S.R., Green, A.L., Karma, A., Kurz, W., Rappaz, M. and Trivedi, R., "Solidification microstructures: recent developments future applications," *Acta. Mater.* 48, 43 (2000).

de Dennes, P-G, Brochard-Wyart, F. and Quere, D., Capillarity and Wetting Phenomena, Springer-Verlag, N.Y. 2004.

Family, F., Platt, D.E. and Vicsek, T, "Deterministic growth model of pattern formatioin in dendritic solidification," *J. Phys. A: Math Gen.* 20 L1177 (1987).

Glicks,man, M.E. and Koss, M.B., "Dentritic growth velocities in micrgravity," *Phys. Rev. Lett.* 73, 573 (1994).

Golliub, J.P. and Langer, J.S., "Pattern formation in nonequilibrium physics," *Reviews of Modern Physics*, 71, S396 (1999).

Hunter, R.J., Foundations of Colloid Science, Oxford University Press, Oxford, 1995.

Hutter, J.L. and Bechhoefer, J., "Three classes of morphological transitions in the solidification of liquid crystals," *Phys. Rev. Lett.* 79, 4022 (1997).

Karma A. and Sarkissian, A., "Interface dynamics and banding in rapid solidification," *Phys. Rev. E* 47, 513 (1993).

Kurz, W. and Fisher, J.D., "Dendritic growth at the limit of high stability: tip radius and spacing," *Acta. Metallurgica* 29, 11 (1981).

Langer, J.S., "Instabilities and Pattern formation in crystal growth," *Rev. Mod. Phys.* 52, 1 (1980).

Li, Q. and Beckerman, C., "Evaluation of the sidebranch structure in free dendrutuc growth," *Acta. Ater.* 47, 2355 (1999).

Mullins, W.W. and Sekerka, R.F., "Morphological Stability of a Particle Growing by Diffusion or Heat Flow," *J. Appl. Phys.* 34, 323 (1963).

Mullins, W.W. and Sekerka, R.F., "Stability of a planar interface during solidification of a dilute binary alloy," *Journal of Applied Physics,* 35, 444 (1964).

Mullis, A.M. and Cochrane, R.F., "Grain refinement and the stability of dendrites growing into supercooled pure metals and alloys," *Journal of Applied Physics,* 82, 3783 (1997).

Pochrau and Georgelin, M., "Cellular arrays in binary alloys: from geometry to stability," *J. of Crystal Growth,* 250, 100 (2003).

Sekerka, R.F., "Optimum stability conjecture for the role of the interface kinetics in selection of the dendritic operating state," *Journal of Crystal Growth,* 154, 377 (1995).

van Saarloos, W., "Three non-equilibrium issues concerning interface dynamics in non-equilibrium pattern formation," *Physics Reports* 301, 9 (1998).

Van Sarrloos, W., "Front propagation into unstable states," *Physics Reports,* 386, 29 (2003).

Warren, J.A. and Langer, J.S., "Prediction of dendritic spacings in a directional-solidification experiment," *Phys. Rev. E* 47, 2702 (1993).

12

Comments on Instabilities and Pattern Formation in Condensed Matter

12.1 Introduction

In this chapter, qualitative examples involving the flow of liquids are discussed in order to provide a broader context for instabilities and pattern formation in condensed matter. Small amplitude shape fluctuations that develop at interfaces of condensed matter tend to increase the free energy of the system. Attempts by the system to stabilize these fluctuations occur through various mechanisms engendered by effects associated with the surface tension, viscosity, or gravity, for example, where appropriate. If the dynamics are driven by gradients of an external field (e.g., mechanical forces or forces associated with gradients in temperature, gravity, potential energy, etc.) then certain dynamical modes in these fluctuations can become amplified. In other words, the system becomes unstable and its structure may subsequently evolve spatially and temporally into a final state characterized by different organizational patterns. Nature selects a range of patterns depending on the parameters that characterize the instability. Examples of such phenomena were discussed in Chapters 9 and 11. In Chapter 9, spinodal patterns were formed when local compositional fluctuations in an otherwise homogeneous mixture became amplified when the mixture was placed in the unstable region of the phase diagram. In Chapter 11, the moving solid/melt interface of a crystal growing in its supercooled (or in a supersaturated environment) melt may become unstable toward shape fluctuations, forcing the system to exhibit different morphological patterns (e.g., dendrites).

Examples of instabilities are ubiquitous and include flow of liquid films on surfaces and the subsequent development of fingering instabilities, bulk flow of liquid columns and the development of Rayleigh instabilities leading to the breakup and formation of droplets. Other diverse examples range from the growth of bacterial colonies (M. Matsushita et al), the drying of liquids into which small solid particles are dispersed (e.g., coffee rings that develop from coffee drops) to the development of weather patterns. Such phenomena constitute an important area of research, cross-cutting diverse areas, from physics, biology, chemistry, materials science, and chemical engineering to mathematics. The interested reader is referred to the end of the chapter for references to further reading on the topic.

12.2 Instabilities That Arise in Driven Liquid Films

There is a clear distinction between instabilities that occur in thick liquid films of thickness microns or thicker and films in the nanometer thickness range. For the latter, long- and short-range intermolecular forces often play a dominant role in instabilities in films that are thin compared to the capillary length, whereas for thick macroscopic films, effects associated with gravitational forces become critical.

12.2.1 Instabilities in Macroscopically Thick Films

One of the most commonly observed instabilities is associated with a liquid flowing down an incline under the action of gravity (constant driving force), as shown in Fig. 12.1.

When the liquid flows down the incline, a rim develops at the line of contact. The development of fluctuations, protrusions, and depressions at the line of contact (Fig. 12.2a) leads to an increase of the interfacial free energy of the system. The nature of the perturbation, illustrated in Fig. 12.2a, and 12.2b, is such that the height of the liquid at the protrusion (h_p) are higher than at the depressions (h_d), $h_p > h_d$. Consequently, transverse flows from the higher to lower regions would decrease the gradient, $h_p - h_d$, and therefore oppose the protrusions. These pressure gradients provide a mechanism by which the system attempts to stabilize the fluctuations. However, thicker regions of the film move at a faster rate down the incline. If the slope is sufficiently large, creating a large enough driving force ($f = g \cdot \sin\alpha$, where g is the gravitational constant), the liquid will flow downward, forming fingers. Fingers represent another type of organization on the substrate. Note that a constant driving force is not sufficient to create the instability; there needs to be an opposing force. In this case the surface tension is responsible.

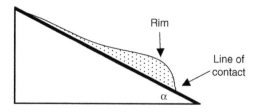

FIG. 12.1
Flow of a liquid film down an incline under the influence of gravity.

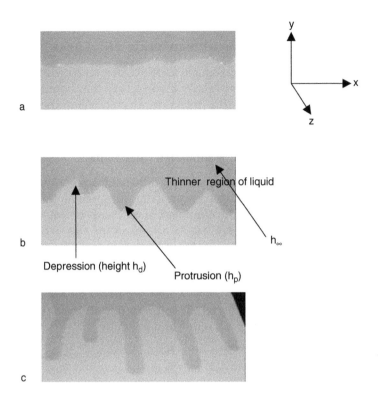

FIG. 12.2
Development of a fingering instability at the edge of a liquid flowing down an incline. a) Local deformations at the line of contact are shown; b) Fingers begin to develop and height at the edge of the moving front is higher than regions far behind the front. c) Fingers are fully developed ($h_p > h_d > h_\infty$).

Other driving forces will lead to fingering instabilities, and they include liquids subjected to centrifugal forces (e.g., photo resist spinning) and gradients in surface tension. Typically when gradients in surface tension exist, the liquid is driven in the direction of higher surface tension. The flow due to surface tension gradients is known as a Marangoni effect. A classic illustration of this effect involves dipping one end of a toothpick in liquid detergent and placing it in a bath of water. The toothpick is propelled in the direction of higher surface tension. The formation of the "tears of wine" observed in a wine glass after the glass of wine is swirled is due to a Marangoni effect. It is associated with the evaporation of alcohol from wine at the sides of the glass to create a gradient in surface tension. The aforementioned are examples involving macroscopic liquid films driven by external forces that induce them to become unstable at the line of contact and form fingers. In each case the driving forces are associated with a gradient of a "field" (surface tension gradients, temperature gradients, gradients in slope under the influence of gravity).

12.3 Instabilities in Films of Nanoscale Thickness

In the foregoing examples we considered films of macroscopic thickness (~*mm*). Another well known set of instabilities involve very thin liquid films in the nanometer thickness range. In this thickness regime, where the thickness is much smaller than the capillary length, gravitational effects are not important.

12.3.1 Pattern Formation in Nanometer-Thick Films

When the film is of nanometer thickness dimensions, long-range van der Waals forces become important. Specifically, long-range intermolecular forces are responsible for the development of an excess pressure, disjoining pressure. The disjoining pressure has its origins in the van der Waals interactions which dictate the nature of the interatomic potential between two particles in a solid. The strength of the interactions between these particles vary as $1/R^6$, where R is the distance of separation between the molecules. In the case of two flat macroscopic surfaces, separated by a distance h, the interactions are more long-ranged. The interaction energy per unit area between the two interfaces is attractive and given by

$$F(h) = -\frac{H_{SS}}{12\pi h^2}$$
                                                                       12.1

where $H_{SS} > 0$ is the Hamaker constant, a measure of the strength of the interactions of all the molecules in the system (J.N. Israelachvili 1985). The Hamaker constant is typically on the order of 10^{-19} J for most systems. If the surfaces are separated by a medium, M, then the form of the free energy function remains the same, $F(h) = -\frac{H_{SMS}}{12\pi h^2}$, but $H_{SMS} = (H_{SS}^{1/2} - H_{MM}^{1/2})^2 > 0$. As an aside, this is an interesting result, because it indicates that, as long as the slabs are made of the same material, the force is attractive! Finally, if a medium, S, is separated from a medium, V, by a third medium, L, then the interaction energy per unit area is

$$P(h) = -\frac{H_{SLV}}{12\pi h^2}$$
                                                                       12.2

This is indeed the case where a liquid is in contact with a substrate. The Hamaker constant can now be negative or positive, depending on the nature of the solid, liquid, and vapor phase. The disjoining pressure, Π, mentioned above is associated with the interaction energy between the L/V and L/S interfaces and is given by

$$\Pi = -\frac{\partial P(h)}{\partial h}$$
                                                                       12.3

One might view this as the excess pressure in the film in relation to that of the bulk. This is believed to be the origin of the destabilizing force in thin liquid films of thickness ~ nm.

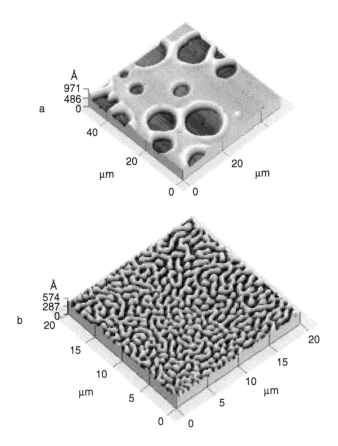

FIG. 12.3
Generic dewetting scanning force microscopy topographies of a thin polymer film a) dewetting via a nucleation and growth mechanism; b) this image is indicative of a spinodal process.

Instabilities that occur in thin liquid films in the nanometer thickness range, include, but are not limited to, spinodal patterns and nucleation and growth patterns, as illustrated in Fig. 12.3.

Physically, any small amplitude fluctuations at the free surface of the liquid film can become amplified by the disjoining pressure, provided the Hamaker constant is positive. Specifically, if the excess interfacial free energy of interaction per unit area, $\Phi(h)$, or equivalently the effective interface potential, between the liquid-substrate interface and the liquid-vapor interface, is $\Phi(h) = -A_{132}/12\pi h^2$, then the disjoining pressure, $\Pi = -\partial\Phi/\partial h = -A_{132}/6\pi h^3$, is created in the film where thinner regions of the film experience greater pressure than thicker regions. A sketch of $\Phi(h)$ is shown in Fig. 12.4. Spinodal dewetting occurs when $\Phi''(h) < 0$.

Hence, a net flow of mass is driven from thinner regions of the film to the thicker regions. The fluctuations produce an increase in the free energy (associated with an increase in the interfacial area A and the related surface tension).

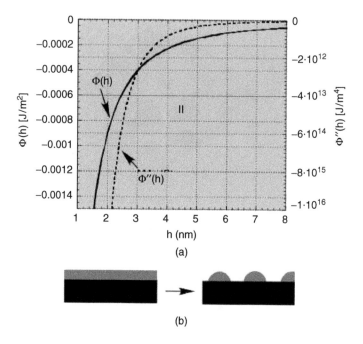

(a)

(b)

FIG. 12.4
A plot of the effective interface potential is shown in part a) for a film that will eventually
dewet the underlying substrate. Spinodal dewetting occurs when $\Phi''(h) < 0$. Thicker films
become unstable via a nucleation and growth process. Part b) illustrates the fact that the initially
stable film will eventually form droplets under these circumstances.

The influence of the surface tension is therefore to oppose these thickness
modulations (Laplace pressure). The competition between the Laplace pressure
and the disjoining pressure dictates a critical wavelength beyond which the
fluctuations will grow. The growth modes are characterized by a dominant
wave vector $q \propto (-\partial^2 \Delta G/\partial h^2)^{1/2}$. This is the process of spinodal dewetting.
Figure 12.3a shows a typical spinodal pattern, where the pattern reflects fluctu-
ations in local film thickness. Such patterns occur in simple liquid films and
liquid films of various materials (Brochard-Wyart et al., 1997, Green, 2003, Reiter
and Sharma, 1993, Seeman et al. 2001, Sharma and Reiter, 1996). Spinodal dewetting is analogous to spinodal decomposition; they are both driven by the
negative curvature of the free energy. In spinodal dewetting the relevant order
parameter is associated with the film thickness, whereas with spinodal decomposition it is the composition.

12.3.2   Fingering in Ultrathin Films

Depending on the molecular weight of the polymer and the film thickness,
the holes in films that are sufficiently thin will spontaneously exhibit fingering instabilities as they grow under the action of the capillary driving
forces (Masson et al 2002; Reiter and Sharma 2001). Figure 12.5 shows

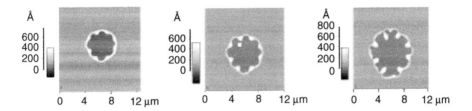

FIG. 12.5
Fingers spontaneously develop at the edge of a growing hole at intervals, from right to left, after 8 minutes, 12 minutes, and 18 minutes at 170°C. The data is for entangled polystyrene thin films supported by a silicon substrate with its native oxide layer.

(schematically) the time-dependent evolution of the growth of fingers at the perimeter of a growing hole.

The fingering phenomenon in ultrathin films is believed to be connected to slip, the nonzero displacement of the polymer at the polymer/substrate interface (deGennes 1985; Léger et al 1997). The extent of slip is characterized by an extrapolation length, b, determined by the viscosity, η, and by the friction coefficient, k $(k = \eta/b)$, between the monomers and the underlying substrate. For long-chain polymers, $b = aN^3/N_e^2$, where a is the monomer size, N is the degree of polymerization and N_e is the number of monomers between entanglements. Clearly, the slip length for polymers is many orders of magnitude larger than for simple liquids.

This mechanism of flow of polymeric films on substrates is distinct from the manner in which simple liquids flow on surfaces. Simple liquids flow along surfaces by a rolling motion and the energy is dissipated within the film due to viscous resistance. It follows that with a polymeric liquid film moving along a surface, dissipation of energy can be due to friction at the polymer/substrate as well as viscous resistance inside the film de Gennes, 1985.

The driving force for growth is provided by the spreading coefficient ($S < 0$). Hence, growth occurs radially in the direction of the "body" of the film. When the hole grows, fluctuations in the shape of the rim develop. Effects due to surface tension act to stabilize the rim, wherein flow of material from thicker to thinner regions occurs. However, because of slip, the growth velocity is proportional to $h^{-1/2}$, indicating that thicker regions of the rim move slower than thinner regions. Under sufficiently large driving forces, fingers eventually form.

12.4 Instabilities Involving Macroscopic or Bulk Flows

The foregoing examples involved liquid films that ranged in thickness from nanometers to millimeters and thicker. We are now interested in common instabilities that have been of interest in bulk systems for some time.

12.4.1   Rayleigh-Bénard Instability

Consider the convection process in a liquid film where one surface is held at temperature T_1 and the other at a higher temperature T_2, see Fig. 12.6. If $\Delta T = T_2 - T_1$ is small then the convective processes are normal and not particularly exciting from the point of view of this subject.

On the other hand, if ΔT is larger than a critical value then flow (upward and downward) occurs locally in regions (cells) throughout the liquid film, as illustrated in the above fig. A description of the basic principle follows. Since the fluid at the top is cold and at the bottom is hot, hot liquid will flow upward and the cold layer will be forced downward. It turns out that the entire layer of liquid at the bottom cannot rise upward simultaneously as the cold layer moves downward. The system accomplishes the transfer of material by partitioning into cells. A natural consequence of the buoyancy force that raises the liquid is the transfer of potential to kinetic energy. The buoyancy force is opposed by the viscous resistance and by the diffusion of heat (conduction). Radiation would be an additional contributor to the dissipation of energy if the temperatures are high. This flow, therefore, is due to a competition between the buoyancy force and the viscous resistance and thermal conduction. The growth of the instability occurs when the driving force, temperature difference $T_2 - T_1$, is sufficiently large. This phenomenon is typically quantified in terms of a Rayleigh number,

$$Ra = \frac{g\alpha(T_2 - T_1)d^3}{\nu D}$$
12.4

where α is the thermal expansion, $\nu = \eta/\rho$ is the kinematic viscosity (ρ is the density and η is the viscosity), and D is the thermal diffusivity of the liquid. The thermal diffusivity is $D = \kappa/\rho c$, where κ is the thermal conductivity and c is the heat capacity. When the Rayleigh number exceeds a critical value, ~1,700, the system becomes unstable. This is the Rayleigh-Bénard instability. A range of patterns including hexagons, and lamellae are possible.

It is interesting to note that if the top surface of the fluid is free, then surface tension effects become important (indicating that a Marangoni effect is also

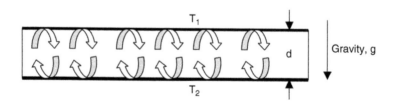

FIG. 12.6
Schematic of a Rayleigh-Bénard convection.

present) and the nature of the patterns that form are different. This, incidentally, was the problem Lord Rayleigh first solved to gain some insight into the experiments Bénard had conducted. A more relevant prediction remained elusive for over a century. Finally, we note that more complex patterns develop if the system is inclined at various angles ("Pattern formation in inclined layer convection," (Daniels, Plapp, and Bodenschatz 2003).

12.4.2   Rayleigh Instability

Imagine the situation in which a column of liquid emerges from a circular opening (e.g., a water faucet). In principle, the liquid emerging from the opening is held together by the surface tension and, over time, fluctuations of a characteristic wavelength develop in the shape of the column (Fig. 12.7a). Eventually the column breaks up into droplets (Fig. 12.7b).

While the cross-section of the column remains circular, the curvature associated with the fluctuations, or undulations, in shape create a variation in pressure (pressure gradient) along the column. The local pressure in the column increases as the wavelength of the instability decreases (Young-Laplace equation). For the same amplitude, the pressure gradients are larger for the smaller wave-length fluctuations. The surface tension provides as stabilizing force. A critical wavelength is dictated by the competing forces. When the wavelength exceeds this critical value, fluctuations of the shape of the cylinder become unstable and droplets are created.

This phenomenon is often identified as the Rayleigh instability. The formation of droplets occurs to minimize the surface area (Eggers 1997, de Gennes et al 2004). This may be seen by considering a column of liquid of length L, radius R, and surface area A_{column}. Imagine further that the column breaks up into n droplets, each of radius r. The surface area of n such droplets is $A_{drops} = 4n\pi R^2$. If the volume is to be conserved then $\pi R^2 L = (4/3)\pi r^3 n$. The relationship between the initial and final areas is

$$A_{column} = \frac{3R}{2r} A_{drops} \qquad\qquad 12.5$$

This result indicates that when $r > 1.5R$ the surface area of the droplets is smaller than that of the column; hence the breakup.

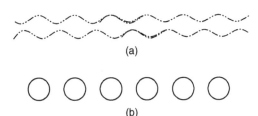

(a)

FIG. 12.7
a) Fluctuations develop in a liquid column that emerges from a circular hole. b) The stream becomes unstable and breaks up into droplets.

(b)

12.5 Final Comments

Interestingly, individual topics in this vast area of pattern formation were first investigated by separate research communities in response to certain technological problems or due to sheer scientific curiosity. In the field of materials science, the Mullins-Sekerka instability was of interest because it plays an important role in understanding microstructural development. In the field of mathematics, the problem of the moving boundary is identified as the Stefan problem. In Chemical Engineering instabilities are an essential component of the study of fluid mechanics. Important connections between these diverse problems (from biology and geology to mathematics and engineering) are now apparent. These topics form the basis of a separate interdisciplinary field known as nonlinear dynamics, or pattern formation, to many researchers. Some representative references are listed at the end of this section. I would not dare suggest that this list is exhaustive by any means.

12.6 References

Brochard-Wyart, F., et al., Dewetting of Supported Viscoelastic Polymer Films: Birth of Rims. *Macromolecules*, 1997. **30**(4): p. 1211–1213.

de Dennes, P-G, Brochard-Wyart, F. and Quere, D., Capillarity and Wetting Phenomena, Springer-Verlag, N.Y. 2004.

de Gennes, P-G.,"Wetting: statics and dynamics," *Rev. Mod. Phys.* 57, 827 (1985).

Edwards, D.A., Brenner, H., and Wasan, D.T., Interfacial Transport Processes and rheology, Butterworth-Heinemann Series in Chemical Engineering, Boston, 1991.

Green, P.F., Wetting and dynamics of structured liquid films. *Journal of Polymer Science, Part B: Polymer Physics*, 2003. **41**(19): p. 2219–2235.

Israelachvili, J.N., Intermolecular and surface forces, Academic Press: London, 1985.

L. Léger, H. Hervet, G. Massey and E. Durliat, "Wall slip in polymer melts," *J. Phys. Cond. Matter*, 9, 7719 (1997).

Masson, J.-L., O. Olufokunbi, and P.F. Green, Flow Instabilities in Entangled Polymer Thin Films. *Macromolecules*, 2002. **35**(18): p. 6992–6996.

Matsushita, M., Wakita, J., Itoh, H., Ráfols, I., Matsuyama, T., Sakaguchi, H., Mimura, M.M., "Interface growth and pattern formation in bacterial colonies," *Physics A*, 249, 517 (1998).

Reiter, G. and Sharma, "Auto-Optimization of Dewetting Rates by Rim Instabilities in Slipping Polymer Films," *Phys. Rev. Lett.*, 87, 166103 (2001).

Reiter, G., Unstable thin polymer films: rupture and dewetting processes. *Langmuir*, 1993. **9**(5): p. 1344–51.

Seemann, R., S. Herminghaus, and K. Jacobs, Dewetting Patterns and Molecular Forces: A Reconciliation. *Physical Review Letters*, 2001. **86**(24): p. 5534–5537.

Sharma, A. and G. Reiter, Instability of thin polymer films on coated substrates: rupture, dewetting, and drop formation. *Journal of Colloid and Interface Science,* 1996. **178**(2): p. 383–99.

12.7 Further Reading

Langer, J.S. "Instabilities and pattern formation in crystal growth," *Rev. Mod. Phys.* 52, 1 (1980).

Oron, A., Davis, S.H. and Bankoff, S.G., "Long-scale evolution of thin liquid films," *Rev. Mod. Phys.* 69, 931 (2997).

van Saarlos, W., "Propogation into unstable states" *Physics Reports,* 386, 29 (2003).

Eggers, J., "nonlinear dynamics and breakup of free-surface flows," *Rev. Mod. Phys.* 69, 865 (1997).

Meakin, P., Fractals, scaling and growth far from equilibrium, Cambridge Nonlinear Science Series 5, Cambridge University Press, 1998.

Gollub, J.P. and Langer, J.S., "Pattern formation in nonequilibrium physics," *Rev. Mod. Phys.* 71, S396 (1999).

Cross, M., and Hohenerg, P.C., "Pattern formation outside of equilibrium," *Rev. Mod. Phys.* 65, 851 (1993).

Eggers, J., "Nonlinear dynamics and breakup of free flows," *Rev. Mod. Phys.* 69, 865 (1997).

Index

A

Activation energy for conductivity, 239
Activation energy for flow, 224
Activity coefficient, 128
Adam-Gibbs model, 220–222
Alkali halide, 123–132
Alkali silicate, 4
Alkali tellurite, 227
alternating copolymers, 155
Anion vacancies, 132
Anionic diffusion, 123
Atomic defects, 86
Atomic diffusion, 79
Atomic packing fraction, 83
Autocorrelation function (of the current), 236
Auto-correlation function, 57, 58
Autocorrelation of the scattered field, 69
Velocity, average component, 13
Energy, average, 8
Pressure, average, 57
Speed, average, 14

B

Binomial distribution function, 52, 53
Body centered cubic, 82, 95
Boltzmann factor, 7, 228, 283,
Boltzmann superposition principle, 175
Branched polymers, 154
Bravis lattice, 81
Brillouin scattering, 67
Brownian motion, 51–70, 262

C

Capillary length, 317–321
Cationic diffusion, 123
Central limit theorem, 56
Characteristic ratio, 158, 191
Chemical potential driving force, 127, 283
Capillarity, 315

partition function, 10, 59
Classical theory of nucleation, 282–284
Coarsening, 269, 319
Complex conductivity, 237
Complex conductivity, 240
Complex dielectric permittivity, 235
Complex liquids, 151
Complex modulus, 174, 176, 177, 202–203
Compliance, 167, 173, 175, 187, 191
Concentration fluctuations, 278
Concentration of impurities in dilute alloy, 109
Concentration of vacancies in dilute alloy, 109
Conductance, 235
Conductivity (composition dependence), 240
Conductivity Relaxation time(temperature dependence), 242
Configuration integral, 59
Conformation, 159
Constitutive model, 175
Constraint release, 180, 199–201
Contour length fluctuation, 180
contour length, 155, 159
Coordination number, 83
copolymers, 154
correlation factor, 90, 97, 111, 112, 113
Correlation function, 51, 56, 60, 259
Correlation length, 280
Creep, 166
Critical energy for formation of a nucleus, 283
Critical point, 273, 280
Critical radius for nucleation, 283
Cross-correlation function, 59
Crystal direction, 82
Crystal planes, 82
Crystal symmetry, 80
Crystal systems, 82
Cubic system (directions, planes), 84, 85

D

Darken equation, 300
Debye frequency, 92, 93, 95, 159
Debye scattering function, 305